Wastewater Treatment Engineering

Editor:

Jaspreet Banga, Ph.D.

ArclerPress

www.arclerpress.com

Wastewater Treatment Engineering

Editor: Jaspreet Banga, Ph.D.

© 2017 Arcler Press LLC
708 3rd Avenue, 6th Floor
New York
NY 10017
United States of America
www.arclerpress.com

ISBN: 978-1-68094-410-5

Library of Congress Control Number: 2016951117

About the Editor

Jaspreet Banga, Ph.D.

Jaspreet Banga has completed PhD from Central Drug Research Institute, India and Master of Technology from Indian Institute of Technology. Jaspreet's interests range from Drug development, Purification, Fermentation technology, Plant Biotechnology and Oncology.

Contents

List of Contributors

Mohamed Samer
Cairo University, Faculty of Agriculture, Department of Agricultural Engineering, Egypt

Parneet Paul
Department of Civil Engineering, School of Natural and Built Environments, Faculty of Science, Engineering and Computing, Kingston University, Penrhyn Road, Kingston upon Thames, Surrey KT1 2EE, UK

Franck Anderson Jones
Department of Mechanical, Aerospace and Civil Engineering, College of Engineering, Design and Physical Sciences, Brunel University, Uxbridge, Middlesex UB8 3PH, UK

Miyoshi Kato
Graduate School of Biosphere Science, Hiroshima University, 1-4-4, Kagamiyama, Higashihiroshima, Hiroshima 739-8527, Japan

Haruyuki Iefuji
Graduate School of Biosphere Science, Hiroshima University, 1-4-4, Kagamiyama, Higashihiroshima, Hiroshima 739-8527, Japan
National Research Institute of Brewing, 3-7-1, Kagamiyama, Higashihiroshima, Hiroshima 739-0046, Japan

Abdullah Erdal Tümer
Computer Engineering Department, University of Necmettin Erbakan, Konya, Turkey

Serpil Edebali
Chemical Engineering Department, University of Selçuk, Konya, Turkey

Yalda Basim
Department of Environmental Engineering, Sazabpardazan Consultant Engineering, Ahvaz, Iran

Mahdi Farzadkia
Department of Environmental Health Engineering, School of Public Health, Tehran University of Medical Sciences, Tehran, Iran

Nematollah Jaafarzadeh
Department of Environmental Health Engineering, School of Health, Ahvaz Jondishapour University of Medical Sciences, Ahvaz, Iran

Tim Hendrickx
Department of Environmental Technology, Wageningen University, Wageningen, The Netherlands

JunfengWan
School of Civil and Environmental Engineering, Nanyang Technological University, 50 Nanyang Avenue, 639798, Singapore
Advanced Environmental Biotechnology Centre, Nanyang Environment and Water Research Institute, Nanyang Technological University, 1 Cleantech Loop, CleanTech One, 637141, Singapore
School of Chemical Engineering and Energy, Zhengzhou University, 100 Science Avenue, 450001, P. R. China

JunGu
Advanced Environmental Biotechnology Centre, Nanyang Environment and Water Research Institute, Nanyang Technological University, 1 Cleantech Loop, CleanTech One, 637141, Singapore

Qian Zhao
School of Civil and Environmental Engineering, Nanyang Technological University, 50 Nanyang Avenue, 639798, Singapore
Advanced Environmental Biotechnology Centre, Nanyang Environment and Water Research Institute, Nanyang Technological University, 1 Cleantech Loop, CleanTech One, 637141, Singapore

Yu Liu
School of Civil and Environmental Engineering, Nanyang Technological University, 50 Nanyang Avenue, 639798, Singapore
Advanced Environmental Biotechnology Centre, Nanyang Environment and Water Research Institute, Nanyang Technological University, 1 Cleantech Loop, CleanTech One, 637141, Singapore

J. L. Campos
Facultad de Ingeniería y Ciencias, Universidad Adolfo Ibáñez, Avenida Padre Hurtado 750, 2520000 Viña del Mar, Chile

D. Valenzuela-Heredia
Facultad de Ingeniería y Ciencias, Universidad Adolfo Ibáñez, Avenida Padre Hurtado 750, 2520000 Viña del Mar, Chile

A. Pedrouso
Department of Chemical Engineering, School of Engineering, University of Santiago de Compostela, Rua Lope Gómez de Marzoa s/n, 15782 Santiago de Compostela, Spain

A. Val del Río
Department of Chemical Engineering, School of Engineering, University of Santiago de Compostela, Rua Lope Gómez de Marzoa s/n, 15782 Santiago de Compostela, Spain

M. Belmonte
Department of Environment, Faculty of Engineering, University of Playa Ancha, Avenida Leopoldo Carvallo 270, 2340000 Valparaíso, Chile
School of Biochemical Engineering, Pontifical Catholic University of Valparaíso, Avenida Brasil 2085, 2340000 Valparaíso, Chile

A. Mosquera-Corral
Department of Chemical Engineering, School of Engineering, University of Santiago de Compostela, Rua Lope Gómez de Marzoa s/n, 15782 Santiago de Compostela, Spain

Nancy Vásquez Sarria
Faculty of Engineering, Universidad del Valle, Colombia

Jenny Rodríguez Victoria
Faculty of Engineering, Universidad del Valle, Colombia

Patricia Torres Lozada
Faculty of Engineering, Universidad del Valle, Colombia

Carlos Madera Parra
Faculty of Engineering, Universidad del Valle, Colombia

Mayamin Razali
School of Chemical Engineering & Analytical Science, The University of Manchester, The Mill, Sackville Street, Manchester M13 9PL, UK

Jeong F. Kim
WCU Department of Energy Engineering, Hanyang University, Republic of Korea

Martin Attfield
School of Chemistry, The University of Manchester, M13 9PL, UK

Peter M. Budd
School of Chemistry, The University of Manchester, M13 9PL, UK

Enrico Drioli
WCU Department of Energy Engineering, Hanyang University, Republic of Korea
Institute on Membrane Technology (ITM-CNR), Via P. Bucci 17/C, 1-87030 Rende, CS, Italy

Young Moo Lee
WCU Department of Energy Engineering, Hanyang University, Republic of Korea

Gyorgy Szekely
School of Chemical Engineering & Analytical Science, The University of Manchester, The Mill, Sackville Street, Manchester M13 9PL, UK

Hassimi Abu Hasan
Department of Chemical and Process Engineering, Faculty of Engineering and Built Environment, 43600 UKM Bangi, Selangor Darul Ehsan Malaysia

Siti Rozaimah Sheikh Abdullah
Department of Chemical and Process Engineering, Faculty of Engineering and Built Environment, 43600 UKM Bangi, Selangor Darul Ehsan Malaysia

Nor Ashikin Shafinaz Binti Whayab
Department of Chemical and Process Engineering, Faculty of Engineering and Built Environment, 43600 UKM Bangi, Selangor Darul Ehsan Malaysia

Fathiyyah Mohd Zainudin
Department of Chemical and Process Engineering, Faculty of Engineering and Built Environment, 43600 UKM Bangi, Selangor Darul Ehsan Malaysia

Muhammad Farhan Md Yusoff
Department of Chemical and Process Engineering, Faculty of Engineering and Built Environment, 43600 UKM Bangi, Selangor Darul Ehsan Malaysia

Nastee Kornochalert
Department of Microbiology, Faculty of Science, Prince of Songkla University, Hat Yai, Thailand

Duangporn Kantachote
Department of Microbiology, Faculty of Science, Prince of Songkla University, Hat Yai, Thailand
Center of Excellence on Hazardous Substance Management, Hat Yai, Thailand

Sumate Chaiprapat
Center of Excellence on Hazardous Substance Management, Hat Yai, Thailand
Department of Civil Engineering, Faculty of Engineering, Prince of Songkla University, Hat Yai, Thailand

Somkiet Techkarnjanaruk
Biochemical Engineering and Pilot Plant Research and Development Unit, National Center for Genetics Engineering and Biotechnology, Bangkok, Thailand

EFC Chaúque
Department of Applied Chemistry, University of Johannesburg, Doornfontein 2028 Johannesburg, South Africa

JN Zvimba
Water Research Commission, Gezina, 0031, Pretoria, South Africa

JC Ngila
Department of Applied Chemistry, University of Johannesburg, Doornfontein 2028 Johannesburg, South Africa

N Musee
Department of Chemical Engineering, University of Pretoria, Hatfield 0028, Pretoria, South Africa

Dilafroza Wani
Centre of Research for Development, University of Kashmir, Srinagar-190006 Jammu and Kashmir, India

Ashok K. Pandit
Centre of Research for Development, University of Kashmir, Srinagar-190006 Jammu and Kashmir, India

Azra N Kamili
Centre of Research for Development, University of Kashmir, Srinagar-190006 Jammu and Kashmir, India

Na Wang
School of Environmental Science and Engineering, Qilu University of Technology, Jinan, Shandong, China

Wenqiang Jiang
School of Environmental Science and Engineering, Qilu University of Technology, Jinan, Shandong, China

Libin Fang
Water Technology Research Institute, A Agrochemical Company of Binzhou, Binzhou, Shandong, China

Andrea G. Capodaglio
University of Pavia (UNIPV), Pavia, Italy, Department of Civil Engineering and Architecture

Petr Hlavínek
Brno University of Technology (BUT), Brno, Czech Republic, Institute of Municipal Water Management

Massimo Raboni
Università Carlo Cattaneo (LIUC), Castellanza, Italy

Khaled Zaher Abdalla
Assistant Professor, Sanitary and Environmental Engineering Division, Public works Engineering Department, Faculty of Engineering, Cairo University, Giza

Gina Hammam
Environmental Engineer, graduate student, Public works Engineering Department, Faculty of Engineering, Cairo university, Giza

Haider M. Zwain
School of Civil Engineering, Universiti Sains Malaysia, Engineering Campus, Seri Ampangan, 14300 Nibong Tebal, Pulau Pinang, Malaysia

Nastaein Qamaruz Zaman
School of Civil Engineering, Universiti Sains Malaysia, Engineering Campus, Seri Ampangan, 14300 Nibong Tebal, Pulau Pinang, Malaysia

Hamidi Abdul Aziz
School of Civil Engineering, Universiti Sains Malaysia, Engineering Campus, Seri Ampangan, 14300 Nibong Tebal, Pulau Pinang, Malaysia

Irvan Dahlan
School of Civil Engineering, Universiti Sains Malaysia, Engineering Campus, Seri Ampangan, 14300 Nibong Tebal, Pulau Pinang, Malaysia

Mohammed Wedyan
Biological Sciences and Biotechnology Department, the Hashemite University, JORDAN

Ahmed Al Harahsheh
Biological Sciences and Biotechnology Department, the Hashemite University, JORDAN

Esam Qnaisb
Biological Sciences and Biotechnology Department, the Hashemite University, JORDAN

Moh Faiqun Ni'am
Dept. of Environmental Eng., Fac. of Engineering, Universitas Islam Sultan Agung, Semarang 50112, Indonesia

Fadil Othman
Dept. of Environmental Eng., Fac. Of Civil Eng., Universiti Teknologi Malaysia, Johor 81310, Malaysia

H. Doan
Department of Chemical Engineering, Ryerson University, 350 Victoria Street, Toronto, On., Canada

A. Lohi
Department of Chemical Engineering, Ryerson University, 350 Victoria Street, Toronto, On., Canada

Preface

Water is becoming a progressively scarce resource on our planet. Although we have vast amount of water on our planet, but fresh water makes up only about 1% of all the planet's water bodies. It is reasonable to say at this point, that any sources of water which might be used economically and effectively be promoted and further developed in the overall benefit of human kind. A large proportion of fresh water is used for human consumption, followed by irrigation use, and whenever fresh water is scarce, marginally good quality or water treated after wastewater treatment should be used for agricultural purposes. Domestic wastewater is comprised of 99.9% water and the 0.1% remaining part consists or organic/inorganic suspended and dissolved solids, making it unfit for human use and placing it in the wastewater category, which may be made suitable for agricultural use. So, the composition of wastewater is dependent upon the functions to which the water was subjected and its treatment is also dependent upon the initial use of water. This book deals with the various treatments to which the waste water is subjected to make reuse of water a possibility.

Editor
Jaspreet Banga, Ph.D.

INTRODUCTION

Wastewater engineering is not usually its own degree course but a specialization from degrees such as civil engineering, environmental engineering or chemical engineering. Wastewater engineering deals with the transportation and cleaning of black water, greater, and irrigation water. Wastewater treatment and water reclamation are areas of concern in this field. Wastewater engineers map out topographical and geographical features of Earth to determine the best means of collection. Wastewater treatment is the process of converting wastewater – water that is no longer needed or is no longer suitable for use – into bilge water that can be discharged back into the environment. It's formed by a number of activities including bathing, washing, using the toilet, and rainwater runoff. Wastewater is full of contaminants including bacteria, chemicals and other toxins. Its treatment aims at reducing the contaminants to acceptable levels to make the water safe for discharge back into the environment. There are two wastewater treatment plants namely chemical or physical treatment plant, and biological wastewater treatment plant. Biological waste treatment plants use biological matter and bacteria to break down waste matter. Physical waste treatment plants use chemical reactions as well as physical processes to treat wastewater. Biological treatment systems are ideal for treating wastewater from households and business premises. Physical wastewater treatment plants are mostly used to treat wastewater from industries, factories and manufacturing firms. This is because most of the wastewater from these industries contains chemicals and other toxins that can largely.

Early history

Irrigation systems were invented five to seven thousand years ago as a means of supplying water to agriculture-based societies. Aqueducts and irrigation systems were among the first forms of wastewater engineering. As population centers became denser, they were used to remove sewage from settlements. The Romans were among the first to demonstrate the effectiveness of the aqueduct. The Dark Ages marked a period where progress in water management came to a halt

Modern history

As populations grew, the management of human waste became a growing concern and a public health threat. By the 1850s in London more than 400,000 tons of sewage were flushed into the Thames River each day - around 150 million tons a year. Diseases such as smallpox, diphtheria, measles, scarlet fever, typhus, cholera and typhoid were spread via the contaminated water supply. During the 19th century major cities started building sewage systems to remove human waste out of cities and into rivers. During the 1900s the activated sludge process was invented. The activated sludge process is a form of water purification that uses bacteria to consume human feces. Chlorine is used to kill off the bacteria.

Over the centuries much has changed in the field of wastewater engineering. Advancements in microbiology, chemistry, and engineering have drastically changed the field. Today wastewater engineers work on the collection of clean water for drinking, chemically treating it, and using UV light to kill off micro-organisms. They also treat Water Pollution in wastewater (blackwater and greywater) so that this water may be made safe for use without endangering the population and environment around it. Today the wastewater managers confront new challenges and the need for new technology as water levels decrease due to increasingly frequent and extended droughts. Technologies such as sonar mapping are being used in wells to determine the volume of water that they can hold. For example the U.S. Geological Survey (USGS) and the State of New York worked together to map underground aquifers since the 1980s. Today they have thorough maps of these aquifers to assist in water management. Wastewater treatment contributes to global warming in many ways. One of the factors that contributes to global warming is wastewater treatment facilities and their emissions of greenhouse gases. Some of those gases are carbon dioxide, methane, and nitrous oxide. These gases occur because of the decomposition of organic material from the anaerobic bacteria. These bacteria clean the leftover waste. Even if the anaerobic bacteria decomposition produces these gases, the percentage of greenhouse gases that other equipment produce is still greater than the contribution of the anaerobic bacteria. Also, the power usage from those machinery is very high. That is why many facilities are undergoing renovation to use higher levels of anaerobic bacteria compared to other types of equipment. The effects of global warming on waste water treatment facilities vary from location to location. In the Northeast there has been increased rainfall causing treatment plants to be too small relative to the amount coming in. In the West there is a lack of rainfall and water sources are being depleted. This means the facilities will need to be updated to meet modern and future environmental changes such as the increased use of recycled and reclaimed water Climate change also affects piping and increases the rate of corrosion, adding to facility cost.

Desalination plants may be required in the future for those regions hardest hit by water scarcity. Desalination is a process of cleaning water by means of evaporation. Water is evaporated and it passes through membranes. The water is then cooled and condenses allowing it to flow either back into the main water line or out to sea. Desalination is not completely "green" since the plants require energy to run and due to the fact that the brine waste must managed to avoid ecological damage to the ocean

Education necessary for wastewater

Formal education for wastewater engineers begins in high school with students taking classes such as chemistry, biology, physics, and higher mathematics including calculus. After high school most jobs require certification from a state agency. Those wanting to advance in the industry should pursue a civil engineering, mechanical engineering, environmental engineering, or a facilities engineering degree. Gaining experience through internships and working while in college is a common pathway toward advancement.

Education about waste treatment requires course work in systems design, machinery design principles, water chemistry, and similar coursework. Other classes may include Chemistry of Plant Processes, and various plant operations courses.

Initial employment in wastewater engineering can be obtained by those with and without advanced formal education. The California State Water Resources Control Board (SWRCB), for example, shows how individuals can advance through a progression of certifications as Waste Water Treatment Operators. The Board uses a five level classification system to classify water treatment facilities into categories I-V according to the population served and the complexity of the treatment system.

The Operator Certification requirements for water treatment operators and waste water treatment operators are described in detail by State law. To meet certification requirements, operators must submit an application to SWRCB, have the necessary work experience, meet the educational requirements, and pass an examination based on the knowledge, skill, and abilities described in the regulations. Operators are required to renew their certificates every three years. To be eligible for renewal, certified operators must complete a specified number of continuing education hours after the previous issuance of a certificate.

Wastewater engineers may advance in their careers through additional education and experience. With additional knowledge and experience one can become the manager of an entire plant. The accreditation body certifying the education for the degree and license is the Accreditation Board for Engineering and Technology (ABET). Over time some companies may require the wastewater engineer to continue their education to keep up with any changes in technology. Obtaining one's master's degree is encouraged since many companies list it as a preference in selection.

In this field 76 percent of those employed have a bachelor's degree, 17 percent have a master's degree and three percent have a post-doctoral degree as of 2013. The average annual salary is approximately $83,360. Wastewater engineers use a variety of skills and must have knowledge of mechanical and environmental engineering. They are required to perform tasks and demonstrate knowledge in design, mathematics, English, construction, physics, chemistry, biology, management, and personnel. Wastewater engineers must have skills in complex problem solving, critical thinking, mathematics, active listening, judgment, reading comprehension, speaking, writing, science, and system analysis. Typical work activities include problem solving, communication with management and staff, gathering information, analyzing data, evaluating standards and complying with them, and communicating with others in the field. Wastewater engineers perform these activities by combining their knowledge and skills to perform tasks. These tasks are to understand computer-aided design programs, and to conduct studies for the construction of facilities, water supply systems and collection systems. They may design systems for wastewater collection machinery, as well as system components. They may perform water flow analysis, then select designs and equipment based on government and industry standards. Wastewater engineers are responsible for planning and designing all processes, structures and facilities involved in the treatment, distribution, and collection of wastewater. Waste treatment engineers work with government agencies and private industry on projects of all sizes and scales, including designing small package plants as well as large municipal and industrial treatment facilities, pumping stations, reservoirs, distribution systems, and sewers.

Wastewater Treatment Processes

Wastewater treatment is closely related to the standards and/or expectations set for the effluent quality. Wastewater treatment processes are designed to achieve improvements in the quality of the wastewater. The various treatment processes may reduce:

- **Suspended solids** (physical particles that can clog rivers or channels as they settle under gravity)
- **Biodegradable organics** (e.g. BOD) which can serve as "food" for microorganisms in the receiving body. Microorganisms combine this matter with oxygen from the water to yield the energy they need to thrive and multiply; unfortunately, this oxygen is also needed by fish and other organisms in the river. Heavy organic pollution can lead to "dead zones" where no fish can be found; sudden releases of heavy organic loads can lead to dramatic "fishkills".
- **Pathogenic bacteria** and other disease causing organisms: These are most relevant where the receiving water is used for drinking, or where people would otherwise be in close contact with it; and
- **Nutrients**, including nitrates and phosphates. These nutrients can lead to high concentrations of unwanted algae, which can themselves become heavy loads of biodegradable organic load Treatment processes may also neutralize or removing

industrial wastes and toxic chemicals. This type of treatment should ideally take place at the industrial plant itself, before discharge of their effluent in municipal sewers or water courses.

- **Primary (mechanical) treatment** is designed to remove gross, suspended and floating solids from raw sewage. It includes screening to trap solid objects and sedimentation by gravity to remove suspended solids. This level is sometimes referred to as "mechanical treatment", although chemicals are often used to accelerate the sedimentation process. Primary treatment can reduce the BOD of the incoming wastewater by 20-30% and the total suspended solids by some 50-60%. Primary treatment is usually the first stage of wastewater treatment. Many advanced wastewater treatment plants in industrialized countries have started with primary treatment, and have then added other treatment stages as wastewater load has grown, as the need for treatment has increased, and as resources have become available.
- **Secondary (biological) treatment** removes the dissolved organic matter that escapes primary treatment. This is achieved by microbes consuming the organic matter as food, and converting it to carbon dioxide, water, and energy for their own growth and reproduction. The biological process is then followed by additional settling tanks ("secondary sedimentation", see photo) to remove more of the suspended solids. About 85% of the suspended solids and BOD can be removed by a well running plant with secondary treatment. Secondary treatment technologies include the basic activated sludge process, the variants of pond and constructed wetland systems, trickling filters and other forms of treatment which use biological activity to break down organic matter.
- **Tertiary treatment** is simply additional treatment beyond secondary! Tertiary treatment can remove more than 99 percent of all the impurities from sewage, producing an effluent of almost drinking-water quality. The related technology can be very expensive, requiring a high level of technical know-how and well trained treatment plant operators, a steady energy supply, and chemicals and specific equipment which may not be readily available. An example of a typical tertiary treatment process is the modification of a conventional secondary treatment plant to remove additional phosphorus and nitrogen.

Disinfection, typically with chlorine, can be the final step before discharge of the effluent. However, some environmental authorities are concerned that chlorine residuals in the effluent can be a problem in their own right, and have moved away from this process. Disinfection is frequently built into treatment plant design, but not effectively practiced, because of the high cost of chlorine, or the reduced effectiveness of ultraviolet radiation where the water is not sufficiently clear or free of particles.

Wastewater Treatment Water Use

We consider wastewater treatment as a water use because it is so interconnected with the other uses of water. Much of the water used by homes, industries, and businesses must be treated before it is released back to the environment.

If the term "wastewater treatment" is confusing to you, you might think of it as "sewage treatment." Nature has an amazing ability to cope with small amounts of water wastes and pollution, but it would be overwhelmed if we didn't treat the billions of gallons of wastewater and sewage produced every day before releasing it back to the environment. Treatment plants reduce pollutants in wastewater to a level nature can handle.

Wastewater is used water. It includes substances such as human waste, food scraps, oils, soaps and chemicals. In homes, this includes water from sinks, showers, bathtubs, toilets, washing machines and dishwashers. Businesses and industries also contribute their share of used water that must be cleaned.

Wastewater also includes storm runoff. Although some people assume that the rain that runs down the street during a storm is fairly clean, it isn't. Harmful substances that wash off roads, parking lots, and rooftops can harm our rivers and lakes.

Why Treat Wastewater?

It's a matter of caring for our environment and for our own health. There are a lot of good reasons why keeping our water clean is an important priority:

Fisheries

Clean water is critical to plants and animals that live in water. This is important to the fishing industry, sport fishing enthusiasts, and future generations.

Wildlife Habitats

Our rivers and ocean waters teem with life that depends on shoreline, beaches and marshes. They are critical habitats for hundreds of species of fish and other aquatic life. Migratory water birds use the areas for resting and feeding.

Recreation and Quality of Life

Water is a great playground Goes to page with picture plus text for us all. The scenic and recreational values of our waters are reasons many people choose to live where they do. Visitors are drawn to water activities such as swimming, fishing, boating and picnicking.

Health Concerns

If it is not properly cleaned, water can carry disease. Since we live, work and play so close to water, harmful bacteria have to be removed to make water safe.

Effects of wastewater pollutants

If wastewater is not properly treated, then the environment and human health can be negatively impacted. These impacts can include harm to fish and wildlife populations, oxygen depletion, beach closures and other restrictions on recreational water use, restrictions on fish and shellfish harvesting and contamination of drinking water. Environment Canada provides some examples of pollutants that can be found in wastewater and the potentially harmful effects these substances can have on ecosystems and human health:

- decaying organic matter and debris can use up the dissolved oxygen in a lake so fish and other aquatic biota cannot survive;
- excessive nutrients, such as phosphorus and nitrogen (including ammonia), can cause eutrophication, or over-fertilization of receiving waters, which can be toxic to aquatic organisms, promote excessive plant growth, reduce available oxygen, harm spawning grounds, alter habitat and lead to a decline in certain species;
- chlorine compounds and inorganic chloramines can be toxic to aquatic invertebrates, algae and fish;
- bacteria, viruses and disease-causing pathogens can pollute beaches and contaminate shellfish populations, leading to restrictions on human recreation, drinking water consumption and shellfish consumption;
- metals, such as mercury, lead, cadmium, chromium and arsenic can have acute and chronic toxic effects on species.
- other substances such as some pharmaceutical and personal care products, primarily entering the environment in wastewater effluents, may also pose threats to human health, aquatic life and wildlife.

Wastewater treatment

The major aim of wastewater treatment is to remove as much of the suspended solids as possible before the remaining water, called effluent, is discharged back to the environment. As solid material decays, it uses up oxygen, which is needed by the plants and animals living in the water.

"Primary treatment" removes about 60 percent of suspended solids from wastewater. This treatment also involves aerating (stirring up) the wastewater, to put oxygen back in. Secondary treatment removes more than 90 percent of suspended solids.

Wastewater treatment problem

The principal objective of wastewater treatment is generally to allow human and industrial effluents to be disposed of without danger to human health or unacceptable damage to the natural environment. Irrigation with wastewater is both disposal and utilization and indeed is an effective form of wastewater disposal (as in slow-rate land treatment). However, some degree of treatment must normally be provided to raw municipal wastewater before it can be used for agricultural or landscape irrigation or for aquaculture. The quality of treated effluent used in agriculture has a great influence on the operation and performance of the wastewater-soil-plant or aquaculture system. In the case of irrigation, the required quality of effluent will depend on the crop or crops to be irrigated, the soil conditions and the system of effluent distribution adopted. Through crop restriction and selection of irrigation systems which minimize health risk, the degree of pre-application wastewater treatment can be reduced. A similar approach is not feasible in aquaculture systems and more reliance will have to be placed on control through wastewater treatment.

The most appropriate wastewater treatment to be applied before effluent use in agriculture is that which will produce an effluent meeting the recommended microbiological and chemical quality guidelines both at low cost and with minimal operational and maintenance requirements. Adopting as low a level of treatment as possible is especially desirable in developing countries, not only from the point of view of cost but also in acknowledgement of the difficulty of operating complex systems reliably. In many locations it will be better to design the reuse system to accept a low-grade of effluent rather than to rely on advanced treatment processes producing a reclaimed effluent which continuously meets a stringent quality standard.

Nevertheless, there are locations where a higher-grade effluent will be necessary and it is essential that information on the performance of a wide range of wastewater treatment technology should be available. The design of wastewater treatment plants is usually based on the need to reduce organic and suspended solids loads to limit pollution of the environment. Pathogen removal has very rarely been considered an objective but, for reuse of effluents in agriculture, this must now be of primary concern and processes should be selected and designed accordingly. Treatment to remove wastewater constituents that may be toxic or harmful to crops, aquatic plants (macrophytes) and fish is technically possible but is not normally economically feasible. Unfortunately, few performance data on wastewater treatment plants in developing countries are available and even then they do not normally include effluent quality parameters of importance in agricultural use.

The short-term variations in wastewater flows observed at municipal wastewater treatment plants follow a diurnal pattern. Flow is typically low during the early morning hours, when water consumption is lowest and when the base flow consists of infiltration-inflow and small quantities of sanitary wastewater. A first peak of flow generally occurs in the late morning, when wastewater from the peak morning water use reaches the treatment plant, and a second peak flow usually occurs in the evening.

The relative magnitude of the peaks and the times at which they occur vary from country to country and with the size of the community and the length of the sewers. Small communities with small sewer systems have a much higher ratio of peak flow to average flow than do large communities. Although the magnitude of peaks is attenuated as wastewater passes through a treatment plant, the daily variations in flow from a municipal treatment plant make it impracticable, in most cases, to irrigate with effluent directly from the treatment plant. Some form of flow equalization or short-term storage of treated effluent is necessary to provide a relatively constant supply of reclaimed water for efficient irrigation, although additional benefits result from storage.

Conventional wastewater treatment processes

Conventional wastewater treatment consists of a combination of physical, chemical, and biological processes and operations to remove solids, organic matter and, sometimes, nutrients from wastewater. General terms used to describe different degrees of treatment, in order of increasing treatment level, are preliminary, primary, secondary, and tertiary and/or advanced wastewater treatment. In some countries, disinfection to remove pathogens sometimes follows the last treatment step. A generalized wastewater treatment diagram is shown in Figure.

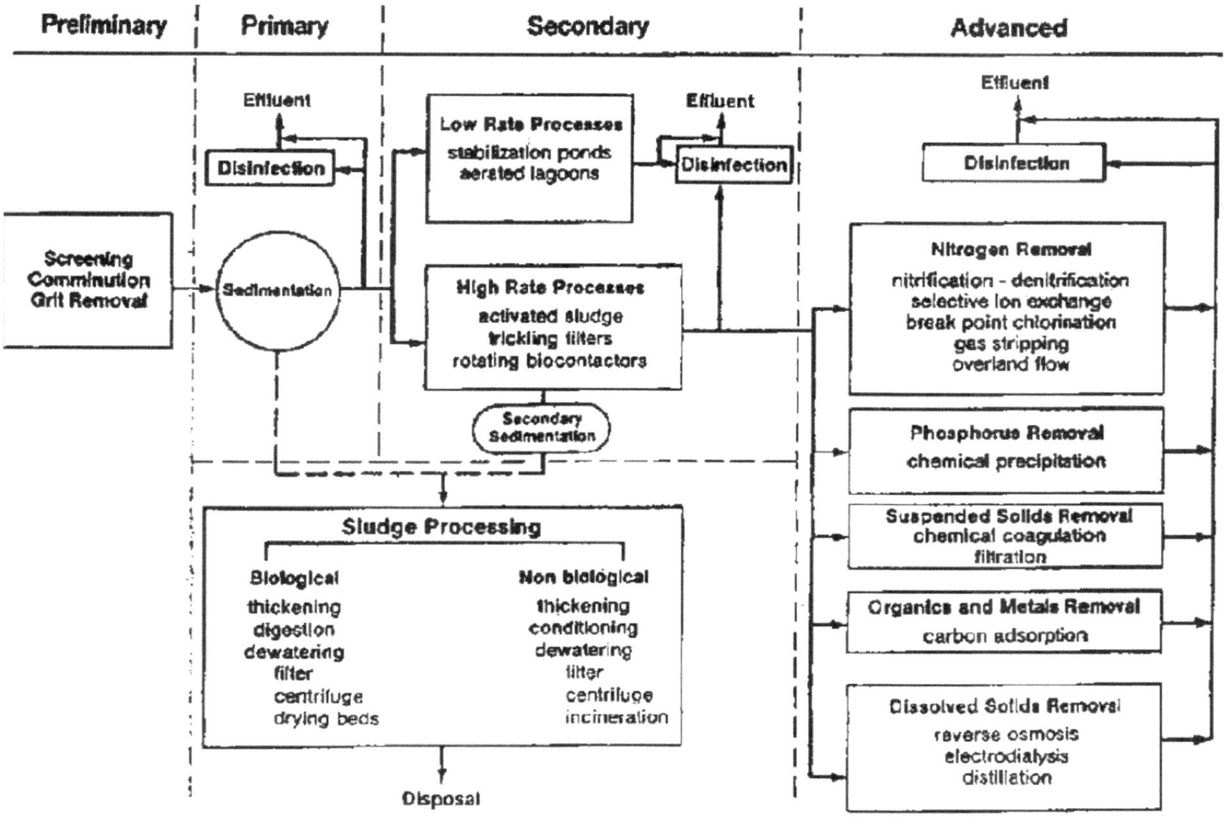

Figure 1: Generalized flow diagram for municipal wastewater treatment.

Preliminary treatment

The objective of preliminary treatment is the removal of coarse solids and other large materials often found in raw wastewater. Removal of these materials is necessary to enhance the operation and maintenance of subsequent treatment units. Preliminary treatment operations typically include coarse screening, grit removal and, in some cases, commination of large objects. In grit chambers, the velocity of the water through the chamber is maintained sufficiently high, or air is used, so as to prevent the settling of most organic solids. Grit removal is not included as a preliminary treatment step in most small wastewater treatment plants. Comminutes are sometimes adopted to supplement coarse screening and serve to reduce the size of large particles so that they will be removed in the form of a sludge in subsequent treatment processes. Flow measurement devices, often standing-wave flumes, are always included at the preliminary treatment stage.

Primary treatment

The objective of primary treatment is the removal of settleable organic and inorganic solids by sedimentation, and the removal of materials that will float (scum) by skimming. Approximately 25 to 50% of the incoming biochemical oxygen demand (BOD5), 50 to 70% of the total suspended solids (SS), and 65% of the oil and grease are removed during primary treatment. Some organic nitrogen, organic phosphorus, and heavy metals associated with solids are also removed during primary sedimentation but colloidal and dissolved constituents are not affected. The effluent from primary sedimentation units is referred to as primary effluent.

Table 2. Quality of Raw Wastewater and Primary Effluent at Selected Treatment Plants

Quality parameters (mg/l, except as otherwise indicated)	**City of Davis**		**San Diego**		**Los Angeles County Joint Plant**	
Raw wastewater	**Primary effluent**	**Raw wastewater**	**Primary effluent**	**Raw wastewater**	**Primary effluent**	
Biochemical oxygen demand, BOD_5	112	73	184	134	-	204
Total organic carbon	63.8	40.6	64.8	52.3	-	-
Suspended solids	185	72	200	109	-	219
Total nitrogen	43.4	34.7	-	-	-	-
NH_3-N	35.6	26.2	21.0	20.0	-	39.5
NO-N	0	0	-	-	-	-
Org-N	7.8	8.5	-	-	-	14.9
Total phosphorus	-	7.5	-	10.2	-	11.2
Ortho-P	-	7.5	11.2		-	
pH (unit)	7.7	-	7.3	7.3	-	-
Cations:						
Ca	-	-	-	-	78.8	-
Mg	-	-	-	-	25.6	-
Na	-	-	-	-	357	359
K	-	-	-	-	19	19
Anions:						
SO_4	-		160		270	
Cl	-		120		397	
Electrical conductivity, dS/m	2.52	2.34			2.19	-
Total dissolved solids	-	-	829	821	1404	1406
Soluble sodium percentage, %	-		-		70.3	
Sodium adsorption ratio	-	-	-	-	8.85	6.8
Boron (B)	-	-	-	-	1.68	1.5
Alkalinity $(CaCO_3)$	-	-	-		322	332
Hardness $(CaCO_3)$	-		-		265	

In many industrialized countries, primary treatment is the minimum level of preapplication treatment required for wastewater irrigation. It may be considered sufficient treatment if the wastewater is used to irrigate crops that are not consumed by humans or to irrigate orchards, vineyards, and some processed food crops. However, to prevent potential nuisance conditions in storage or flow-equalizing reservoirs, some form of secondary treatment is normally required in these countries, even in the case of non-food crop irrigation. It may be possible to use at least a portion of primary effluent for irrigation if off-line storage is provided.

Primary sedimentation tanks or clarifiers may be round or rectangular basins, typically 3 to 5 m deep, with hydraulic retention time between 2 and 3 hours. Settled solids (primary sludge) are normally removed from the bottom of tanks by sludge rakes that scrape the sludge to a central well from which it is pumped to sludge processing units. Scum is swept across the tank surface by water jets or mechanical means from which it is also pumped to sludge processing units.

In large sewage treatment plants (> 7600 m3/d in the US), primary sludge is most commonly processed biologically by anaerobic digestion. In the digestion process, anaerobic and facultative bacteria metabolize the organic material in sludge, thereby reducing the volume requiring ultimate disposal, making the sludge stable (nonputrescible) and improving its dewatering characteristics. Digestion is carried out in covered tanks (anaerobic digesters), typically 7 to 14 m deep. The residence time in a digester may vary from a minimum of about 10 days for high-rate digesters (well-mixed and heated) to 60 days or more in standard-rate digesters. Gas containing about 60 to 65% methane is produced during digestion and can be recovered as an energy source. In small sewage treatment plants, sludge is processed in a variety of ways including: aerobic digestion, storage in sludge lagoons, direct application to sludge drying beds, in-process storage (as in stabilization ponds), and land application.

Secondary treatment

The objective of secondary treatment is the further treatment of the effluent from primary treatment to remove the residual organics and suspended solids. In most cases, secondary treatment follows primary treatment and involves the removal of biodegradable dissolved and colloidal organic matter using aerobic biological treatment processes. Aerobic biological treatment is performed in the presence of oxygen by aerobic microorganisms (principally bacteria) that metabolize the organic matter in the wastewater, thereby producing more microorganisms and inorganic end-products (principally CO_2, NH_3, and H_2O). Several aerobic biological processes are used for secondary treatment differing primarily in the manner in which oxygen is supplied to the microorganisms and in the rate at which organisms metabolize the organic matter.

High-rate biological processes are characterized by relatively small reactor volumes and high concentrations of microorganisms compared with low rate processes. Consequently, the growth rate of new organisms is much greater in high-rate systems because of the well-controlled environment. The microorganisms must be separated from the treated wastewater by sedimentation to produce clarified secondary effluent. The sedimentation tanks used in secondary treatment, often referred to as secondary clarifiers, operate in the same basic manner as the primary clarifiers described previously. The biological solids removed during secondary sedimentation, called secondary or biological sludge, are normally combined with primary sludge for sludge processing.

Common high-rate processes include the activated sludge processes, trickling filters or bio filters, oxidation ditches, and rotating biological contactors (RBC). A combination of two of these processes in series (e.g., bio filter followed by activated sludge) is sometimes used to treat municipal wastewater containing a high concentration of organic material from industrial sources.

Activated Sludge

In the activated sludge process, the dispersed-growth reactor is an aeration tank or basin containing a suspension of the wastewater and microorganisms, the mixed liquor. The contents of the aeration tank are mixed vigorously by aeration devices which also supply oxygen to the biological suspension. Aeration devices commonly used include submerged diffusers that release compressed air and mechanical surface aerators that introduce air by agitating the liquid surface. Hydraulic retention time in the aeration tanks usually ranges from 3 to 8 hours but can be higher with high BOD5 wastewaters. Following the aeration step, the microorganisms are separated from the liquid by sedimentation and the clarified liquid is secondary effluent. A portion of the biological sludge is recycled to the aeration basin to maintain a high mixed-liquor suspended solids (MLSS) level. The remainder is removed from the process and sent to sludge processing to maintain a relatively constant concentration of microorganisms in the system. Several variations of the basic activated sludge process, such as extended aeration and oxidation ditches, are in common use, but the principles are similar.

Trickling Filters

A trickling filter or bio filter consists of a basin or tower filled with support media such as stones, plastic shapes, or wooden slats. Wastewater is applied intermittently, or sometimes continuously, over the media. Microorganisms become attached to the media and form a biological layer or fixed film. Organic matter in the wastewater diffuses into the film, where it is metabolized. Oxygen is normally supplied to the film by the natural flow of air either up or down through the media, depending on the relative temperatures of the wastewater and ambient air. Forced air can also be supplied by blowers but this is rarely necessary. The thickness of the biofilm increases as new organisms grow. Periodically, portions of the film 'slough off the media. The sloughed material is separated from the liquid in a secondary clarifier and discharged to sludge processing. Clarified liquid from the secondary clarifier is the secondary effluent and a portion is often recycled to the biofilter to improve hydraulic distribution of the wastewater over the filter.

Rotating Biological Contactors

Rotating biological contactors (RBCs) are fixed-film reactors similar to biofilters in that organisms are attached to support media. In the case of the RBC, the support media are slowly rotating discs that are partially submerged in flowing wastewater in the reactor. Oxygen is supplied to the attached biofilm from the air when the film is out of the water and from the liquid when submerged, since oxygen is transferred to the wastewater by surface turbulence created by the discs' rotation. Sloughed pieces of biofilm are removed in the same manner described for biofilters.

High-rate biological treatment processes, in combination with primary sedimentation, typically remove 85 % of the BOD5 and SS originally present in the raw wastewater and some of the heavy metals. Activated sludge generally produces an effluent of slightly higher quality, in terms of these constituents, than biofilters or RBCs. When coupled with a disinfection step, these processes can provide substantial but not complete removal of bacteria and virus. However, they remove very little phosphorus, nitrogen, non-biodegradable organics, or dissolved minerals.

Tertiary and/or advanced treatment

Tertiary and/or advanced wastewater treatment is employed when specific wastewater constituents which cannot be removed by secondary treatment must be removed. As shown in Figure 2, individual treatment processes are necessary to remove nitrogen, phosphorus, additional suspended solids, refractory organics, and heavy metals and dissolved solids. Because advanced treatment usually follows high-rate secondary treatment, it is sometimes referred to as tertiary treatment. However, advanced treatment processes are sometimes combined with primary or secondary treatment (e.g., chemical addition to primary clarifiers or aeration basins to remove phosphorus) or used in place of secondary treatment (e.g., overland flow treatment of primary effluent).

An adaptation of the activated sludge process is often used to remove nitrogen and phosphorus and an example of this approach is the 23 Ml/d treatment plant commissioned in 1982 in British Columbia, Canada (World Water). The Bardenpho Process adopted is shown in simplified form in Figure 2. Effluent from primary clarifiers flows to the biological reactor, which is physically divided into five zones by baffles and weirs. In sequence these zones are: (i) anaerobic fermentation zone (characterized by very low dissolved oxygen levels and the absence of nitrates); (ii) anoxic zone (low dissolved oxygen levels but nitrates present); (iii) aerobic zone (aerated); (iv) secondary anoxic zone; and (v) final aeration zone. The function of the first zone is to condition the group of bacteria responsible for phosphorus removal by stressing them under low oxidation-reduction conditions, which results in a release of phosphorus equilibrium in the cells of the bacteria. On subsequent exposure to an adequate supply of oxygen and phosphorus in the aerated zones, these cells rapidly accumulate phosphorus considerably in excess of their normal metabolic requirements. Phosphorus is removed from the system with the waste activated sludge.

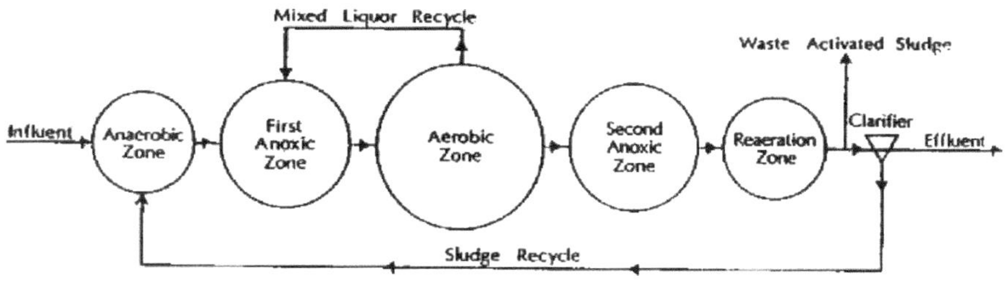

Above: Figure 1 - Bardenpho process train

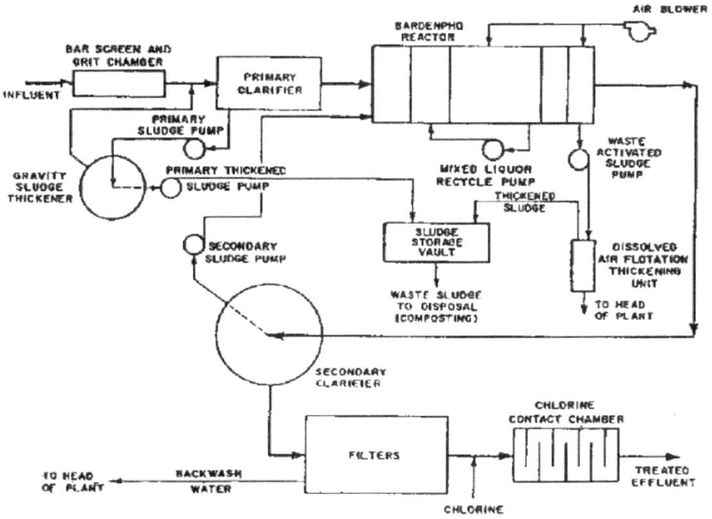

Figure 2: Simplified flow diagram of Bardenpho-plant.

Most of the nitrogen in the influent is in the ammonia form, and this passes through the first two zones virtually unaltered. In the third aerobic zone, the sludge age is such that almost complete nitrification takes place, and the ammonia nitrogen is converted to nitrites and then to nitrates. The nitrate-rich mixed liquor is then recycled from the aerobic zone back to the first anoxic zone. Here denitrification occurs, where the recycled nitrates, in the absence of dissolved oxygen, are reduced by facultative bacteria to nitrogen gas, using the influent organic carbon compounds as hydrogen donors. The nitrogen gas merely escapes to atmosphere. In the second anoxic zone, those nitrates which were not recycled are reduced by the endogenous respiration of bacteria. In the final re-aeration zone, dissolved oxygen levels are again raised to prevent further denitrification, which would impair settling in the secondary clarifiers to which the mixed liquor then flows.

An experimentation programmer on this plant demonstrated the importance of the addition of volatile fatty acids to the anaerobic fermentation zone to achieve good phosphorus removal. These essential short-chain organics (mainly acetates) are produced by the controlled fermentation of primary sludge in a gravity thickener and are released into the thickener supernatant, which can be fed to the head of the biological reactor. Without this supernatant return flow, overall phosphorus removal quickly dropped to levels found in conventional activated sludge plants. Performance data over three years have proved that, with thickener supernatant recycle, effluent quality median values of 0.5-1.38 mg/l Ortho-P, 1.4-1.6 mg/l Total nitrogen and 1.4-2.0 mg/l nitrate-N are achievable. This advanced biological wastewater treatment plant cost only marginally more than a conventional activated sludge plant but nevertheless involved considerable investment. Furthermore, the complexity of the process and the skilled operation required to achieve consistent results make this approach unsuitable for developing countries.

CHAPTER 1

BIOLOGICAL AND CHEMICAL WASTEWATER TREATMENT PROCESSES

Mohamed Samer

Keywords: Wastewater treatment, biological treatment, chemical treatment, bioremediation, phytoremediation, mycoremediation, vermifiltration, treatment plant

INTRODUCTION

The chapter concerns with wastewater treatment engineering, with focus on the biological and chemical treatment processes. It aims at providing a brief and obvious description of the treatment methods, designs, schematics, and specifications. The chapter also answers an important question on how the different processes are interrelated and the correct order of these processes in relation to each other. The main objective of this work was to summarize the work of the eminent scientists in this field in order to provide a clear but concise chapter that can be used as a quick reference for environmental engineers and researchers, and to be effectively implemented in higher education teaching undergraduate and graduate students, as well as extension and outreach.

CHAPTER DESCRIPTION AND CONTENTS OVERVIEW

The chapter describes the biological and chemical wastewater treatment processes that include:

a. Bioremediation of wastewater using oxidation ponds, aeration lagoons, anaerobic lagoons, aerobic and anaerobic bioreactors, activated sludge, percolating or trickling filters, biological filters, rotating biological contactors, and biological removal of nutrients;

b. Mycoremediation of wastewater using bioreactors;

c. Phytoremediation of wastewater that includes: constructed wetlands, rhizofiltration, rhizodegradation, phytodegradation, phytoaccumulation, phytotransformation, and hyperaccumulators;

d. Vermifiltration and vermicomposting;

e. Microbial fuel cells for electricity production from wastewater;

f. Chemical wastewater treatment processes that include: chemical precipitation, ion exchange, neutralization, adsorption and disinfection (chlorination/dechlorination, ozone, ultraviolet radiation);

g. Wastewater treatment plants. The chapter elucidates and illustrates the plant sizing, plant layout, plant design, and plant location.

OVERVIEW

Wastewater Treatment Techniques

Wastewater, or sewage, originates from human and home wastewaters, industrial wastes, animal wastes, rain runoff, and groundwater infiltration. Generally, wastewater is the flow of used water from a neighborhood. The wastewater consists of 99.9% water by weight, where the remaining 0.1% is suspended or dissolved material. This solid material is a mixture of excrements, detergents, food leftovers, grease, oils, salts, plastics, heavy metals, sands, and grits [1, 2]. Types of wastewaters include: municipal wastewater, industrial wastewaters, mixtures of industrial/domestic wastewaters, and agricultural wastewaters. Typical agricultural industries include: dairy processing industries, meat processing factories, juice and beverage industries, slaughterhouses, vegetable processing facilities, rendering plants, and drainage water of irrigation systems.

Citation: Mohamed Samer (2015). Biological and Chemical Wastewater Treatment Processes, Wastewater Treatment Engineering, Associate Prof. Mohamed Samer (Ed.), InTech, DOI: 10.5772/61250.

Subsequent to primary treatment of wastewater, i.e., physical treatment of wastewater, it still contains large amounts of dissolved and colloidal material that must be removed before discharge. The issue is how to transform the dissolved materials or particulate matters that are too little for sedimentation into larger particles to allow the separation processes to eliminate them. This can be accomplished by secondary treatment, i.e., biological treatment. The treatment of wastewater subsequent to the removal of suspended solids by microorganisms such as algae, fungi, or bacteria under aerobic or anaerobic conditions during which organic matter in wastewater is oxidized or incorporated into cells that can be eliminated by removal process or sedimentation is termed biological treatment. Biological treatment is termed secondary treatment. Chemical treatment, or tertiary treatment, using chemical materials will react with a portion of the undesired chemicals and heavy metals, but a portion of the polluting material will remain unaffected. Additionally, the cost of chemical additives and the environmental problem of disposing large amounts of chemical sludge make this treatment process deficient [1]. Alternatively, the biological treatment must be implemented. This treatment process implements naturally occurring microorganisms to transform the dissolved organic matter into a dense biomass that can be separated from the treated wastewater by the sedimentation process. In fact, the microorganisms utilize the dissolved organic matter as food for themselves, where the generated sludge will be far less for chemical treatment. In practice, therefore, secondary treatment tends to be a biological process with chemical treatment implemented for the removal of toxic compounds.

Aims of Wastewater Treatment

The goals of treating the wastewaters are:

a. Transforming the materials available in the wastewater into secure end products that are able to be safely disposed off into domestic water devoid of any negative environmental effects;

b. Protecting public health;

c. Ensuring that wastewaters are efficiently handled on a trustworthy basis without annoyance or offense;

d. Recycling and recovering the valuable components available in wastewaters;

e. Affording feasible treatment processes and disposal techniques;

f. Complying with the legislations, acts and legal standards, and approval conditions of discharge and disposal.

Biological Treatment Processes

The secondary treatment can be defined as "treatment of wastewater by a process involving biological treatment with a secondary sedimentation". In other words, the secondary treatment is a biological process. The settled wastewater is introduced into a specially designed bioreactor where under aerobic or anaerobic conditions the organic matter is utilized by microorganisms such as bacteria (aerobically or anaerobically), algae, and fungi (aerobically). The bioreactor affords appropriate bioenvironmental conditions for the microorganisms to reproduce and use the dissolved organic matter as energy for themselves. Provided that oxygen and food, in the form of settled wastewater, are supplied to the microorganisms, the biological oxidation process of dissolved organic matter will be maintained. The biological process is mostly carried out bacteria that form the basic trophic level (the level of an organism is the position it occupies in a food chain) of the food chain inside the bioreactor. The bioconversion of dissolved organic matter into thick bacterial biomass can fundamentally purify the wastewater. Subsequently, it is crucial to separate the microbial biomass from the treated wastewater though sedimentation. This secondary sedimentation is basically similar to primary sedimentation except that the sludge contains bacterial cells rather than fecal solids. The biological removal of organic matter from settled wastewater is conducted by microorganisms, mainly heterotrophic bacteria but also occasionally fungi. The microorganisms are able to decompose the organic matter through two different biological processes: biological oxidation and biosynthesis [1]. The biological oxidation forms some end-products, such as minerals, that remain in the solution and are discharged with the effluent (Eq. 1). The biosynthesis transforms the colloidal and dissolved organic matter into new cells that form in turn the dense biomass that can be then removed by sedimentation (Eq. 2). Figure 1 summarizes these processes. On the other hand, algal photosynthesis plays an important role in some cases (Figure 2).

Oxidation:

$$\underset{\text{(Organic matter)}}{COHNS} + O_2 + \text{Bacteria} \rightarrow CO_2 + \underset{\text{+Energy}}{NH_3} + \text{Other end products} \quad (1)$$

Biosynthesis:

$$\underset{\text{(Organic matter)}}{COHNS} + O_2 + \text{Bacteria} \rightarrow \underset{\text{(New cells)}}{C_5H_7NO_2} \quad (2)$$

Useful Terms

The following terms are the most used in biological treatment processes [2]:

a. DO: Dissolved Oxygen (mg L^{-1})

b. BOD: Biochemical Oxygen Demand (mg L^{-1})

c. BOD_5: BOD (mg L^{-1}), incubation at 15°C for 5 days

d. COD: Chemical Oxygen Demand (mg L^{-1})

e. CBOD: Carbonaceous BOD (mg L^{-1})

f. NBOD: Nitrogenous (mg L^{-1})

g. SOD: Sediment Oxygen Demand (mg L^{-1})

h. TBOD: Total BOD (mg L^{-1})

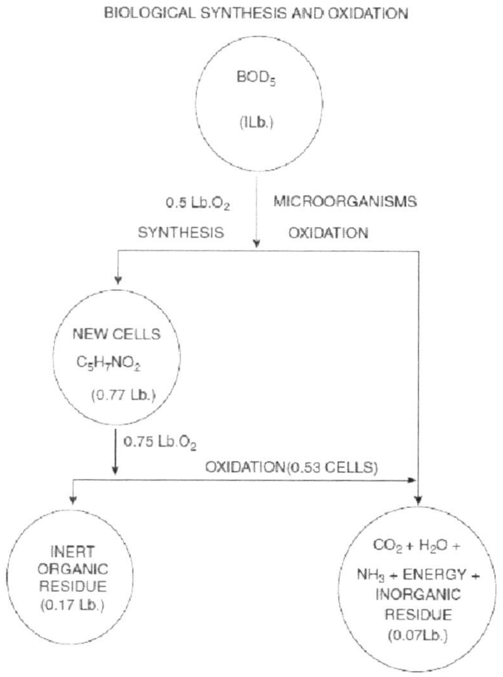

Figure 1. Biological synthesis and oxidation [3].

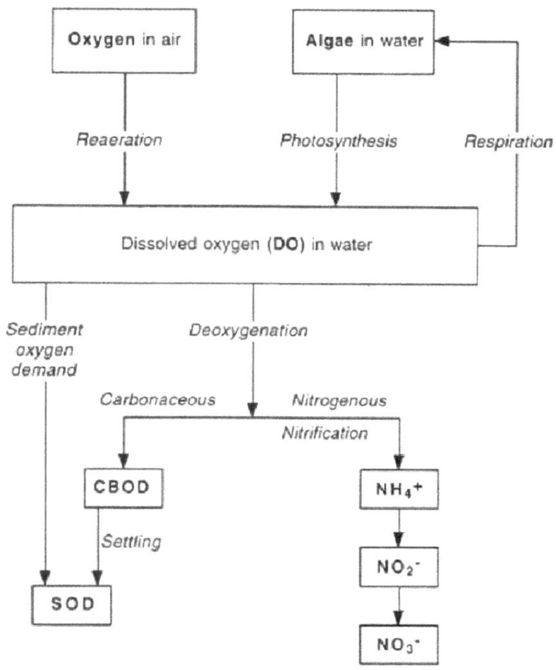

Figure 2. Photosynthesis and oxidation [2].

Chemical Treatment Processes

In early wastewater treatment technologies, chemical treatment has preceded biological treatment. Recently, the biological treatment precedes chemical treatment in the treatment process. Chemical treatment is now considered as a tertiary treatment that can be more broadly defined as "treatment of wastewater by a process involving chemical treatment". The mostly implemented chemical treatment processes are: chemical precipitation, neutralization, adsorption, disinfection (chlorine, ozone, ultraviolet light), and ion exchange.

BIOLOGICAL TREATMENT OF WASTEWATER

Biological Growth Equation

The biological growth can be described according to the Monod equation:

$$\mu = (\lambda S) / (K_S + S)$$

Where, μ is the specific growth rate coefficient; λ is the maximum growth rate coefficient that occurs at 0.5 μ_{max}; S is the concentration of limiting nutrient, that is BOD and COD; and KS is the Monod coefficient [3].

Generally, the bacterial growth can be explained by the following simplified figure:

$$\text{Organics+Bacteria+Nutrients+Oxygen}$$
$$\rightarrow \text{New Bacteria+}CO_2\text{+}H_2O\text{+Residual Organics+Inorganics}$$

Several bioenvironmental factors affect the activity of bacteria and the rate of biochemical reactions. The most important factors are: temperature, pH, dissolved oxygen, nutrient concentration, and toxic materials. All these factors can be controlled within a biological treatment system and/or a bioreactor in order to ensure that the microbial growth is maintained under optimum bioenvironmental conditions. The majority of biological treatment systems operate in the mesophilic temperature range, where the optimal temperature ranges from 20°C to 40°C. Aeration tanks and percolating filters operate at the temperature of the wastewater that ranges from 12°C to 25°C; although in percolating filters, the air temperature and the ventilation rate may have a significant effect on heat loss. The higher temperatures increase the biological activity and metabolism, which result in increasing the substrate removal rate. However, the increased metabolism at the higher temperatures may lead to problems of oxygen limitations.

Bacterial Kinetics

The bacterial kinetics can be shown in Figures 3 and 4. The microbial growth curve that shows bacterial density and specific growth rate at the different growth phases is shown in Figure 3. The microbial growth curves that compare the total biomass and the variable biomass are shown in Figure 4.

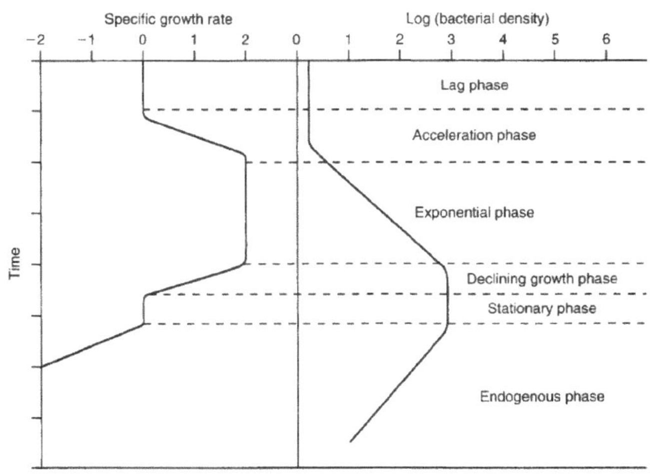

Figure 3. Microbial growth curve [1].

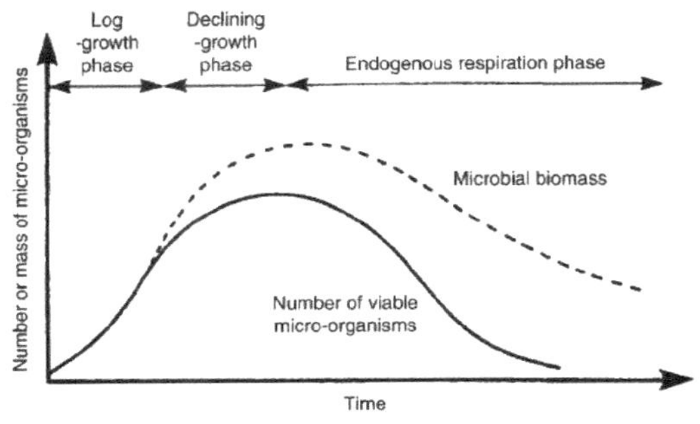

Figure 4. Microbial growth curves [1].

Principles of Biological Treatment

The principles of biological treatment of wastewater were stated by [3]. The following is a summary of the principles:

- The biological systems are very sensitive for extreme variations in hydraulic loads. Diurnal variations of greater than 250% are problematic because they will create biomass loss in the clarifiers.
- The growth rate of microorganisms is highly dependent on temperature. A 10°C reduction in wastewater temperature dramatically decreases the biological reaction rates to half.
- BOD is efficiently treated in the range of 60 to 500 mg L^{-1}. Wastewaters in excess of 500 mg L^{-1}BODs have been treated successfully if sufficient dilution is applied in the treatment process, or if an anaerobic process was implemented as a pretreatment process.
- The biological treatment is effective in removing up to 95% of the BOD. Large tanks are required in order to eliminate the entire BOD, which is not feasible.
- The biological treatment systems are unable to handle "shock loads" efficiently. Equalization is necessary if the variation in strength of the wastewater is more than 150% or if that wastewater at its peak concentration is in excess of 1,000 mg L^{-1} BOD.
- The carbon:nitrogen:phosphorus (C:N:P) ratio of wastewater is usually ideal. The C:N:P ratio of industrial wastewaters should range from 100:20:1 to 100:5:1 for a most advantageous biological process.
- If the C:N:P ratio of the wastewater is strong in an element in comparison to the other elements, then poor treatment will result. This is especially true if the wastewater is very strong in carbon. The wastewater should also be neither very weak nor very strong in an element; although very weak is acceptable, it is difficult to treat.
- Oils and solids cannot be handled in a biological treatment system because they negatively affect the treatment process. These wastes should be pretreated to remove solids and oils.
- Toxic and biological-resistant materials require special consideration and may require pretreatment before being introduced into a biological treatment system.
- Although the capacity of the wastewater to utilize oxygen is unlimited, the capacity of any aeration system is limited in terms of oxygen transfer.

Bioremediation of Wastewater

Bioremediation is a treatment process that involves the implementation of microorganisms to remove pollutants from a contaminated setting. Bioremediation can be defined as "treatment that implements natural organisms to decompose hazardous materials into less toxic or nontoxic materials". Some examples of bioremediation-related technologies are phytoremediation, bioaugmentation, rhizofiltration, and biostimulation. The microorganisms implemented to carry out the bioremediation are called bioremediators. However, some pollutants are not easily removed or decomposed by bioremediation. For example, heavy metals such as lead and cadmium are not eagerly captured by bioremediators. Example of bioremediation: fish bone char has been shown to bioremediate small amounts of cadmium, copper, and zinc.

The bioremediation of wastewater can be achieved by autotrophs or heterotrophs. A heterotroph is an organism that is unable to fix carbon and utilizes organic carbon for its growth. Heterotrophs are divided based on their source of energy. If the heterotroph utilizes light as its source of energy, then it is considered a photoheterotroph. If the heterotroph utilizes organic and/or inorganic compounds as energy sources, it is then considered a chemoheterotroph. Autotrophs, such as plants and algae, that are able to utilize energy from sunlight are called photoautotrophs. Autotrophs that utilize inorganic compounds to produce organic compounds such as carbohydrates, fats, and proteins from inorganic carbon dioxide are called lithoautotrophs. These reduced carbon compounds can be utilized as energy sources by autotrophs and provide the energy in food consumed by heterotrophs. Over 95% of all organisms are heterotrophic.

Aerobic Treatment

Aeration has been used to remove trace organic volatile compounds (VOCs) in water. It has also been employed to transfer a substance, such as oxygen, from air or a gas phase into water in a process called "gas adsorption" or "oxidation", i.e., to oxidize iron and/or manganese. Aeration also provides the escape of dissolved gases, such as CO_2 and H_2S. Air stripping has been also utilized effectively to remove NH_3 from wastewater and to remove volatile tastes and other such substances in water [2]. Samer [4] and Samer et al. [5] mentioned that aerobic treatment with biowastes is effective in reducing harmful gaseous emissions as greenhouse gases (CH_4 and N_2O) and ammonia.

Oxidation Ponds

Oxidation ponds (Figure 5) are aerobic systems where the oxygen required by the heterotrophic bacteria (a heterotroph is an organism that cannot fix carbon and uses organic carbon for growth) is provided not only by transfer from the atmosphere but also by photosynthetic algae. The algae are restricted to the euphotic zone (sunlight zone), which is often only a few centimeters deep. Ponds are constructed to a depth of between 1.2 and 1.8 m to ensure maximum penetration of sunlight, and appear dark green in color due to dense algal development. Samer [6] and Samer et al. [7] illustrated the structures and constructions of the aerobic treatment tanks and the used building materials.

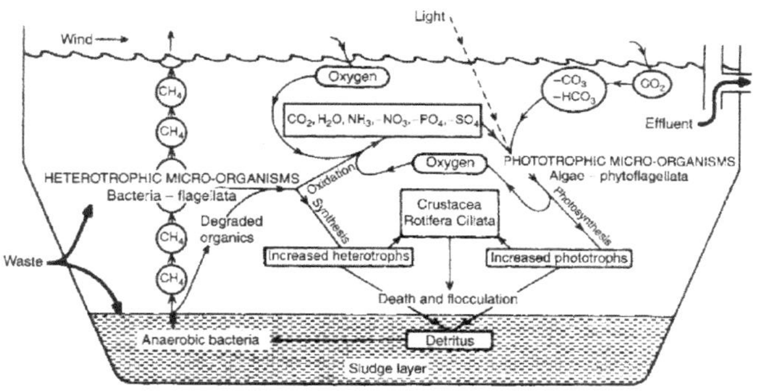

Figure 5. Aerobic system/oxidation pond [1].

In oxidation ponds, the algae use the inorganic compounds (N, P, CO_2) released by aerobic bacteria for growth using sunlight for energy. They release oxygen into the solution that in turn is utilized by the bacteria, completing the symbiotic cycle. There are two distinct zones in facultative ponds: the upper aerobic zone where bacterial (facultative) activity occurs and a lower anaerobic zone where solids settle out of suspension to form a sludge that is degraded anaerobically.

Aeration Lagoons

Aeration lagoons are profound (3–4 m) compared to oxidation ponds, where oxygen is provided by aerators but not by the photosynthetic activity of algae as in the oxidation ponds. The aerators keep the microbial biomass suspended and provide sufficient dissolved oxygen that allows maximal aerobic activity. On the other hand, bubble aeration is commonly used where the bubbles are generated by compressed air pumped through plastic tubing laid through the base of the lagoon. A predominately bacterial biomass develops and, whereas there is neither sedimentation nor sludge return, this procedure counts on adequate mixed liquor formed in the tank/lagoon. Therefore, the aeration lagoons are suitable for strong but degradable wastewater such as wastewaters of food industries. The hydraulic retention time (HRT) ranges from 3 to 8 days based on treatment level, strength, and temperature of the influent. Generally, HRT of about 5 days at 20°C achieves 85% removal of BOD in household wastewater. However, if the temperature falls by 10°C, then the BOD removal will decrease to 65% [1].

Anaerobic Treatment

The anaerobic treatments are implemented to treat wastewaters rich in biodegradable organic matter (BOD >500 mg L^{-1}) and for further treatment of sedimentation sludges. Strong organic wastewaters containing large amounts of biodegradable materials are discharged mainly by agricultural and food processing industries. These wastewaters are difficult to be treated aerobically due to the troubles and expenses of fulfillment of the elevated oxygen demand to preserve the aerobic conditions [1]. In contrast, anaerobic degradation occurs in the absence of oxygen. Although the anaerobic treatment is time-consuming, it has a multitude of advantages in treating strong organic wastewaters. These advantages include elevated levels of purification, aptitude to handle high organic loads, generating small amounts of sludges that are usually very stable, and production of methane (inert combustible gas) as end-product. Anaerobic digestion is a complex multistep process in terms of chemistry and microbiology. Organic materials are degraded into basic constituents, finally to methane gas under the absence of an electron acceptor such as oxygen [8]. The basic metabolic pathway of anaerobic digestion is shown in Figures 6and 7. To achieve this pathway, the presence of very different and closely dependent microbial population is required.

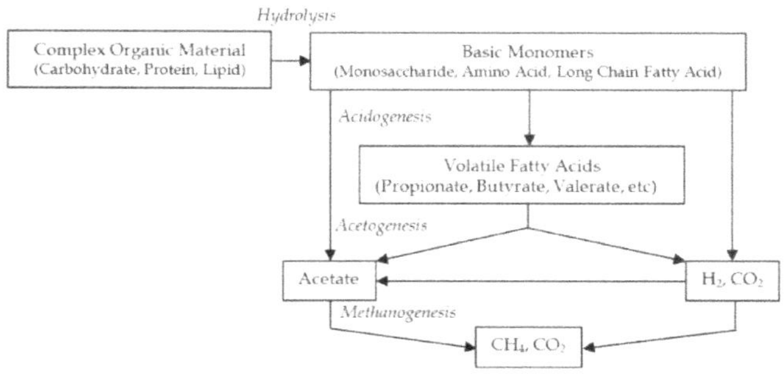

Figure 6. Steps of the anaerobic digestion process [8].

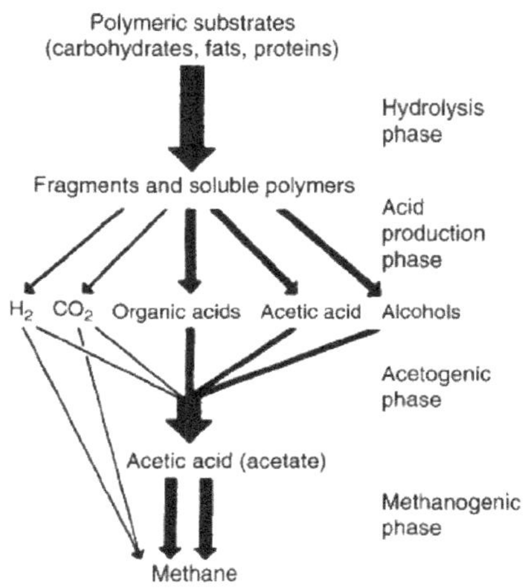

Figure 7. Major steps in anaerobic decomposition [1].

Suitable wastewaters include livestock manure, food processing effluents, petroleum wastes (if the toxicity is controlled), and canning and dyestuff wastes where soluble organic matters are implemented in the treatment. Most anaerobic processes (solids fermentation) occur in two predetermined temperature ranges: mesophilic or thermophilic. The temperature ranges are 30–38°C and 38–50°C, respectively

[3]. In contrast to aerobic systems, absolute stabilization of organic matter is not achievable under anaerobic conditions. Therefore, subsequent aerobic treatment of the anaerobic effluents is usually essential. The final waste matter discharged by the anaerobic treatment includes solubilized organic matter that is acquiescent to aerobic treatment demonstrating the possibility of installing collective anaerobic and aerobic units in series [1].

Anaerobic Digesters

Samer [9] elucidated and illustrated the structures and constructions of the anaerobic digesters and the used building materials. Samer [10] developed an expert system for planning and designing biogas plants. Figures 8 to 13 show different types of anaerobic digesters. While Figures 14 and 15 show some industrial applications. Table 1 shows the advantages and disadvantages of anaerobic treatment compared to aerobic treatment.

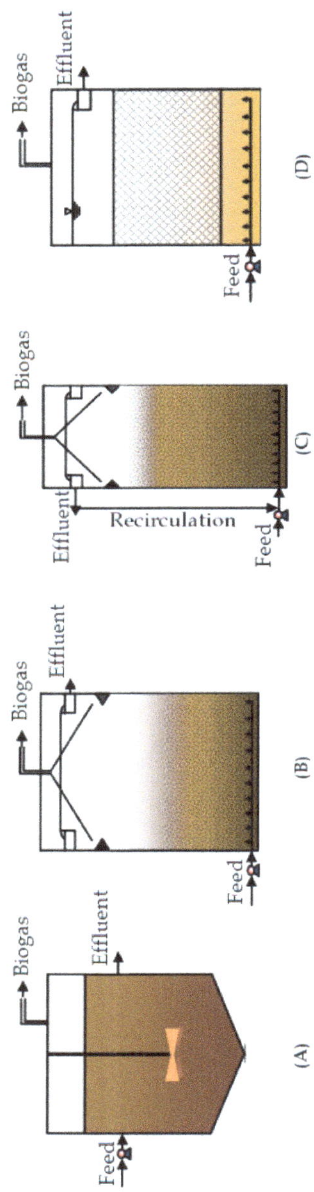

Figure 8. Most commonly used anaerobic reactor types: (A) Completely mixed anaerobic digester, (B) UASB reactor, (C) AFB or EGSB reactor, and (D) Upflow AF [8].

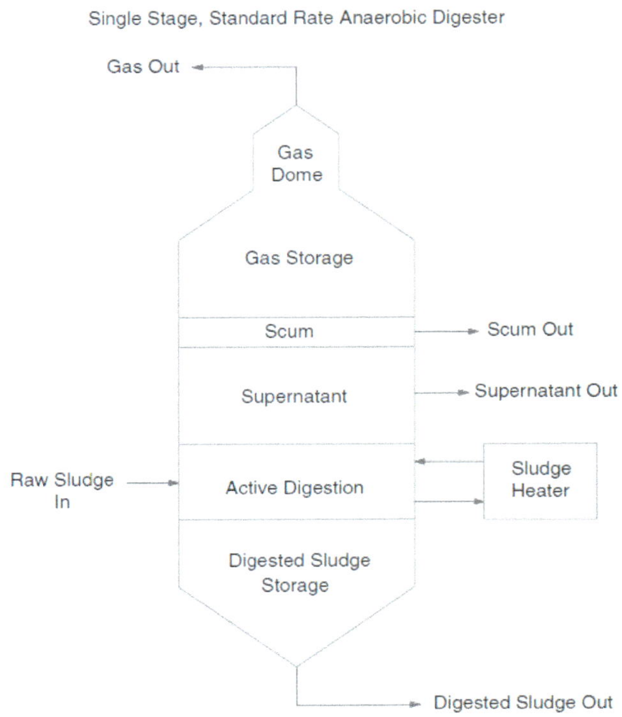

Figure 9. Single-stage conventional anaerobic digester [3].

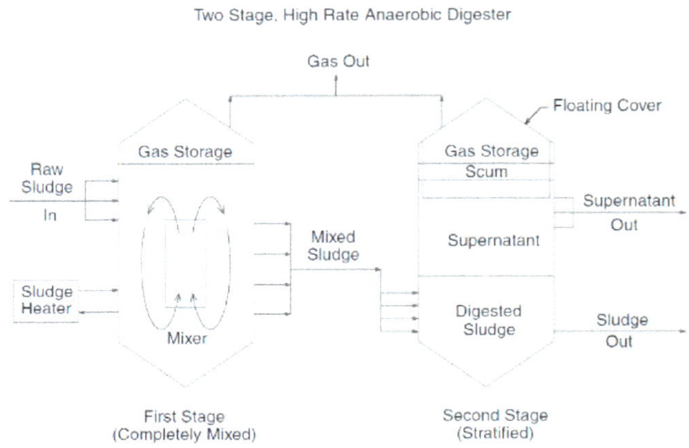

Figure 10. Dual-stage high rate digester [3].

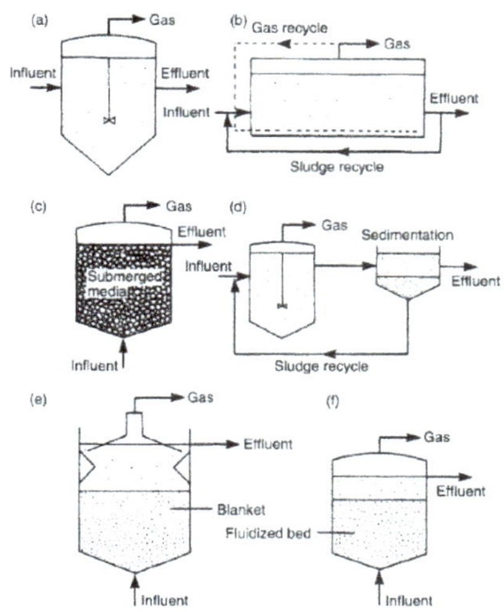

Figure 11. Schematic representation of digester types. Flow-through (A–B) and contact systems (C–F) [1].

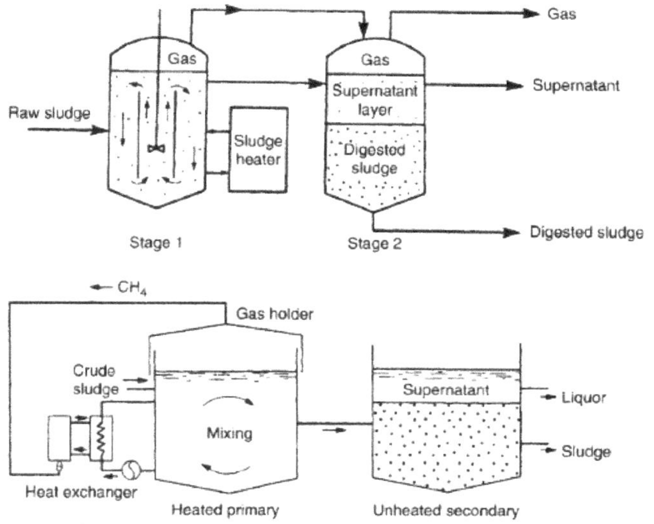

Figure 12. The upper scheme shows a two-stage anaerobic sludge digester, while the lower scheme shows the conventional sludge digestion plant [1].

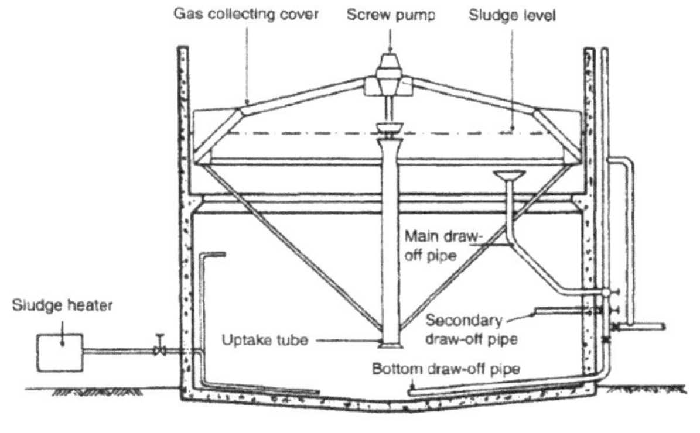

Figure 13. Primary digestion tank with screw mixing pump and external heater [1].

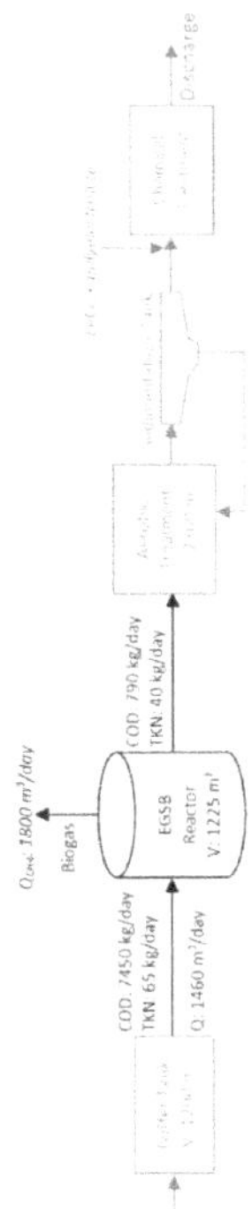

Figure 14. Wastewater treatment plant for corn processing industry [8].

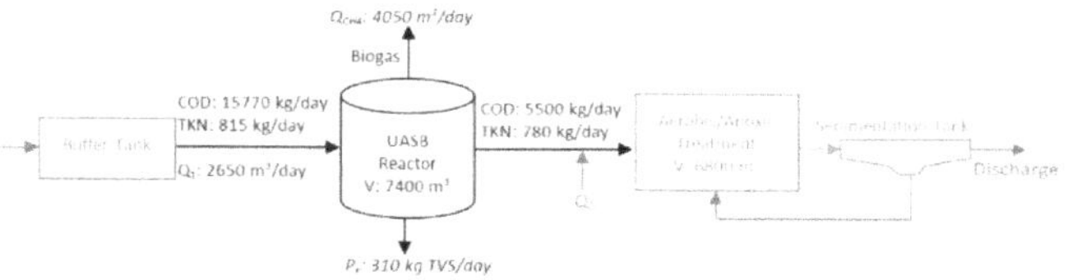

Figure 15. Mass balance study for a wastewater treatment plant of the baker's yeast industry [8].

By definition, the anaerobic treatment is conducted without oxygen. It is different from an anoxic process, which is a reduced environment in contrast to an environment without oxygen. Both processes are anoxic, but anaerobic is an environment beyond anoxic where the oxidation reduction potential (ORP) values are highly negative. In the anaerobic process, nitrate is reduced to ammonia and nitrogen gas, and sulfate (SO_3^{2-}) is reduced to hydrogen sulfide (H_2S). Phosphate is also reduced because it is often transformed through the ADP–ATP chain [3].

Table 1. The advantages and disadvantages of anaerobic treatment compared to aerobic treatment [1]

The advantages and disadvantages of anaerobic treatment compared to aerobic treatment

Advantages	*Disadvantages*
Low operational costs	High capital costs Generally require heating
Low sludge production	Low retention times required (>24 h)
Reactors sealed giving no odour or aerosols	Corrosive and malodorous compounds produced during anaerobiosis
Sludge is highly stabilized	Not as effective as aerobic stabilization for pathogen destruction
Methane gas produced as end product	Hydrogen sulphide also produced
Low nutrient requirement due to lower growth rate of anaerobes	Reactor may require additional alkalinity
Can be operated seasonally	Slow growth rate of anaerobes can result in long initial start-up of reactors and recovery periods
Rapid start-up possible after acclimation	Only used as pre-treatment for liquid wastes

Anaerobic Lagoons

An anaerobic lagoon is a deep lagoon, fundamentally without dissolved oxygen, that enforces anaerobic conditions. The anaerobic process occurs in deep ground ponds, and such basins are implemented for anaerobic pretreatment. The anaerobic lagoons are not aerated, heated, or mixed. The depth of an anaerobic lagoon should be typically deeper than 2.5 m, where deeper lagoons are more efficient. Such depths diminish the amount of oxygen diffused from the surface, allowing anaerobic conditions to prevail (U.S. EPA, 2002). Figures 16 to 18 show different types of anaerobic lagoons.

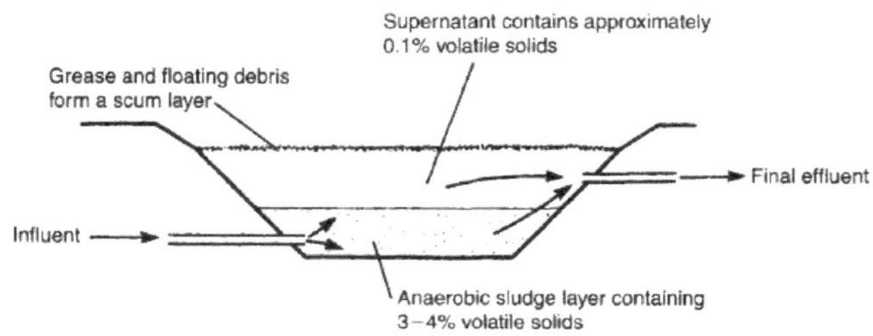

Figure 16. Anaerobic lagoon for strong wastewater treatment, such as meat processing wastewater [1].

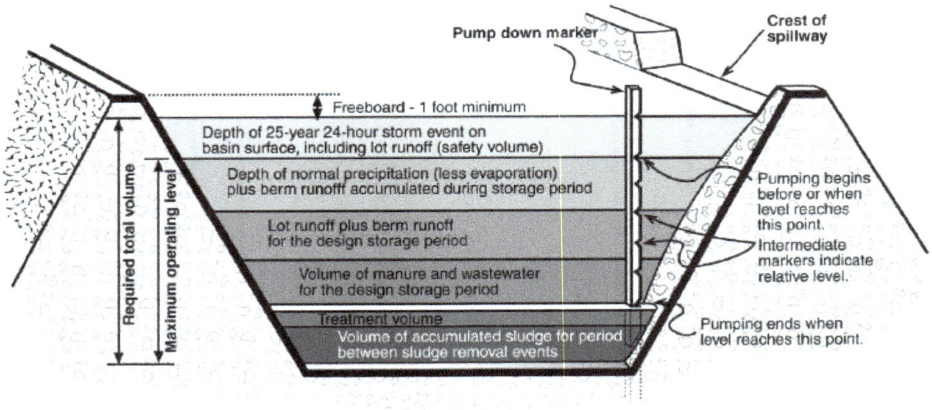

Figure 17. Schematic of volume fractions in anaerobic lagoon design [11].

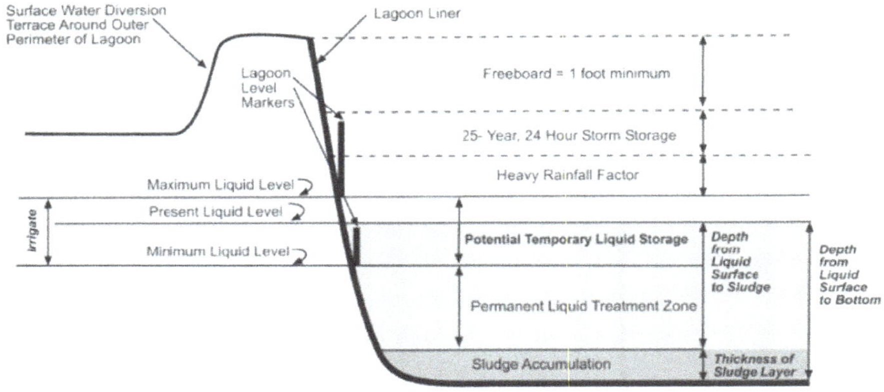

Figure 18. Anaerobic wastewater treatment lagoon [12].

Bioreactors

A bioreactor can be defined as "engineered or manufactured apparatus or system that controls the embraced or encompassed bioenvironment". Precisely, the bioreactor is a vessel in which a biochemical process is conducted, where it involves microorganisms (e.g., bacteria, algae, fungi) or biochemical substances (e.g., enzymes) derived from such microorganisms. The treatment can be conducted under either aerobic or anaerobic conditions. The bioreactors are commonly made of stainless steel, usually cylindrical in shape and range in size from liters to cubic meters. The bioreactors are classified as batch, plug, or continuous flow reactors (e.g., continuous stirred-tank bioreactor).

Mycoremediation is a type of bioremediation where fungi are implemented to break down the contaminants. The term "mycoremediation" refers particularly to the implementation of fungal "mycelia" in bioremediation. The principal role of fungi in the ecological system is the breakdown of pollutants, which is performed by the mycelium. The mycelium, the vegetative part of a fungus, secretes enzymes and acids that biodegrade lignin and cellulose that are the main components of vegetative fibers. Lignin and cellulose are organic compounds composed of long chains of carbon and hydrogen, and therefore they are structurally similar to several organic pollutants. One key issue is specifying the right fungus to break down a determined pollutant. Similarly, mycofiltration is a process that uses fungal mycelia to filter toxic compounds from wastewater.

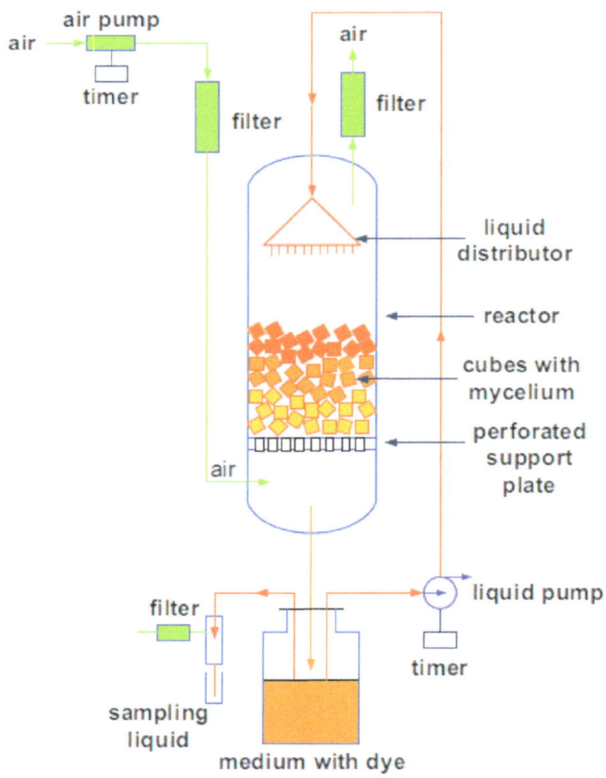

Figure 19.A bioreactor for fungal degradation: trickle bed bio-reactor [13].

In an experiment, wastewater contaminated with diesel oil was inoculated with mycelia of oyster mushrooms. One month later, more than 93% of many of the polycyclic aromatic hydrocarbons (PAH) had been reduced to non-toxic components in the mycelial-inoculated samples. The natural microbial community participates with the fungi to break down contaminants, eventually into CO_2 and H_2O. Wood-degrading fungi are particularly effective in breaking down aromatic pollutants (toxic components of petroleum), as well as chlorinated compounds (certain persistent pesticides). Figures 19 to 22 show different types and designs of bioreactors.

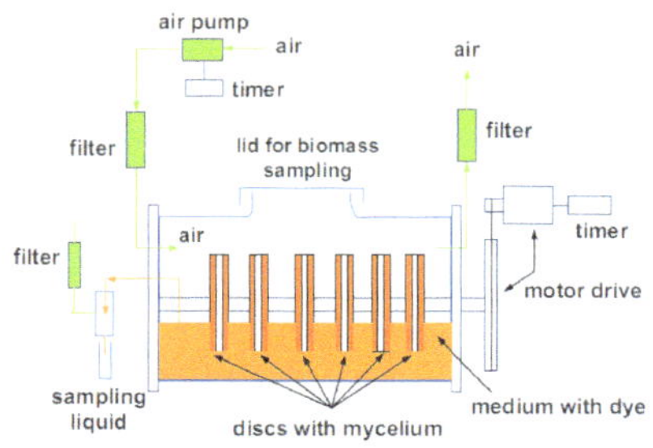

Figure 20.A bioreactor for fungal degradation: rotating disc bioreactor [13].

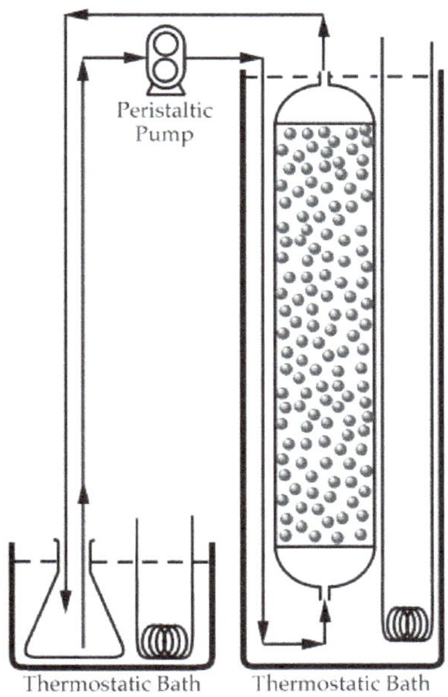

Figure 21. Fluidized bed bioreactor [14].

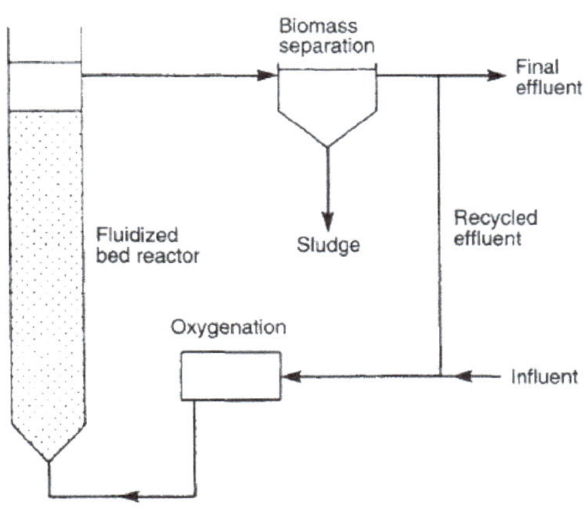

Figure 22. Typical design of fluidized bed reactor system [1].

Activated Sludge

The activated sludge process is based on a mixture of thick bacterial population suspended in the wastewater under aerobic conditions. With unlimited nutrients and oxygen, high rates of bacterial growth and respiration can be attained, which results in the consumption of the available organic matter to either oxidized end-products (e.g., CO_2, NO_3^-, SO_4^{2-}, and PO_4^{3-}) or biosynthesis of new microorganisms. The activated sludge process is based on five interdependent elements, which are: bioreactor, activated sludge, aeration and mixing system, sedimentation tank, and returned sludge [1]. The biological process using activated sludge is a commonly used method for the treatment of wastewater, where the running costs are inexpensive (Figure 23). However, a huge quantity of surplus sludge is produced in wastewater treatment plants (WWTPs) which is an enormous burden in both economical and environmental aspects. The excess sludge contains a lot of moisture and is not easy to treat. The byproducts of WWTPs are dewatered, dried, and finally burnt into ashes. Some are used in farm lands as compost fertilizer [15]. However, it is suggested that the dried byproducts of WWTPs are fed into the pyrolysis process rather than the burning process.

The sludge volume index (SVI) is an estimation that specifies the tendency of aerated solids, i.e., activated sludge solids, to become dense or concentrated through the thickening process. SVI can be computed as follows: (a) allowing a mixed liquor sample from the aeration tank to sediment in 30 min; (b) determining the concentration of the suspended solids for a sample of the same mixed liquor; (c) SVI is then computed as ratio of the measured wet volume (mL/L) of the settled sludge to the dry weight concentration of MLSS in g/L (Source: Office of Water Programs, Sacramento State, USA).

During the treatment of wastewater in aeration tanks through the activated sludge process (Table 2) there are suspended solids, where the concentration of the suspended solids is termed as mixed liquor suspended solids (MLSS), which is measured in milligrams per liter (mg L^{-1}). Mixed liquor is a mixture of raw wastewater and activated sludge in an aeration tank. MLSS consists mainly of microorganisms and non-biodegradable suspended solids. MLSS is the effective and active portion of the activated sludge process that ensures that there is adequate quantity of viable biomass available to degrade the supplied quantity of organic pollutants at any time. This is termed as Food to Microorganism Ratio (F/M Ratio) or food to mass ratio. If this ratio is kept at the suitable level, then the biomass will be able to consume high quantities of the food, which reduces the loss of residual food in the discharge. In other words, the more the biomass consumes food the lower the BOD will be in the treated effluent. It is important that MLSS eliminates BOD in order to purify the wastewater for further usage and hygiene. Raw sewage is introduced into the wastewater treatment process with a concentration of several hundred mg L^{-1} of BOD. The concentration of BOD in wastewater is reduced to less than 2 mg L^{-1} after being treated with MLSS and other treatment methods, which is considered to be safe water to use.

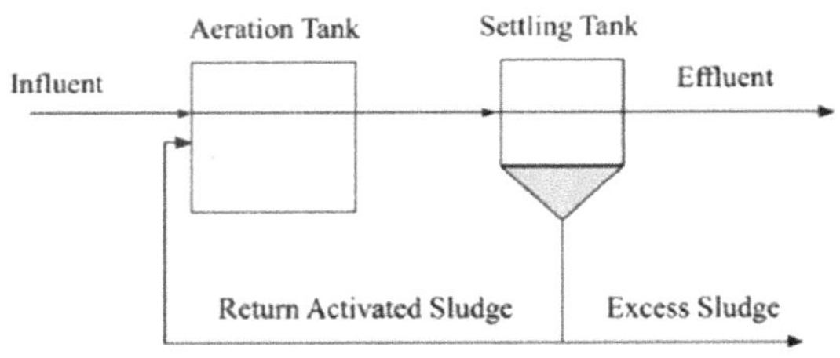

Figure 23. Activated sludge [15].

Table 2. Conventional activated sludge [15].

Specification	Value	Unit
BOD-Sludge Loading	0.40	mg $^{L-}$1
BOD-Volume Loading	0.20	mg $^{L-}$1
MLSS	2000	mg $^{L-}$1
COD of Influent	300	mg $^{L-}$1
Amount of Influent	4.48	L $^{d-}$1
Aeration Rate	3.00	L mi^{n-}1

The biological treatment process is the most commonly implemented method for the treatment of domestic sewage. This method implements bacterial populations that possess superior sedimentation characteristics. The living microorganisms break down the organic matter in the wastewater and consequently purify the wastewater from biological waste [15].

According to [1], the main components of all activated sludge systems are:

- The bioreactor: it can be a lagoon, tank, or ditch. The main characteristic of a bioreactor is that it contains sufficiently aerated and mixed contents. The bioreactor is also known as the aeration tank.
- Activated sludge: it is the bacterial biomass inside the bioreactor that consists mostly of bacteria and other flora and microfauna. The sludge is a flocculent suspension of these microorganisms and is usually termed as the mixed liquor suspended solids (MLSS) that ranges between 2,000 and 5,000 mg L^{-1}.
- Aeration and mixing system: the aeration and mixing of the activated sludge and the raw influent are necessary. While these processes can be accomplished separately, they are usually conducted using a single system of either surface aeration or diffused air.
- Sedimentation tank: clarification or settlement of the activated sludge discharged from the aeration tank is essential. This separates the bacterial biomass from the treated wastewater.
- Returned sludge: the settled activated sludge in the sedimentation tank is returned to the bioreactor to maintain the microbial population at a required concentration to guarantee persistence of treatment process.

Several parameters should be considered while operating activated sludge plants. The most important parameters are: (1) biomass control, (2) plant loading, (3) sludge settleability, and (4) sludge activity. The main operational variable is the aeration, where its major functions are: (1) ensuring a sufficient and continuous supply of dissolved oxygen (DO) for the bacterial population, (2) keeping the bacteria and the biomass suspended, and (3) mixing the influent wastewater with the biomass and removing from the solution the excessive CO_2 resulting from oxidation of organic matter [1].

There are several types of activated sludge processes, e.g., conventional activated sludge plant (Figure 24), complete mix plant (Figure 25), contact stabilization plant (Figure 26), and step aeration plant (Figure 27). Figure 28 shows the food pyramid that represents the feeding relationships within the activated sludge process.

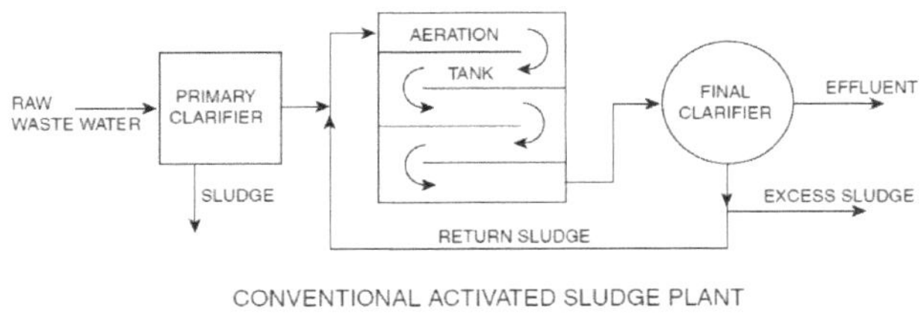

Figure 24. Conventional activated sludge plant [3].

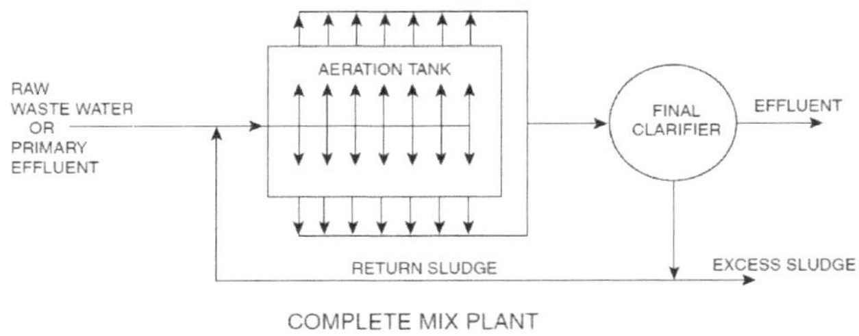

Figure 25. Complete mix plant [3].

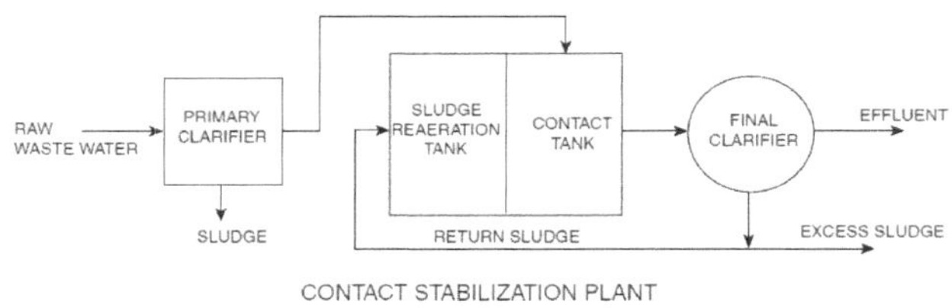

Figure 26. Contact stabilization plant [3].

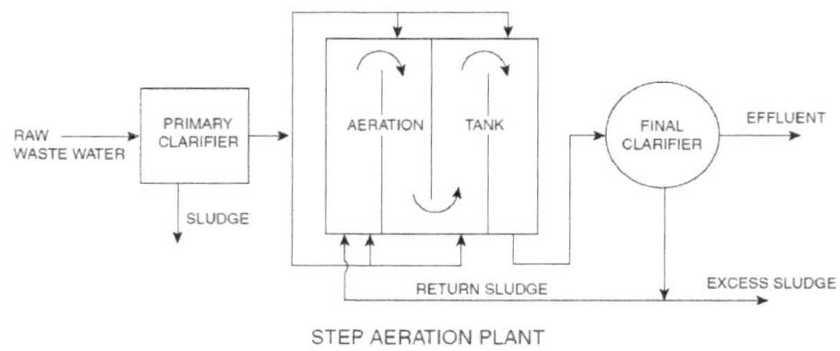

Figure 27. Step aeration plant [3].

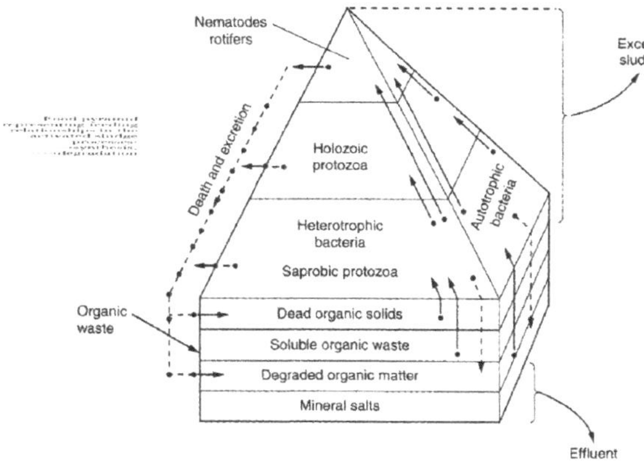

Figure 28. Food pyramid illustrating the feeding relationships within the activated sludge process [1].

Biological Filters

The main systems of operation of biological filters are: (a) single filtration, (b) recirculation, (c) ADF, and (d) two-stage filtration with high-rate primary biotower (Figure 29). There are several types of biological filters, for example, submerged aerated filters that are widely known as biological aerated filters (BAFs) and are the commonly implemented design (Figure 30), and the percolating (trickling) filters (Figure 31). The BAFs implement either the sunken granular media with upward (Figure 30a) or download (Figure 30b) flows, or floating granular media with upward flow (Figure 30c), which is the most common design of BAFs. In order to compare the biological filters and the activated sludge systems (Figures 31 and 32), the comparison is based on the oxidation that can be accomplished by three processes:

- Spreading the wastewater into a thin film of liquid with a large surface area, consequently the required oxygen can be supplied by gaseous diffusion, which is the case of the percolating filters.
- Aerating the wastewater by pumping air in the form of bubbles or stirring forcefully, which is the case of the activated sludge process.
- Implementing algae to produce oxygen by photosynthesis, which is the case of the stabilization ponds.

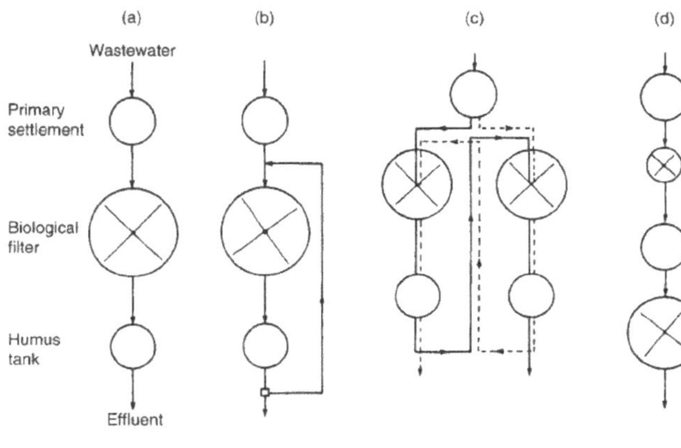

Figure 29. The main systems of operation of biological filters [1].

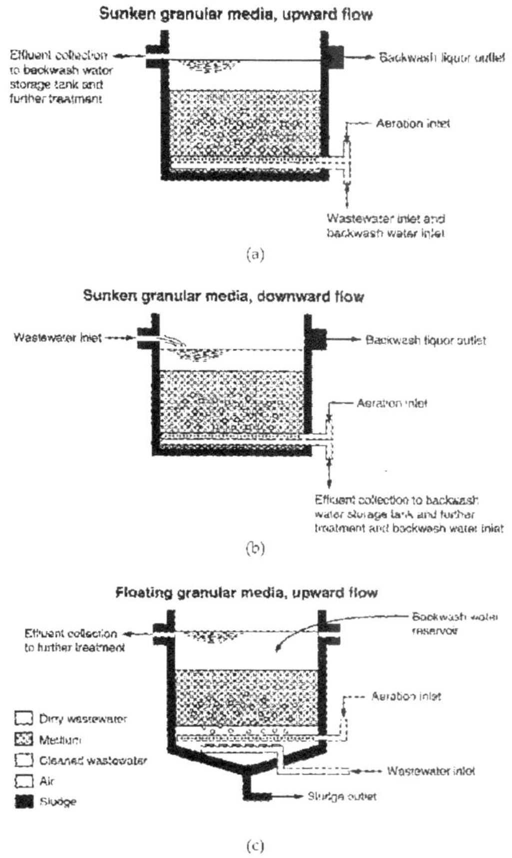

Figure 30. Biological aerated filters [1].

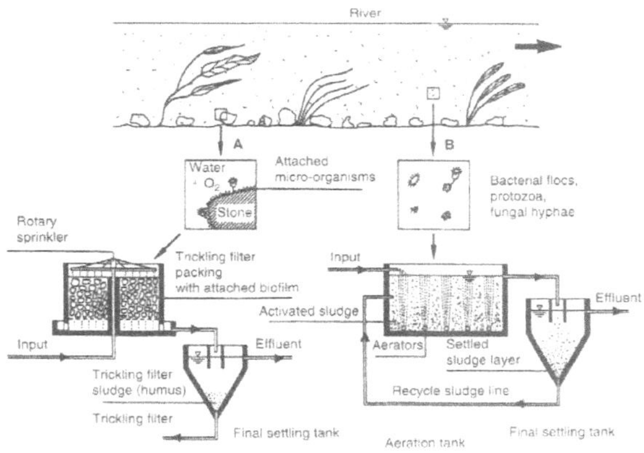

Figure 31. Relationship between the natural bacterial populations in rivers and the development of (A) trickling (percolating) filter and (B) activated sludge system [1].

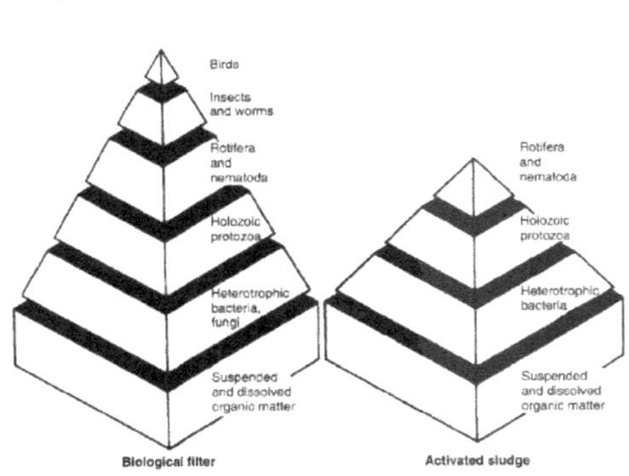

Figure 32. Comparison of the food chain pyramids for biological filters and activated sludge systems [1].

Rotating Biological Contactors

The rotating biological contactors (RBC) system (Figure 33) can be implemented to amend and improve the available treatment processes as the secondary or tertiary treatment processes. The RBC is successfully implemented in all three steps of the biological treatment, which are BOD_5 removal, nitrification, and denitrification. The process is a fixed-biofilm of either aerobic or anaerobic biological treatment system for removal of nitrogenous and carbonaceous compounds from wastewater (Figure 34). The RBC installations (Figure 35) were designed for removal of BOD_5 or ammonia nitrogen (NH_3-N), or both, from wastewater [1, 2].

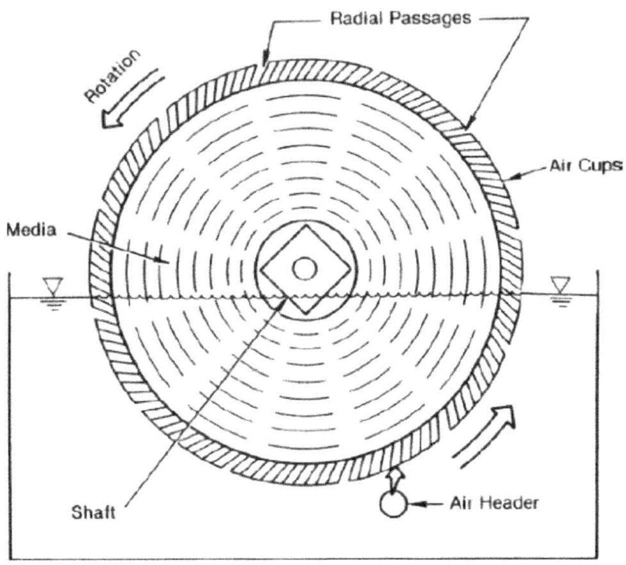

Figure 33. Schematic diagram of air-drive RBC [2].

The RBC consists of media, shaft, drive, bearings, and cover (Figure 34). The RBC hardware consists of a large diameter and closely spaced circular plastic media that is mounted on a horizontal shaft supported by bearings and is slowly rotated by an electric motor. The plastic media are made of corrugated polystyrene or polyethylene material with different designs, dimensions, and densities. The model designs are based on increasing surface area and firmness, allowing a winding wastewater flow path and stimulating air turbulence [1, 2].

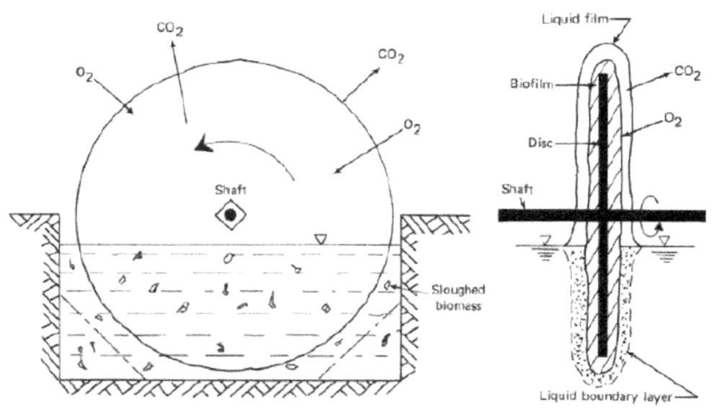

Figure 34. Mechanism of attached growth media in an RBC system [2].

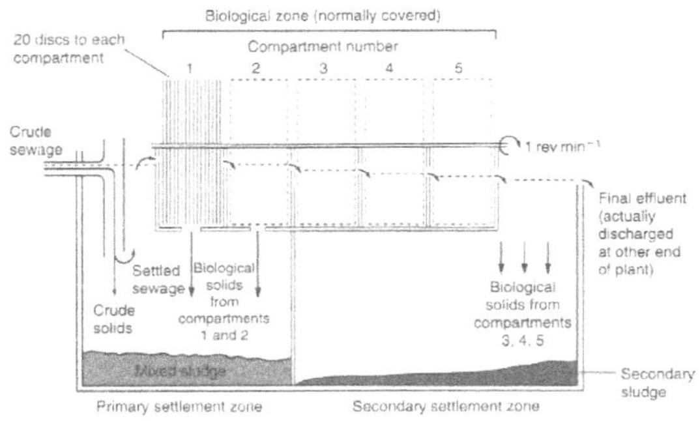

Figure 35.RBC system [1].

Biological Removal of Nutrients

Biological Phosphorous Removal

It is widely agreed that microorganisms utilize acetate and fatty acids to accumulate polyphosphates as poly-β-hydroxybutyrate, which is an acid polymer. The precise mechanism is based on the production and regeneration of adenosine diphosphate (ADP) within the bacteria, and it involves the adenosine triphosphate (ATP). Phosphate removal requires true anaerobic conditions, which occur only when there is no other oxygen donor [3]. Figure 36 shows a phosphate removal process. This process needs long narrow tanks for maintenance of plug flow.

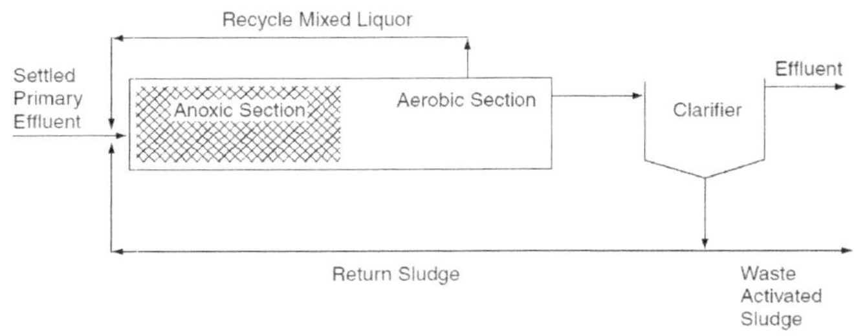

Figure 36. Phosphate removal process [3].

Biological Removal of Nitrogen

The nitrification and denitrification processes are responsible for N_2O production (Figure 37). Figure 38shows a nitrification/ denitrification system for biological removal of nitrogen.

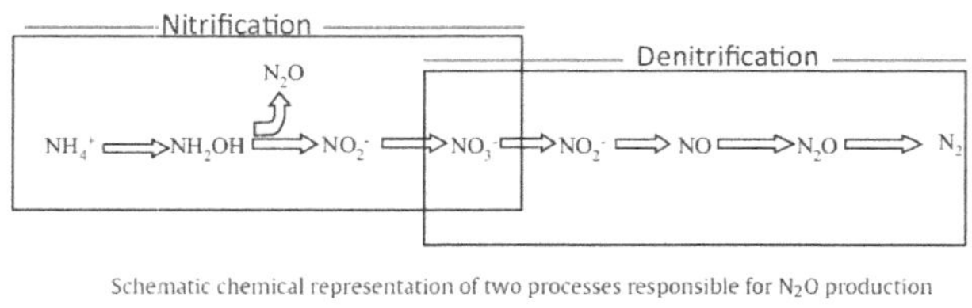

Figure 37. Schematic illustration of nitrification and denitrification processes that are responsible for N_2O release [16].

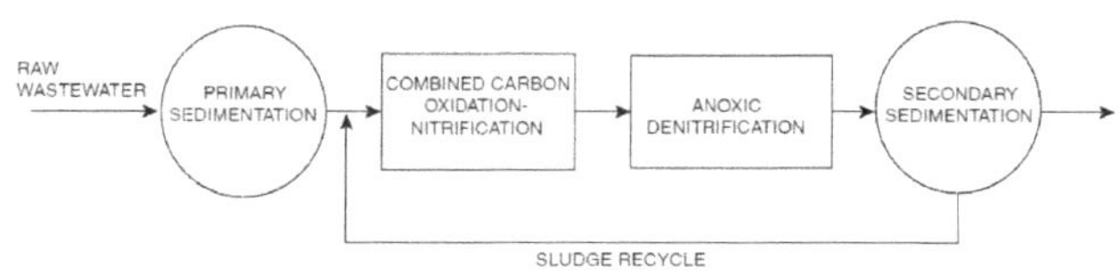

Figure 38. Nitrification/denitrification system for biological removal of nitrogen [3].

Phytoremediation

Phytoremediation is a treatment process that solves environmental problems by implementing plants that abate environmental pollution without excavating the pollutants and disposing them elsewhere. Phytoremediation is the abatement of pollutant concentrations in contaminated soils or water using plants that are able to accumulate, degrade, or eliminate heavy metals, pesticides, solvents, explosives, crude oils and its derivatives, and a multitude of other contaminants and pollutants from water and soils. Figures 39 through 44 show the designs of constructed wetlands where the phytoremediation takes place.

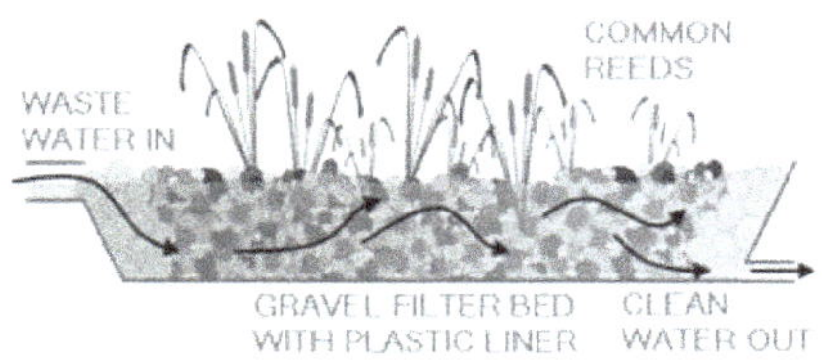

Figure 39. Cross-sectional view of a typical subsurface flow constructed wetland [17].

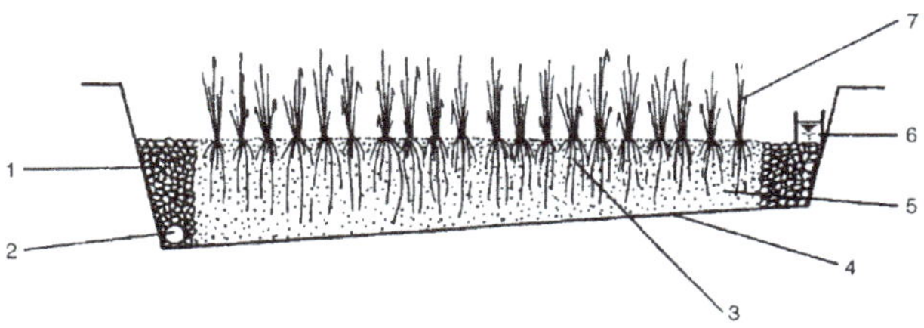

Figure 40. Components of a horizontal flow reed bed: (1) drainage zone consisting of large rocks, (2) drainage tube of treated effluent, (3) root zone, (4) impermeable liner, (5) soil or gravel, (6) wastewater distribution system, and (7) reeds [1].

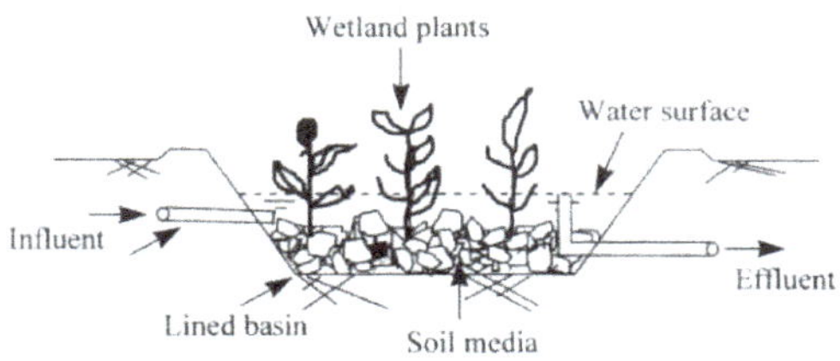

Figure 41. Free water surface system [18].

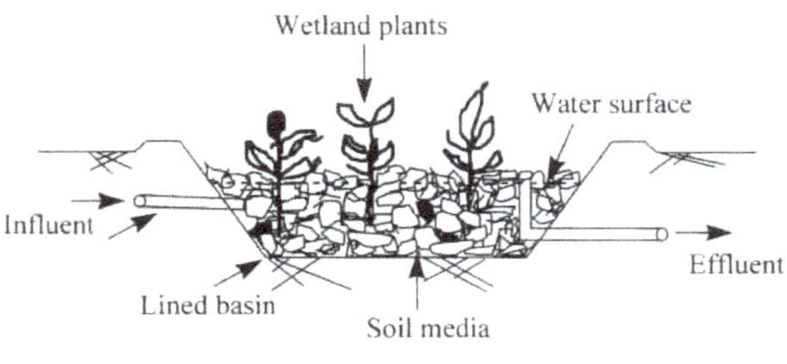

Figure 42. Sub-surface flow system [18].

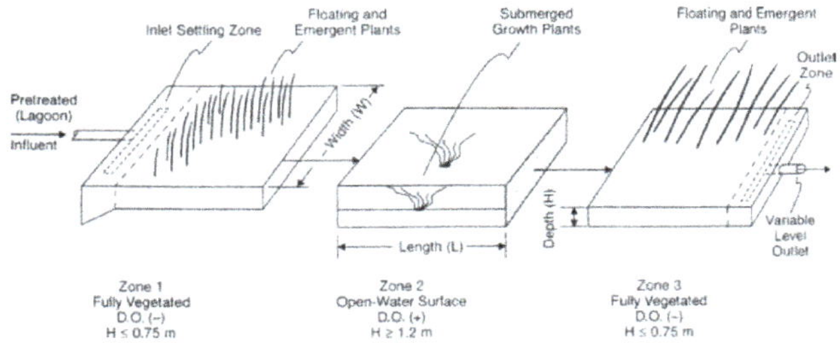

Figure 43. Components of a free water surface constructed wetland [2].

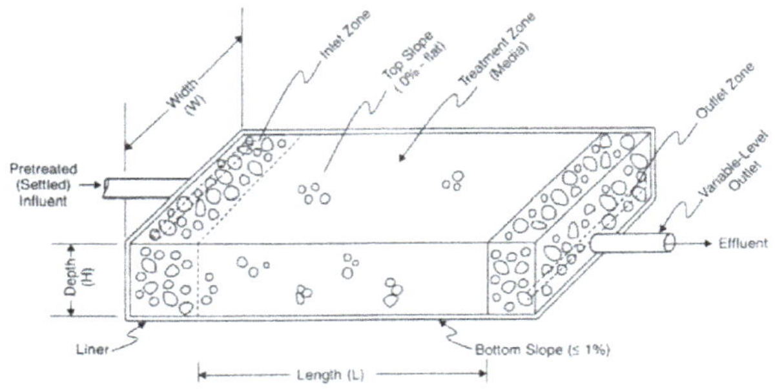

Figure 44. Components of a vegetated submerged bed system [2].

The incorporation of heavy metals, such as mercury, into the food chain may be a deteriorating matter. Phytoremediation is useful in these situations, where natural plants or transgenic plants are able to phytodegrade and phytoaccumulate these toxic contaminants in their above-ground parts, which will be then harvested for extraction. The heavy metals in the harvested biomass can be further concentrated by incineration and recycled for industrial implementation. Rhizofiltration is a sort of phytoremediation that involves filtering wastewater through a mass of roots to remove toxic substances or excess nutrients. Phytoaccumulation or phytoextraction implements plants or algae to remove pollutants and contaminants from wastewater into plant biomass that can be harvested. Organisms that accumulate over than usual amounts of pollutants from soils are termed hyperaccumulators, where a multitude of tables that show the different hyperaccumulators are available and should be referred to.

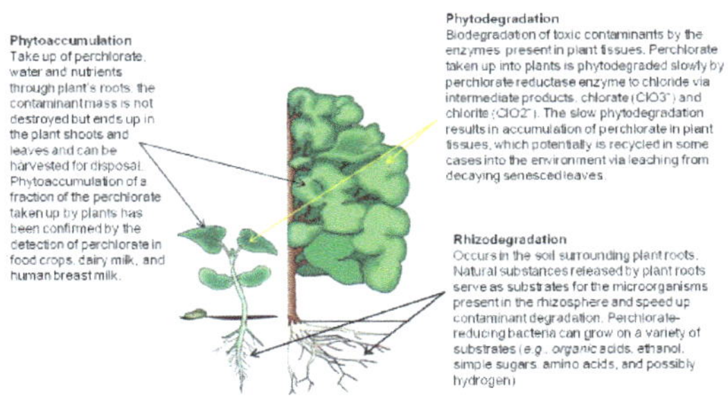

Figure 45. Rhizofiltration, phytodegradation, and phytoaccumulation [19].

In the case of organic pollutants, such as pesticides, explosives, solvents, industrial chemicals, and other xenobiotic substances, certain plants render these substances non-toxic by their metabolism and this process is called phytotransformation. In other cases, microorganisms that live in symbiosis with plant roots are able to metabolize these pollutants in wastewater. Figure 45 shows the tissues where the rhizofiltration, phytodegradation, and phytoaccumulation take place.

Vermifiltration

Vermiculture, or worm farming, is the implementation of some species of earthworm, such as Eisenia fetida (known as red wiggler, brandling, or manure worm) and Lumbricus rubellus, to make vermicompost, also known as worm compost, vermicast, worm castings, worm humus, or worm manure, which is the end-product of the breakdown of organic matter and considered to be a nutrient-rich biofertilizer and soil conditioner. Vermiculture can be implemented to transform livestock manure, food leftovers, and organic matters into a nutrient-rich biofertilizer.

The potential use of earthworms to break down and manage sewage sludge began in the late 1970s [20] and was termed vermicomposting. The introduction of earthworms to the filtration systems, termed vermifiltration systems, was advocated by José Toha in 1992 [21]. Vermifilter is widely used to treat wastewater, and appeared to have high treatment efficiency, including synchronous stabilization of wastewater and sludge [22, 23, 24]. Vermifiltration is a feasible treatment method to reduce and stabilize liquid-state sewage sludge under optimal conditions [24, 25, 26]. Vermicomposting involves the joint action of earthworms and microorganisms [24, 27, 28], and significantly enhances the breakdown of sludge. Earthworms operate as mechanical blenders and by comminuting the organic matter they modify its physical and chemical composition, steadily decreasing the C:N ratio, increasing the surface area exposed to microorganisms, and making it much more suitable for bacterial activity and further breakdown. Throughout the passageway is the earthworm gut, they move fragments and bacteria-rich excrements, consequently homogenizing the organic matter [29]. An intensified bacterial diversity was found in vermifilter, compared with conventional biofilter without earthworms [25]. The principle of using earthworms to treat sewage sludge is based on the perception that there is a net loss of biomass and energy when the food chain is extended [25]. Compared to other technologies of liquid-state sludge stabilization, such as anaerobic digestion and aerobic digestion [30], vermifiltration is a low-cost and an ecologically sound technique, and more suitable for sewage sludge treatment of small or developing-countries' WWTPs [23, 24, 25, 26, 31]. Figure 46 illustrates schematic diagram of a vermifilter, where the earthworms are in the filter bed.

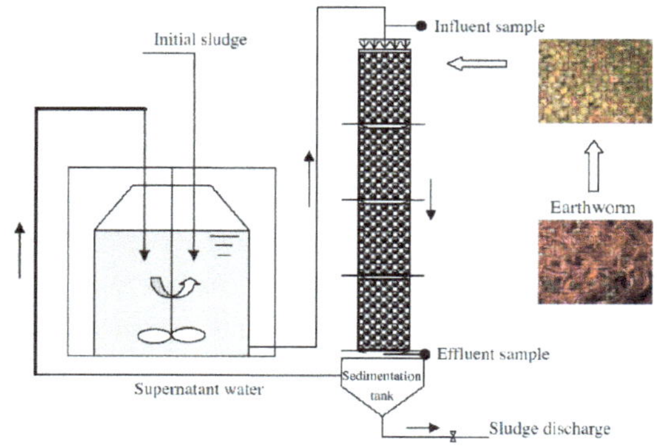

Figure 46. Schematic diagram of a vermifilter [24].

An important application is in livestock manure treatment as shown in Figure 47, where manure is flushed out from the livestock building to a raw effluent tank then the raw effluent is screened to separate the solid waste from manure. The screened effluent is then introduced to the vermifilter to produce the vermicompost. The vermifiltered effluent is then stored in a sedimentation tank. Afterwards, the vermifiltered effluent is introduced to constructed wetlands where the phytoremediation process takes place. The purified water can be then used to flush the water from the livestock building.

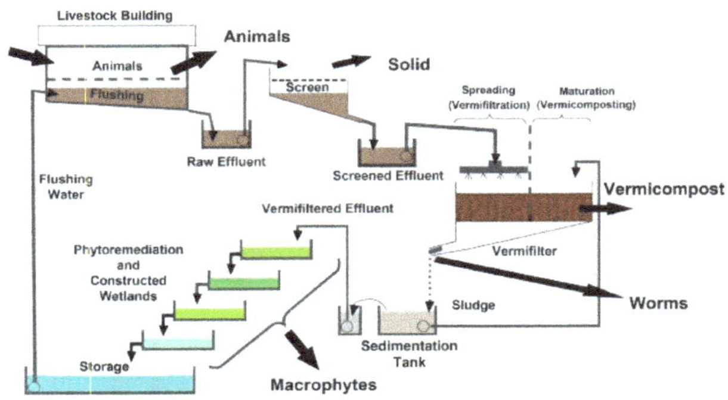

Figure 47. Schematic diagram of a manure treatment system containing vermifiltration and phytoremediation processes (Amended and redrawn from Morand et al. [32]).

Microbial Fuel Cells

The microbial fuel cells (MFCs) allow bacteria to grow on the anode by oxidizing the organic matter that result in releasing electrons. The cathode is sparked with air to provide dissolved oxygen for the reaction of electrons, protons, and oxygen on the cathode, which result in completing the electrical circuit and producing electrical energy (Figure 48).

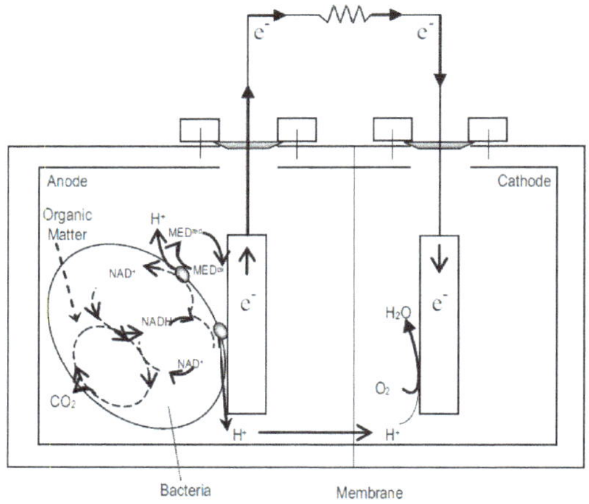

Figure 48. Schematic diagram of the essential components of an MFC [33].

CHEMICAL TREATMENT OF WASTEWATER

Chemical Precipitation

The dissolved inorganic components can be removed by adding an acid or alkali, by changing the temperature, or by precipitation as a solid. The precipitate can be removed by sedimentation, flotation, or other solid removal processes [1].

Although chemical precipitation (coagulation, flocculation) is still implemented, it is highly recommended to substitute the chemical precipitation process by phytoremediation (see previous section), where the trend is to ramp up the implementation of bioremediation and phytoremediation to reduce the use of chemicals, which is in line with the "Green Development".

Neutralization

Neutralization is controlling the pH of the wastewater whether it is acidic or alkaline to keep the pH around 7. The lack of sufficient alkalinity will require the addition of a base (Table 3) to adjust the pH to the acceptable range. Lime (CaO), calcium hydroxide ($Ca(OH)_2$), sodium hydroxide (NaOH), and sodium carbonate (Na_2CO_3), also known as soda ash, are the most common chemicals used to adjust the pH [34]. The lack of sufficient acidity will require the addition of an acid to adjust the pH to the acceptable range. Sulfuric acid (H_2SO_4) and carbonic acid (H_2CO_3) are the most common chemicals used to adjust the pH.

Table 3. Neutralization: Case of acidic wastewater [34].

Neutralization reactions	
To neutralize sulfuric acid with	
Lime:	$H_2SO_4 + CaO \rightleftharpoons CaSO_4 + H_2O$
Calcium hydroxide:	$H_2SO_4 + Ca(OH)_2 \rightleftharpoons CaSO_4 + H_2O$
Sodium hydroxide:	$H_2SO_4 + NaOH \rightleftharpoons NaSO_4 + 2H_2O$
Soda ash:	$H_2SO_4 + Na_2CO_3 \rightleftharpoons Na_2SO_4 + H_2O + CO_2$
To neutralize hydrochloric acid with	
Lime:	$2HCl + CaO \rightleftharpoons CaCl_2 + H_2O$
Calcium hydroxide:	$HCl + Ca(OH)_2 \rightleftharpoons CaCl_2 + H_2O$
Sodium hydroxide:	$HCl + NaOH \rightleftharpoons NaCl + H_2O$
Soda ash:	$HCl + Na_2CO_3 \rightleftharpoons NaCl + H_2O + CO_2$

Note: a stoichiometric reaction will yield a pH of 7.0.

Adsorption

Adsorption is a physical process where soluble molecules (adsorbate) are removed by attachment to the surface of a solid substrate (adsorbent). Adsorbents should have an extremely high specific surface area. Examples of adsorbents include activated alumina, clay colloids, hydroxides, resins, and activated carbon. The surface of the adsorbent should be free of adsorbate. Therefore, the adsorbent should be activated before use. A wide range of organic materials can be removed by adsorption, including detergents and toxic compounds. The most widely used adsorbent is activated carbon, which can be produced by pyrolytic carbonization of biomass [1]. Figure 49 illustrates the difference between absorption and adsorption. Activated carbon is the most implemented adsorbent and is a sort of carbon processed to be riddled with small, low-volume pores that enlarge the surface area available for adsorption. Owing to its high level of microporosity, 1 g of activated carbon has a surface area larger than 500 m^2, which was determined by gas adsorption. Figure 50 shows a bed carbon adsorption unit. Note that the carbon can be regenerated by thermal oxidation or steam oxidation and reused. The adsorption capacity, one of the most important characteristics of an adsorbent, can be calculated as follows:

$$\underset{mg/g}{\text{Adsorption Capacity}} = \underset{mg}{\text{Adsorbate}} / \underset{g}{\text{Adsorbent}}$$

The factors that affect adsorption are [3]:

1. Particle diameter: the adsorption is inversely proportional to the particle size of the adsorbent, and directly proportional to surface area.
2. Adsorbate concentration: the adsorption is directly proportional to adsorbate concentration.
3. Temperature: the adsorption is directly proportional to temperature.
4. Molecular weight: generally, the adsorption is inversely proportional to molecular weight depending upon the compound weight and configuration of pores diffusion control.
5. pH: the adsorption is inversely proportional to pH due to surface charge.
6. Individual properties of adsorbate and adsorbent are difficult to compare.
7. Iodine number: is the mass of iodine (g) that is consumed by 100 g of a substance.

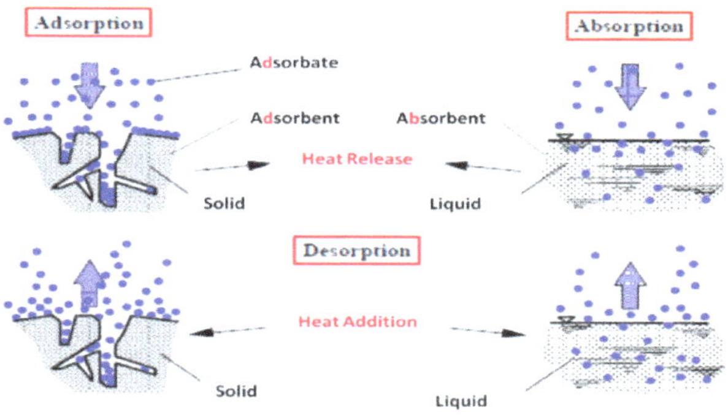

Figure 49. A comparison between absorption and adsorption.

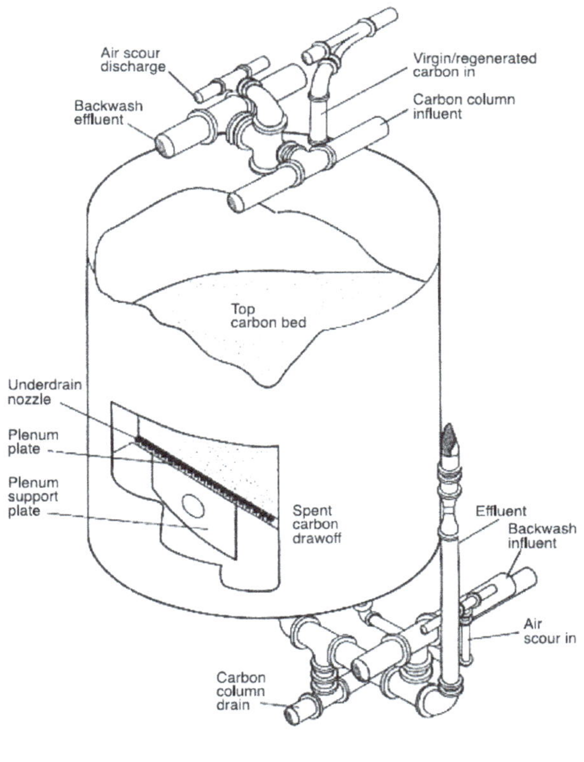

Figure 50. A bed carbon adsorption unit [35].

Disinfection

The disinfection of wastewater is the last treatment step of the tertiary treatment process. Disinfection is a chemical treatment process conducted by treating the effluent with the selected disinfectant to exterminate or at least inactivate the pathogens. The rationales behind effluent disinfection are to protect public health by exterminating or inactivating the pathogens such as microbes, viruses, and protozoan, and to meet the wastewater discharge standards. The purpose of disinfection is the protection of the microbial wastewater quality. The ideal disinfectant should have bacterial toxicity, is inexpensive, not dangerous to handle, and should have reliable means of detecting the presence of a residual. The chemical disinfection agents include chlorine, ozone, ultraviolet radiation, chlorine dioxide, and bromine [3].

Chlorine

Chlorine is one of the oldest disinfection agents used, which is one of the safest and most reliable. It has extremely good properties, which conform to the aspects of the ideal disinfectant. Effective chlorine disinfection depends upon its chemical form in wastewater. The influencing factors are pH, temperature, and organic content in the wastewater [3]. When chlorine gas is dissolved in wastewater, it rapidly hydrolyzes to hydrochloric acid (HCl) and hypochlorous acid (HOCl) as shown in the following chemical equation:

$$Cl_2+H_2O \leftrightarrow H^++Cl^-+HOCl$$

Free ammonia combines with the HOCl form of chlorine to form chloramines in a three-step reaction, as follows:

$$NH_3+HOCl \rightarrow NH_2Cl+H_2O$$
$$NH_2Cl+HOCl \rightarrow NHCl_2+H_2O$$
$$NHCL_2+HOCl \rightarrow NCl_3+H_2O$$

Figure 51 illustrates the chlorination curve, where the formation of chloramines occurs at the breakpoint. The free chlorine residual first rises then falls until the reaction with ammonia has been completed. As additional chlorine is applied and ammonia is consumed, the chlorine residual rises again.

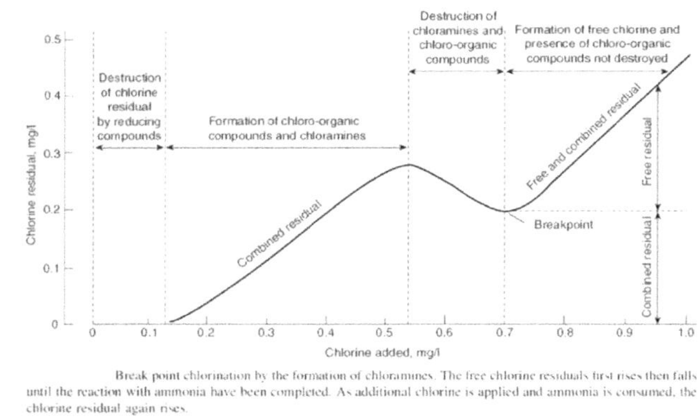

Figure 51. Chlorination curve [3].

Dechlorination is a very important process, where activated carbon, sulfur compounds, hydrogen sulfide, and ammonia can be implemented to minimize the residual chlorine in a disinfected effluent prior to discharge. Activated carbon and sulfur compounds are the most widely used [3]. The commonly used sulfur compounds are sulfur dioxide (SO_2), sodium metabisulfite (NaS_2O_5), sodium bisulfate ($NaHSO_3$), and sodium sulfite (Na_2SO_3). The dechlorination reactions with the abovementioned compounds are described in the following equations:

Ozone

Ozone (O_3) is a very strong oxidant typically used in wastewater treatment. Ozone is able to oxidize a multitude of organic and inorganic compounds in wastewater. These reactions cause an ozone demand in the treated wastewater, which should be fulfilled throughout wastewater ozonation prior to developing an assessable residual. Ozone should be generated at the point of application for use in wastewater treatment as ozone is an unstable molecule [3]. Figure 52 illustrates the corona discharge method for making ozone. Ozone is generally formed by combining an oxygen atom with an oxygen molecule (O_2) as follows:

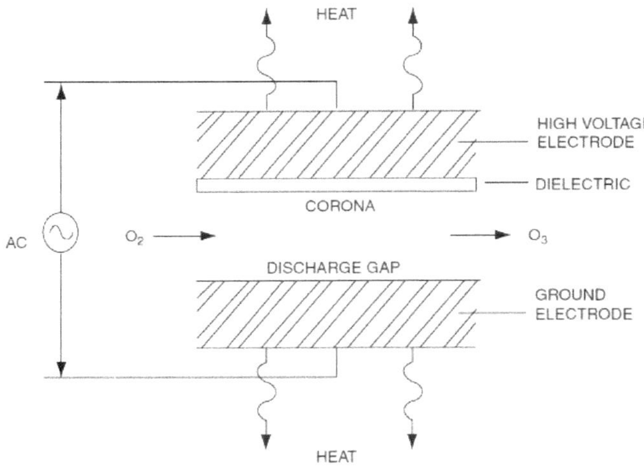

Figure 52. Schematic drawing of corona discharge method for making ozone [3].

Ultraviolet Light

Ultraviolet (UV) radiation is a microbial disinfectant that leaves no residual. It requires clear, un-turbid, and non-colored water for its implementation. The commercial UV disinfection systems use low- to medium-powered UV lamps with a wavelength of 354 nm [3]. The UV dosage can be calculated as follows:

$$D=I\cdot t$$

where, D is the UV dose (mW. s/cm^2); I is the intensity (mW/cm^2); and t is the exposure time (s).

The advantages of UV radiation are: (1) directly effective against the DNA of many microorganisms, (2) not reactive with other forms of carbonaceous demand, and (3) provides superior bactericidal kill values while not leaving any residues. The advantage is often the disadvantage, because power fluctuations, variations in hydraulic flow rates, and color or turbidity can cause the treatment to be ineffective [3]. Additionally, cell recovery and re-growth of the damaged organisms because of the inactivation of their predators and competitors has come to light.

Ion Exchange

Ion exchange (IX) is a reversible reaction in which a charged ion in a solution is exchanged with a similarly charged ion which is electrostatically attached to an immobile solid particle. The most common implementation of ion exchange method in wastewater treatment is for softening, where polyvalent cations (e.g., calcium and magnesium) are exchanged with sodium [36]. Practically, wastewater is introduced into a bed of resin. The resin is manufactured by converting a polymerization of organic compounds into a porous matrix. Typically, sodium is exchanged with cations in the solution [34]. The bed is shut down when it becomes saturated with the exchanged ions, where it should be regenerated by passing a concentrated solution of sodium back through the bed. Figure 53 shows the schematic illustration of organic cation-exchange bead. Figure 54 shows a typical ion exchange resin column. Table 4 shows the ion preference and affinity for some selected compounds.

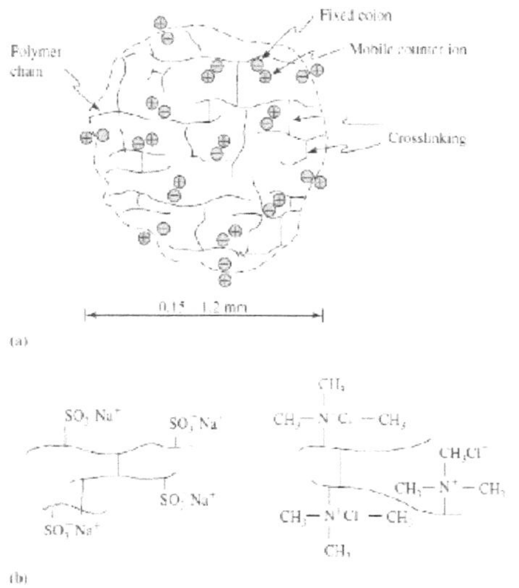

Schematic of organic cation-exchange bead:

(*a*) *The bead is shown as a polystyrene polymer cross-linked with divinyl benzene* with fixed coions (minus charge) balanced by mobile positively charged counterions (positive charge).

(*b*) *strong cation exchange* resin on left (Na form) and strong-base on right (Cl form).

Figure 53. Schematic illustration of organic cation-exchange bead [34].

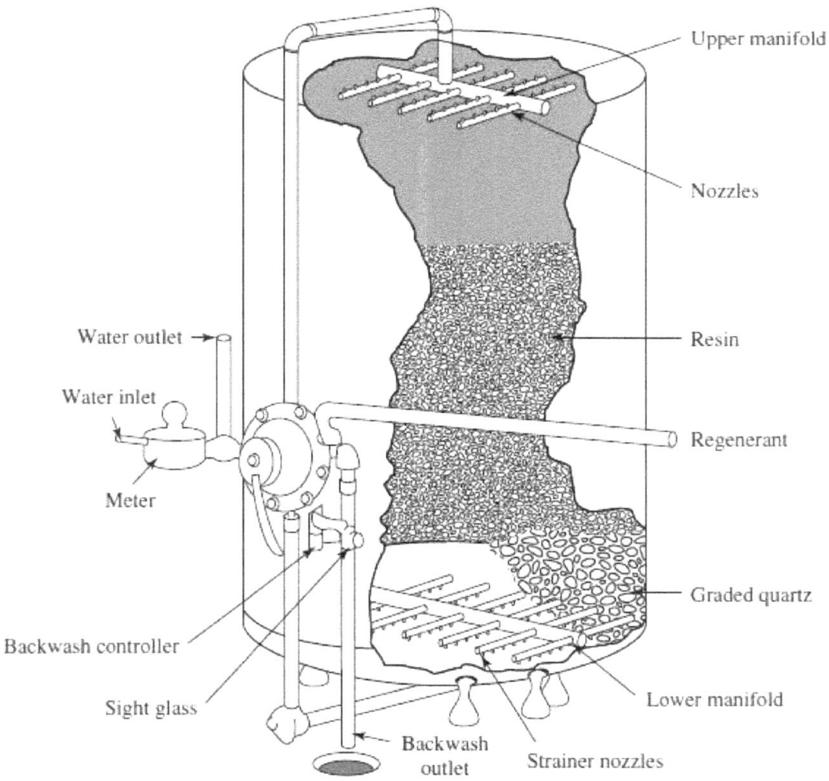

Typical ion exchange resin column.

(*Source:* U.S. EPA, 1981.)

Figure 54. Typical ion exchange resin column [37].

Table 4. Ion preference and affinity for some selected compounds [3].

Strong Acid Cation Exchanger	Strong Base Anion Exchanger	Weak Acid Cation Exchanger	Weak Base Anion Exchanger	Weak Acid Chelate Exchanger
Barium (2+)	Iodide (1−)	Hydrogen (1+)	Hydroxide (1−)	Copper (2+)
Lead (2+)	Nitrate (1−)	Copper (2+)	Sulfate (2−)	Iron (2+)
Mercury (2+)	Bisulfite (1−)	Cobalt (2+)	Chromate (2−)	Nickel (2+)
Copper (1+)	Chloride (1−)	Nickel (2+)	Phosphate (2−)	Lead (2+)
Calcium (2+)	Cyanide (1−)	Calcium (2+)	Chloride (1−)	Manganese (2+)
Nickel (2+)	Bicarbonate (1−)	Magnesium (2+)		Calcium (2+)
Cadmium (2+)	Hydroxide (1−)	Sodium (1+)		Magnesium (2+)
Copper (2+)	Fluoride (1−)			Sodium (1+)
Cobalt (2+)	Sulfate (2−)			
Zinc (2+)				
Cesium (1+)				
Iron (2+)				
Magnesium (2+)				
Potassium (1+)				
Manganese (2+)				
Ammonia (1+)				
Sodium (1+)				
Hydrogen (1+)				
Lithium (1+)				

Physicochemical Treatment Processes

The principal advanced physicochemical wastewater treatment processes are elucidated in Table 5.

Table 5. Principal advanced physicochemical wastewater treatment processes [1].

Principal advanced physico-chemical wastewater treatment processes

Process	*Removal function*
Filtration	Suspended solids
Air stripping	Ammonia
Breakpoint chlorination	Ammonia
Ion exchange	Nitrate, dissolved inorganic solids
Chemical precipitation	Phosphorus, dissolved inorganic solids
Carbon adsorption	Toxic compounds, refractory organics
Chemical oxidation	Toxic compounds, refractory organics
Ultrafiltration	Dissolved inorganic solids
Reverse osmosis	Dissolved inorganic solids
Electrodialysis	Dissolved inorganic solids
Volatilization and gas stripping	Volatile organic compounds

WASTEWATER TREATMENT PLANTS

This section shows some examples of WWTPs as shown in Figure 55 (a, b, and c) and Figure 56. On the other hand, there are some computer programs for planning and designing WWTPs (Figures 57, 58, and 59).

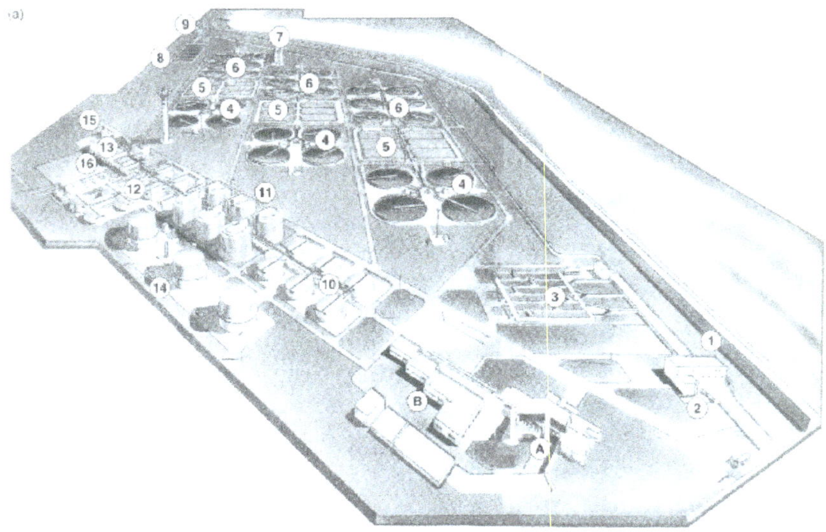

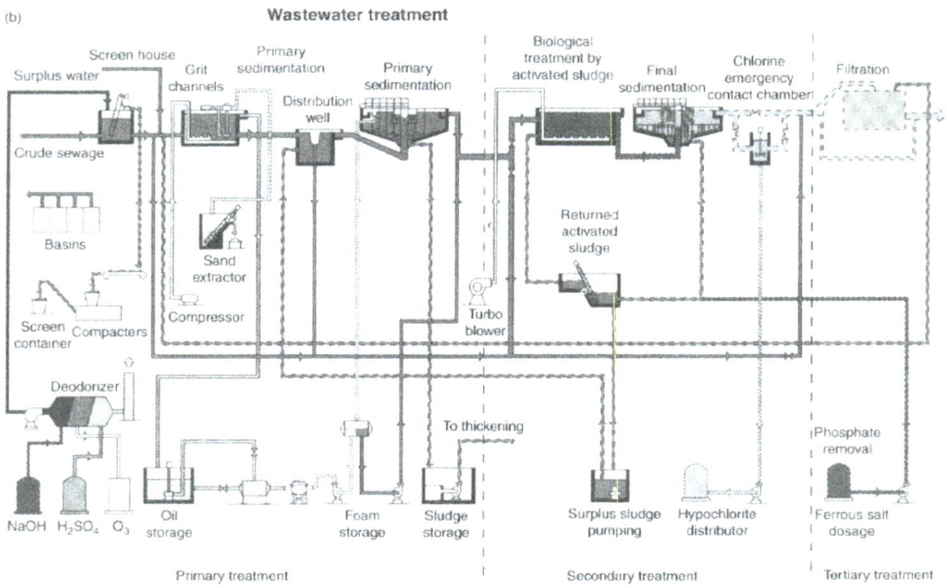

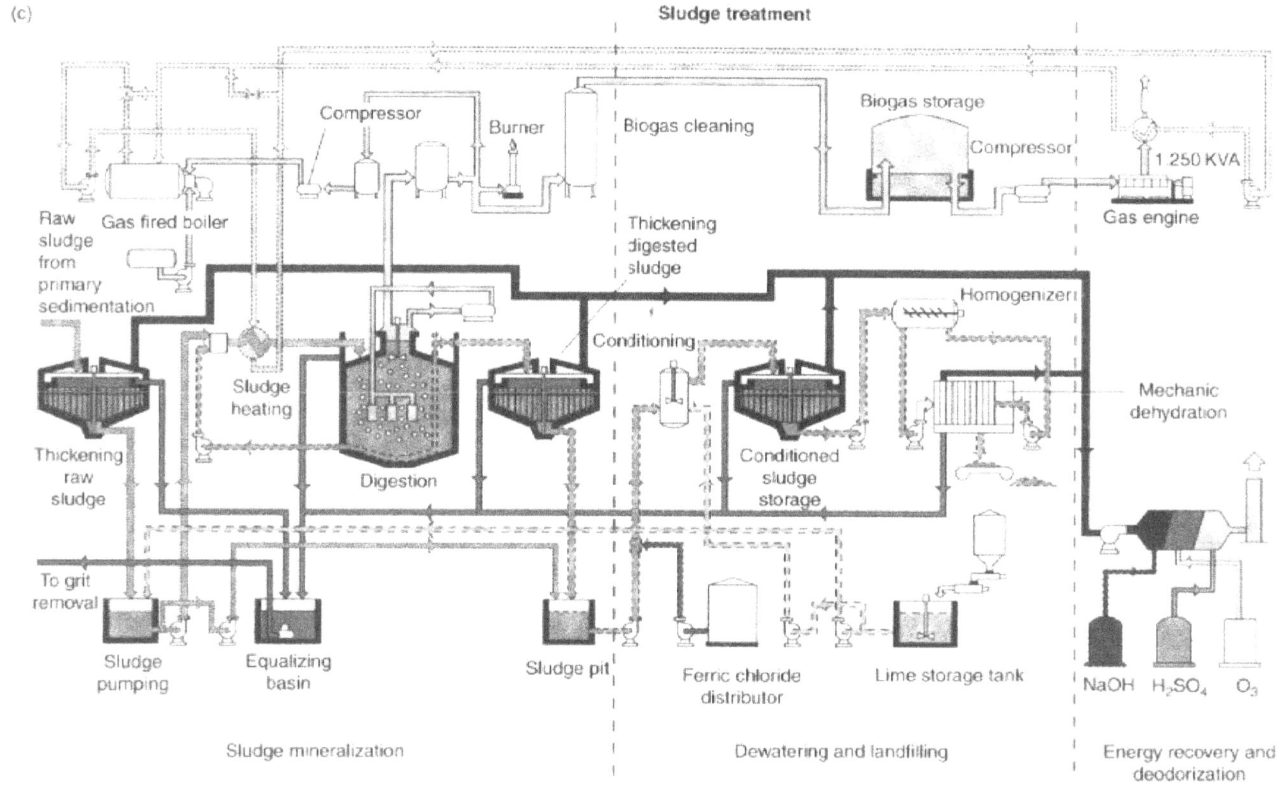

Figure 55. WWTP showing: (a) layout of the plant, (b) wastewater process flow diagrams, and (c) sludge process flow diagram. Wastewater treatment: 1. Storm water overflow; 2. screening; 3. grit removal; 4. primary sedimentation; 5. aeration tanks; 6. Secondary sedimentation; 7. emergency chlorination; 8. filtration; 9. effluent outfall. Sludge treatment: 10. raw sludge thickeners; 11. digestion tanks; 12. digested sludge thickeners; 13. power house; 14. biogas storage; 15. filter press house; 16. transformer station. A and B are administrative areas [1].

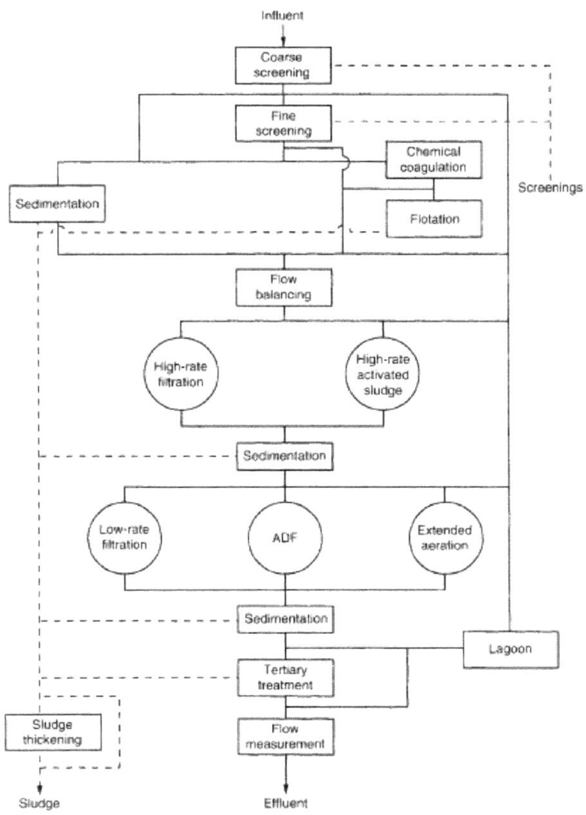

Figure 56. Summary of the main process options commonly employed at both domestic and industrial WWTPs. Not all of these unit processes may be selected, but the order of their use remains the same [1].

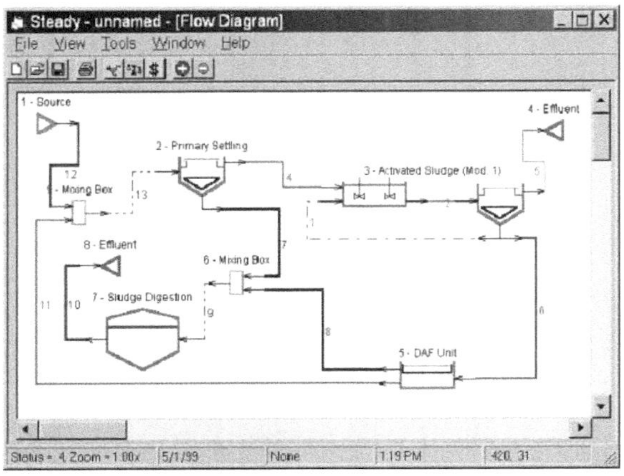

Figure 57. Screenshot of the STEADY program [3].

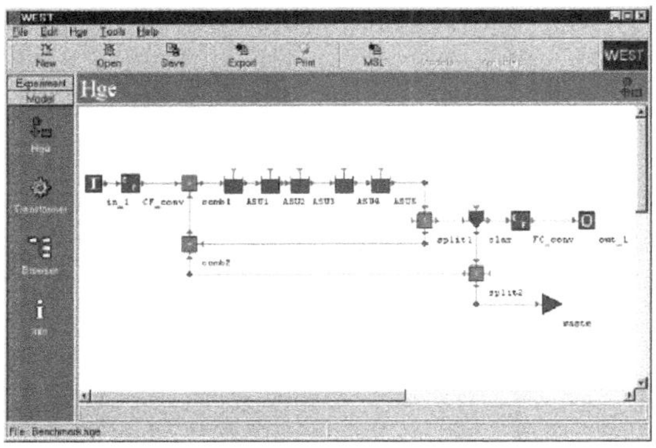

Figure 58.WEST software typical plant configuration [3].

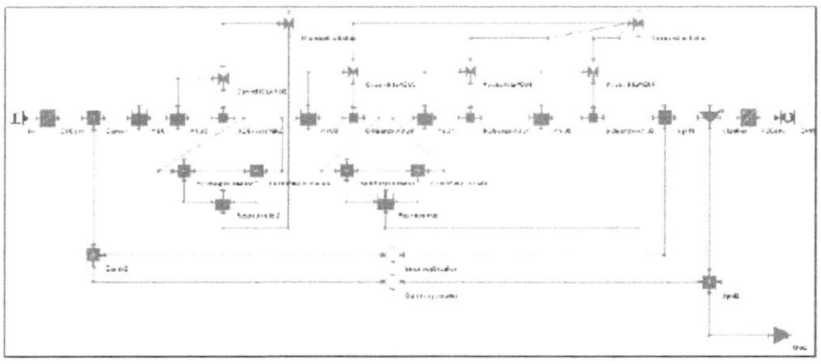

Figure 59.WEST configuration for multitank system [3].

CONCLUSIONS

According to this study, it can be conclude that:

1. The trend is to ramp up the implementation of bioremediation, phytoremediation, and mycoremediation to reduce the use of chemicals, which is in line with the "Green Development".
2. The recent developments elucidate that subsequent to the physical treatment processes (the primary treatment) the biological treatment processes come in turn as secondary treatment and precede the chemical treatment processes, which constitute the tertiary treatment.
3. Microbial fuel cells, phytoremediation, and mycoremediation are the focus of the future development in this field.

REFERENCES

1. Gray N. F. (2005). Water Technology: An Introduction for Environmental Scientists and Engineers (2nd Edition), Elsevier Science & Technology Books, ISBN 0750666331, Amsterdam, The Netherlands.
2. Lin, S. D. (2007). Water and Wastewater Calculations Manual (2nd Edition), McGraw-Hill Companies, Inc., ISBN 0-07-154266-3, New York, USA.
3. Russell D. L. (2006). Practical Wastewater Treatment, John Wiley & Sons, Inc., ISBN-13: 978-0-471-78044-1, Hoboken, New Jersey, USA.
4. Samer M. (2015). GHG Emission from Livestock Manure and its Mitigation Strategies, In: Climate Change Impact on Livestock: Adaptation and Mitigation, V. Sejian, J. Gaughan, L. Baumgard & C. Prasad (Eds.), In Press, Springer International, ISBN 978-81-322-2264-4, Germany.
5. Samer M., Mostafa E. & Hassan A. M. (2014). Slurry Treatment with Food Industry Wastes for Reducing Methane, Nitrous Oxide and Ammonia Emissions. Misr Journal of Agricultural Engineering, 31 (4): 1523–1548.
6. Samer M. (2011). How to Construct Manure Storages and Handling Systems? IST Transactions of Biosystems and Agricultural Engineering, Vol. 1, No. 1, pp. 1–7, ISSN 1913-8741.
7. Samer M., Grimm H., Hatem M., Doluschitz R. & Jungbluth T. (2008). Mathematical Modeling and Spark Mapping for Construction of Aerobic Treatment Systems and their Manure Handling System, Proceedings of International Conference on Agricultural Engineering, Book of Abstracts p. 28, EurAgEng, Crete, Greece, June 23-25, 2008.
8. Ersahin M. E., Ozgun H., Dereli R. K. & Ozturk I. (2011). Anaerobic Treatment of Industrial Effluents: An Overview of Applications, In: Waste Water - Treatment and Reutilization, F. Einschlag (Ed.), pp. 3–28, InTech, ISBN 978-978-953-307-249-4, Rijeka, Croatia.
9. Samer M. (2012). Biogas Plant Constructions, In: Biogas, S. Kumar (Ed.), pp. 343–368, InTech, ISBN 978-953-51-0204-5, Rijeka, Croatia.
10. Samer M. (2010). A Software Program for Planning and Designing Biogas Plants. Transactions of the ASABE, Vol. 53, No. 4, pp. 1277–1285, ISSN 2151-0032.
11. Pfost D. L. & Fulhage C. D. (2000). Anaerobic Lagoons for Storage/Treatment of Livestock Manure, EQ 387, MU Extension University of Missouri-Columbia, USA.
12. Westerman P. W., Shaffer K. A. & Rice J. M. (2008). Sludge Survey Methods for Anaerobic Lagoons, AG-639W, North Carolina Cooperative Extension Service, College of Agriculture & Life Sciences, North Carolina State University, USA.
13. Pavko A. (2011). Fungal Decolourization and Degradation of Synthetic Dyes Some Chemical Engineering Aspects, In: Waste Water - Treatment and Reutilization, F. Einschlag (Ed.), pp. 65–88, InTech, ISBN 978-978-953-307-249-4, Rijeka, Croatia.
14. Diano N. & Mita D.G. (2011). Removal of Endocrine Disruptors in Waste Waters by Means of Bioreactors, In:

Waste Water - Treatment and Reutilization, F. Einschlag (Ed.), pp. 29-48, InTech, ISBN 978-978-953-307-249-4, Rijeka, Croatia.

15. Kabir M., Suzuki M. & Yoshimura N. (2011). Excess Sludge Reduction in Waste Water Treatment Plants, In: Waste Water - Treatment and Reutilization, F. Einschlag (Ed.), pp. 133–150, InTech, ISBN 978-978-953-307-249-4, Rijeka, Croatia.
16. Chadwick D., Sommer S., Thorman R., Fangueiro D., Cardenas L., Amon B. & Misselbrook T. (2011). Manure Management: Implications for Greenhouse Gas Emissions. Animal Feed Science and Technology, 166–167(2011): 514– 531, ISSN 0377-8401.
17. Morris R. H. & Knowles P. (2011). Measurement Techniques for Wastewater Filtration Systems, In: Waste Water - Treatment and Reutilization, F. Einschlag (Ed.), pp. 109–132, InTech, ISBN 978-978-953-307-249-4, Rijeka, Croatia
18. ESCWA (2003). Economic and Social Commission for Western Asia.
19. Kucharzyk K. H., Soule T., Paszczynski A. J. & Hess T. F. (2011). Perchlorate: Status and Overview of New Remedial Technologies, In: Waste Water - Treatment and Reutilization, F. Einschlag (Ed.), pp. 171–194, InTech, ISBN 978-978-953-307-249-4, Rijeka, Croatia.
20. Li, X., Xing M., Yang J. & Huang Z. (2011). Compositional and functional features of humic acid-like fractions from vermicomposting of sewage sludge and cow dung, J. Hazard. Mater. 185: 740–748.
21. Li Y., Robin P., Cluzeau D., Bouché M., Qiu J., Laplanche A., Hassouna M., Morand P., Dappelo C. & Callarec J. (2008). Vermifiltration as a stage in reuse of swine wastewater: monitoring methodology on an experimental farm, Ecol. Eng. 32: 301–309.
22. Li Y., Xiao Y., Qiu J., Dai Y. & Robin P. (2009). Continuous village sewage treatment by vermifiltration and activated sludge process, Water Sci. Technol. 60: 3001–3010.
23. Xing M., Yang J., Wang Y., Liu J. & Yu F. (2011). A comparative study of synchronous treatment of sewage and sludge by two vermifiltration using epigeic earthworm Eisenia fetida, J. Hazard. Mater. 185 (2–3): 881–888.
24. Xing M., Li X., Yang J., Lv B. & Lu Y. (2012). Performance and mechanism of vermifiltration system for liquid-state sewage sludge treatment using molecular and stable isotopic techniques. Chemical Engineering Journal 197: 143–150.
25. Zhao L., Wang Y., Yang J., Xing M., Li X., Yi D. & Deng D. (2010). Earthworm-microorganism interactions: a strategy to stabilize domestic wastewater sludge, Water Res. 44: 2572–2582.
26. Xing M., Zhao L., Yang J., Huang Z. & Xu Z. (2011). Distribution and transformation of organic matter during liquid-state vermiconversion of activated sludge using elemental analysis and spectroscopic evaluation, Environ. Eng. Sci. 28 (9): 619–626.
27. Suthar S. (2008). Development of a novel epigeic–anecic-based polyculture vermireactor for efficient treatment of municipal sewage water sludge,Int. J. Environ. Waster Manage. 2: 84–101.
28. Xing M., Li X., Yang J., Huang Z. & Lu Y. (2011). Changes in the chemical characteristics of water-extracted organic matter from vermicomposting of sewage sludge and cow dung, J. Hazard. Mater. http://dx.doi.org/10.1016/j.jhazmat.2011.11.070.
29. Domínguez J. & Edwards C.A. (2004). Vermicomposting organic wastes, in: S.H. Shakir Hanna, W.Z.A. Mikhaïl (Eds.), Soil Zoology for Sustainable Development in the 21st Century, Cairo, pp. 369–395.
30. Fytili D. & Zabaniotou A. (2008). Utilization of sewage sludge in EU application of old and new methods – a review, Renew. Sust. Energ. Rev. 12: 116–140.
31. Sinha R.K., Bharambe G. & Chaudhari U. (2008). Sewage treatment by vermifiltration with synchronous treatment of sludge by earthworms: A low-cost sustainable technology over conventional systems with potential for decentralization, Environmentalist. 28: 409–420.
32. Morand P., Robin P., Qiu J.P., Li Y., Cluzeau D., Hamon G., Amblard C., Fievet S., Oudart D., Pain le Quere C., Pourcher A.-M., Escande A., Picot B. & Landrain B. (2009). Biomass production and water purification from fresh liquid manure by vermiculture, macrophytes ponds and constructed wetlands to recover nutrients and recycle water for flushing in pig housing. Proceedings of the International Congress "Ecological engineering: From concepts to applications Foreword (EECA)", 2-4 December 2009, Paris, France.
33. Logan B. (2008). Microbial Fuel Cells, Wiley. Hoboken, NJ, EEUU. 200 pp.
34. Davis M. L. (2010). Water and Wastewater Engineering: Design Principles and Practice, McGraw-Hill Companies, Inc., ISBN 978-0-07-171385-6, New York, USA.
35. Sincero, A. P. & G. A. Sincero (2003). Physical-Chemical Treatment of Water and Wastewater, CRC Press LLC, ISBN 1-84339-028-0, Boca Raton, Florida, USA.
36. Clifford D. A. (1999). Ion Exchange and Inorganic Adsorption, In: Water Quality and Treatment (5th Edition), R. D. Letterman (Ed.), pp. 9.1–9.91., American Water Works Association, McGraw-Hill, New York, USA.
37. U.S. EPA (1981) Development Document for Effluent Limitations: Guideline and Standards for the Metal Finishing Point Source Category, U.S. Environmental Protection Agency, EPA/440/1-83-091, Washington, D.C., USA.
38. U.S. EPA (2002). Wastewater Technology Fact Sheet: Anaerobic Lagoons, U.S. Environmental Protection Agency, EPA 832-F-02-009, Washington, D.C., USA.

CHAPTER 2

ADVANCED WASTEWATER TREATMENT ENGINEERING

Parneet Paul and Franck Anderson Jones

Keywords: Wastewater; membrane bioreactor (MBR); rotating membranes; fouling; modelling; cake filtration

INTRODUCTION

In the wastewater reuse, recycling and reclamation sectors, considerable research is on-going to create environmental systems engineering processes that address the water scarcity and security issues facing large parts of the world. As part of this work advanced wastewater treatment systems using membrane processes are becoming very popular as alternatives to traditional processes since they allow wastewater reuse and recycling. Thus, static or non-rotating membrane bioreactors (MBRs) are coveted systems widely used in filtration technologies for biomass separation and to remove microorganisms, particles and turbidity [1,2]. This is mainly because they offer many beneficial advantages over conventional treatment processes. These advantages include reduced footprint and sludge production by achieving high biomass concentrations in the bioreactor; high quality of reusable effluent produced as permeate; and, high nitrification and particles' removal rates [3,4]. However well these static MBR systems perform they might be superseded in the future by newly developed rotating MBR systems that potentially have reduced energy consumption and fouling [5]. Rotating MBR systems typically induce high shear effects on the membrane surface thereby reducing associated fouling whilst minimising energy usage. These systems have been shown to yield high permeate flux in the ultra-filtration (UF) range [6], whilst the very high shear rate simultaneously yields a good system performance by preventing cake formation and subsequent increased concentration polarisation [7]. For instance in Jørgensen *et al.* [8] study that used rotating ceramic membrane discs fouled by sludge, the shear directly impacted on the fouling levels, and in consequence a model was developed that linked the shear rate to the limiting flux.

A major defining issue in the application of MBR systems for water processes is the phenomenon known as fouling, which in turn reduces productivity and increases operational and maintenance costs [4,9]. Fouling by non-Newtonian fluids such as activated sludge used in MBR systems is a key limiting factor in UF membrane processes. The true origin of fouling has yet to be fully defined, although many researchers widely acknowledge that SMP (soluble microbial products) and EPS (extracellular polymeric substances) are the most likely fouling agents [9]. This is since the build-up of SMP and EPS can cause reduction in membrane permeability [10,11]. Additional factors affecting fouling mechanisms include: scaling; biofilm formation; operating conditions, such as pH; temperature and flow rates; and, solution properties such as particle size distribution [9]. According to Hermia [12], during constant pressure UF, three major fouling mechanisms are likely to occur, and these are generally known as pore constriction; pore blocking (typically divided into either complete or intermediate); and cake filtration. The aforementioned fouling mechanisms describe the accumulation of particles, solutes, and colloids inside the membrane's pores and on the membrane's surface leading to a reduction in the diameter of open pores (*i.e.*, pore constriction); an obstruction of the pores by particles larger than the membrane's pore size (*i.e.*, pore blockage) and the deposition of layers of particles onto the blocked membrane's surface (*i.e.*, cake filtration). Additionally, depending on the composition of the fluid being filtered and the interactions between the membrane and the bulk liquid, one fouling process may dominate over the other two or conversely all three fouling mechanisms may occur simultaneously during the filtration time [12]. Many fouling studies have been carried out to date using pilot units in order

Citation: Parneet Paul and Franck Anderson Jones, Advanced Wastewater Treatment Engineering, doi:10.3390/ijerph13010100

to determine the best operating conditions of MBR systems, although currently, due to the complexity of the biomass matrix which includes living microorganisms, no definitive theories on membrane fouling have been established [3]. Different approaches have also been developed for modelling the physical and biological aspects of membrane fouling in a normal non-rotational MBR system. For instance Meng *et al.* [13] established a fractal permeation model while Liu *et al.* [14] presented an empirical hydrodynamic model. Duclos-Orsello *et al.* [15] introduced a fouling model that combined all three classical fouling mechanisms [12], which was later used by Paul [16] as a starting point for a greatly refined model for a side-stream MBR that incorporated both hydrodynamics and SMP effects.

Notwithstanding numerous experimental studies, simulation and modelling of rotating MBR systems is still a nascent topic to date due to the poor understanding and great complexity of the system hydrodynamics involved. The dearth of actual commercially available systems on the market that could be studied does not help the situation either. Since mathematical modelling can be used to simulate flux decline and thus potentially afford a greater understanding of the membrane fouling mechanisms involved, the aims of this research work were to create:

- A fully comprehensive fouling model for a static MBR system incorporating all three classical fouling mechanisms.
- A complete fouling model integrating the three conventional fouling processes for a rotating MBR system that would also include the rotational hydrodynamics effect.
- A comparison of both MBR systems and identification of the impact of the rotational effect on the rotating MBR system.

Full data sets would be used to validate both models under short and medium term filtration conditions.

THEORETICAL APPROACH

The fouling model for the rotating MBR system used in this study was developed earlier by Paul and Jones [17] and extensively tested for validity. These equations will once again be reiterated in this section.

Using the power law for non-Newtonian fluids, such as activated sludge, the viscosity (Pa·s) of mixed liquor suspended solids (MLSS) in a MBR is proportional to the shear rate as depicted in Equation (1) [5].

$$\mu = m\dot{\gamma}^{n-1} \tag{1}$$

To calculate the shear rate over the rotating membrane, consideration must be given to the type flow through the membrane module. The shear rate based on different flow regimes is computed using Equation (2) [18].

$$\text{if,}\begin{cases} Re_{NN} \leqslant 2.10^5,\ \text{laminar flow},\ \dot{\gamma} = 1.81.(k_{\omega}.\omega)^{1.5}.r_o.\nu^{-0.5} \\ Re_{NN} \geqslant 2.10^5,\ \text{turbulent flow},\ \dot{\gamma} = 0.057.(k_{\omega}.\omega)^{1.8}.{r_o}^{1.6}.\nu^{-0.8} \end{cases} \tag{2}$$

Firstly, it was assumed that the pores were cylindrical and uniformly distributed throughout the membrane, so that fluid flow could be described by Hagen-Poiseuille flow. Hence, pore constriction occurs through all open pores, and gradually the membrane surface becomes obstructed by aggregates forming a somewhat uneven blocked area. Once the pores are blocked by aggregates pore constriction is stopped. Consequently, a cake layer will form over any blocked area. The resistance of this deposit layer is time dependent with regions of greatest resistance delivering the smallest flux. However, in reality the actual membrane fouling process is extremely complex in nature with usually all effects occurring simultaneously. Nevertheless, to simplify the model the above assumptions are made as well as overlooking the effect of frictional forces and temperature. Figure 1 shows the combined fouling mechanisms in the model.

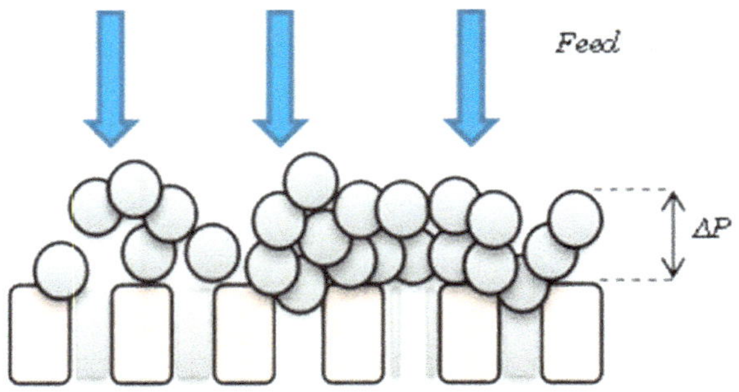

Figure 1. Diagram of the combined fouling mechanisms. Colloids or small particles constrict the pores while larger particles blocked them, forming a cake.

In a similar manner to the reformulation of the Duclos-Orsello *et al.* [15] model undertaken by Paul [16], the bulk concentration C_b (g/L) is replaced by the MLSS concentration, C_{MLSS} (g/L).

Assuming the membrane rotates around a fixed axis (here defined as an imaginary straight line passing through the shaft) with angular velocity, ω, and using the pore constriction model, the unblocked flux, J_u (m·s^{-1}), is defined as a function of time within the unblocked area, Au (m^2), as shown in Equation (3) [15,16].

$$\frac{J_u(t)}{J_0} = \frac{1}{(1+\beta.Q_0.C_{MLSS}.t)^2};\ \text{where}\ \beta = \frac{\sigma_a}{\pi.(r_p)^2.\delta_m}$$
$$\rightarrow$$
$$J_u(t) = \frac{r'_0.k_{\omega}.\omega}{(1+\beta.Q_0.C_{MLSS}.t)^2};\ \text{where}\ J_0 = r'_0.k_{\omega}.\omega \tag{3}$$

As the membrane fouls with time, the unblocked area also decreases at the same rate, and the rate of unblocked area reduction is given in Equation (4).

$$\frac{dA_u}{A_u} = \frac{-\alpha.C_{MLSS}.r'_0.k_{\omega}.\omega}{(1+\beta.Q_0.C_{MLSS}.t)^2}.dt \tag{4}$$

Assuming that at time $t = 0$, the initial unblocked area through the membrane is A_{u0} (m^2), then by integrating Equation (4) between the time filtration boundaries, Equation (5) is derived.

$$A_u(t) = A_{u_0}.e^{\frac{\alpha.r'_0.k_{\omega}.\omega}{\beta.Q_0}.\left(\frac{1}{1+\beta.Q_0.C_{MLSS}.t}-1\right)} \tag{5}$$

By combining Equations (3) and (5), the volumetric flow rate (Q_u, m^3/s) through open pores can be calculated as shown in Equation (6).

$$Q_u(t) = A_u(t).J_u(t) = \frac{A_{u_0}.r_0'.k_\omega.\omega}{(1+\beta.Q_0.C_{MLSS}.t)^2}.e^{\{\frac{\alpha.r_0'.k_\omega.\omega}{\beta.Q_0}.(\frac{1}{1+\beta.Q_0.C_{MLSS}.t}-1)\}} \quad (6)$$

The blocked flux, J_b ($m{\cdot}s^{-1}$), can be calculated from Equation (7) using Darcy's Law and a resistance in-series approach, whilst the trans-membrane pressure (TMP), can be expressed in terms of density and angular velocity in Equation (8) [18].

$$J_b = \frac{TMP}{\mu.(R_{in,b}+R_b)} \quad (7)$$

$$TMP = -PT - \left(\frac{1}{4}.\rho_f.(k_\omega.\omega.r_o)^2\right) \quad (8)$$

Once the pore constriction stops at time, t_b, the time at which a pore was first blocked, $R_{in,b}$ can be calculated from Equation (9) [15].

$$R_{in,b} = R_m(1+\beta.Q_0.C_{MLSS}.t_b)^2 \quad (9)$$

The resistance of the particles deposited increases with time due to the growth in mass (or thickness) of the cake layer, and within the cake filtration model, the resistance R_b is determined in Equation (10).

$$\frac{dR_b}{dt} = f'.R'.J_b.C_{MLSS} \quad (10)$$

Assuming, no loss in area, the blocked area, A_b (m^2), is given by Equation (11), and is directly proportional to the unblocked area A_u (m^2) at time t_b.

$$\frac{dA_b}{dt_b} = -\frac{dA_u}{dt_b} \rightarrow A_b(t_b) = \int_0^t \left(\frac{A_{u0}.\alpha.C_{MLSS}.r'_0.k_\omega.\omega}{(1+\beta.Q_0.C_{MLSS}.t_b)^2}.e^{\{\frac{\alpha.r'_0.k_\omega.\omega}{\beta.Q_0}.(\frac{1}{1+\beta.Q_0.C_{MLSS}.t_b}-1)\}}\right)dt_b \quad (11)$$

At low rotational speeds, the flow is considered laminar and by combining Equations (1), (2), (7)–(9) and (11), the volumetric flow rate (Q_b, m^3/s) through the blocked pores is given by Equation (12).

$$Q_b(t) = \frac{-PT-\left(\frac{1}{4}.\rho_f.(k_\omega.\omega.r_o)^2\right)}{m.\left(1.81.(k_\omega.\omega)^{1.5}.r_o.\nu^{-0.5}\right)^{n-1}.\left(R_m(1+\beta.Q_0.C_{MLSS}.t_b)^2+R_b\right)}.\int_0^t\left(\frac{A_{u0}.\alpha.C_{MLSS}.r'_0.k_\omega.\omega}{(1+\beta.Q_0.C_{MLSS}.t_b)^2}.e^{\{\frac{\alpha.r'_0.k_\omega.\omega}{\beta.Q_0}.(\frac{1}{1+\beta.Q_0.C_{MLSS}.t_b}-1)\}}\right)dt_b \quad (12)$$

Hence the total volumetric flow rate through the membrane is expressed as the summation of the flow rate through the unblocked Q_u and blocked Q_b pores respectively as shown in Equation (13).

$$\begin{aligned}Q(t) &= \frac{A_{u_0}.r'_0.k_\omega.\omega}{(1+\beta.Q_0.C_{MLSS}.t)^2}.e^{\{\frac{\alpha.r'_0.k_\omega.\omega}{\beta.Q_0}.(\frac{1}{1+\beta.Q_0.C_{MLSS}.t}-1)\}}\\ &+ \frac{-PT-\left(\frac{1}{4}.\rho_f.(k_\omega.\omega.r_o)^2\right)}{m.\left(1.81.(k_\omega.\omega)^{1.5}.r_o.\nu^{-0.5}\right)^{n-1}.\left(R_m(1+\beta.Q_0.C_{MLSS}.t_b)^2+R_b\right)}.\int_0^t\left(\frac{A_{u0}.\alpha.C_{MLSS}.r'_0.k_\omega.\omega}{(1+\beta.Q_0.C_{MLSS}.t_b)^2}.e^{\{\frac{\alpha.r'_0.k_\omega.\omega}{\beta.Q_0}.(\frac{1}{1+\beta.Q_0.C_{MLSS}.t_b}-1)\}}\right)dt_b\end{aligned} \quad (13)$$

It is interesting to note that, at $t = 0$, $t_b = 0$, thus, $Q = Q_0$.

When describing the hydrodynamic regime in this model, the air scouring flux, J_{air} (m/s), is a key parameter for the management and prevention of membrane fouling in most submerged MBR systems. As such, cake layer growth rate depends on the scouring energy induced by the aeration. Furthermore, rotation in rotating MBRs produces a torque that induces additional shear effects to reduce fouling on the membrane surface. Rightfully so, since the rotating MBR has a very low rotational speed of 2.09 rad/s (or 20 revolutions per minute), the above-mentioned scenario and ensuing equations will be correct. However, it is worth mentioning that at very high rotational speeds there is a high possibility that the air scouring effect will be significantly less than that induced by rotation. The net total effect responsible for reducing fouling on the membrane can tentatively be calculated by the summation of the air scouring and rotational effects. However in hindsight, at some point during the filtration process, these two effects work in opposing directions. This fact alone ultimately poses a physical limitation to the model since a completely isolated hydrodynamic study of the shear stresses will be required, which is not the scope of this study. The cake's resistance is consequently decreased to allow the system to gain flux due to these membrane cleaning effects. To account for these changes, an additional removal term was added to the rate of membrane blocked area, and was defined as the flux induced by the air scouring flow combined with rotational effects. This additional removal term is also in-line with Liang *et al.* [19] cake's formulation equation that accounted for the change in reversible fouling due to cake build-up. An analogous reformulation is found in Equation (14) but includes air scouring and rotational effects.

$$\frac{dR_b}{dt} = f'.R'.J_b.C_{MLSS} - g_o.(\alpha_v.J_{air} - k_\omega.\omega.r_o).\delta.R_c \qquad (14)$$

Subsequently, the blocked area, A_b (m^2), is mathematically given by Equation (15).

$$\frac{dA_b}{dt} = \alpha.J_u.A_u.C_{MLSS} - k_{A_b}.(\alpha_v.J_{air} - k_\omega.\omega.r_o).\theta_c(t) \qquad (15)$$

Static Square-Shaped MBR System

Equation (13) can be reduced to Equation (16), if the rotational switching function is removed so that the model reverts to a submerged static MBR system that now simply includes the air scouring term.

$$Q(t) = \frac{A_{u_0}.J_0}{(1+\beta.Q_0.C_{MLSS}.t)^2}.e^{\left\{\frac{\alpha.J_0}{\beta.Q_0}.\left(\frac{1}{1+\beta.Q_0.C_{MLSS}.t}-1\right)\right\}} + \frac{-PT}{\mu.\left(R_m(1+\beta.Q_0.C_{MLSS}.t_b)^2+R_b\right)}.\int_0^t\left(\frac{A_{u0}.\alpha.C_{MLSS}.J_0}{(1+\beta.Q_0.C_{MLSS}.t_b)^2}.e^{\left\{\frac{\alpha.J_0}{\beta.Q_0}.\left(\frac{1}{1+\beta.Q_0.C_{MLSS}.t_b}-1\right)\right\}}\right)dt_b \qquad (16)$$

The only prevalent hydrodynamic factor to take into account during operation of the static square-shaped MBR is the coarse bubble air scour that is mainly used to mitigate cake growth and thus hamper fouling. Consequently, the air scouring flux J_{air} ($m\cdot s^{-1}$) also becomes a vital parameter for the management of fouling in this static square-shaped MBR system. Hence, this air scour removal term was added to the rate of fouling resistance build up and rate of increase in blocked area. This is a comparable formulation that is also in-line with Liang *et al.* [19] rate of membrane biomass buildup equation.

In a similar manner for the rotational MBR model in Equations (14) and (15), the hydrodynamics effect in the form of air scour alone can be reduced simply to Equations (17) and (18).

$$\frac{dR_b}{dt} = f'.R'.J_b.C_{MLSS} - g_o.(\alpha_v.J_{air}).\delta.R_c \qquad (17)$$

$$\frac{dA_b}{dt} = \alpha.J_u.A_u.C_{MLSS} - k_{A_b}.(\alpha_v.J_{air}).\theta_c(t) \qquad (18)$$

MATERIALS AND METHOD

Materials

The rotating MBR pilot unit used to generate short and medium term filtration data for testing of the fouling model consisted of an UF module of 36 circular flat membrane sheets as shown inFigure 2 (FUV-185-A15R Flexidisks by Avanti Membrane Technology, Walnut, CA, USA). These circular membrane sheets were attached to a single shaft rotating via an electrical motor with an operational speed of 20 revolutions per minute (RPM). Each membrane sheet in disc form was composed of hydrophilic, low fouling polyvinylidene fluoride (PVDF) polymer with the manifold that collected the permeate flow being located in the single shaft. The measurements of TMP, dissolved oxygen levels, temperature, pH, permeate flux, air scouring flow rates, viscosity and MLSS concentration were all covered in detail in Paul and Jones [17]. Figure 2 shows part of the set-up of this research rig in operation.

Figure 2. Rotating membrane bioreactor (MBR) system RPU-185 in operation. The membrane module is located in the batch tank for filtration purposes.

Comparable to above, Figure 3 shows the set-up and operation of a second research MBR rig that was fabricated at Brunel University using a bespoke static square-shaped membrane module (Flexidisks by Avanti Membrane Technology, Walnut, CA, USA). Again in a similar manner to above, this rig generated results and data that in turn were used to validate and test the static MBR fouling model described in Section 2.1. This UF membrane module comprised 20 static membrane flat sheets. Each membrane sheet in square form was also made of hydrophilic, low fouling PVDF polymer. The viscosity of the fluid was measured daily by the aid of rotating viscometers (Rotary-Viscometer ASTM by PCE Instruments UK Ltd., Southampton, UK; and High Shear CAP-2000+ by Brookfield Viscometers Ltd., Essex, UK), while the MLSS concentration was logged continuously by a MLSS analyser (GE-138 MLSS Suspended Solids Sludge Concentration Meter Analyser Monitor by A. Yite Technology Group, Wanchai, Hong Kong). Table 1 shows this second unit's membrane dimensioning and a general overview of the operating conditions of the MBR system as provided by the manufacturer.

Plant Operational Regime and Filtration Experimentation

Both MBR plants were initially seeded with activated sludge supplied by Thames Water, UK, and were semi-batch fed a synthetic wastewater made up using a standard recipe to mimic an influent wastewater source. MLSS concentrations were kept between the range of 6.32 g/L and 7.24 g/L by periodic excess sludge wasting. The influents had an average pH between the range of 7.8 and 8.6, and experiments were carried out at a constant room temperature of 23 °C.

Figure 3. Static square-shaped membrane bioreactor (MBR) system in operation. The square membrane module is located in the larger tank on the right for filtration purposes.

Table 1. Dimensions and operating conditions of the square-shaped Flexidisks MBR module.

Description	Unit	Values
Individual membrane width	m	0.24
Individual membrane length	m	0.24
Total membrane area	m^2	1.152
Operating temperature	°C	~5–60 °C
Permeate flux	$L \cdot m^2 \cdot h^{-1}$	>30
TSS (Total suspended solids)	$g \cdot L^{-1}$	>8
TMP (trans-membrane pressure)	bar	(>3)

Shear and Viscosity Experiments for the Rotating MBR

Using the Brookfield rotating viscometer [17] that also acted as a rheometer, measurements were taken following protocols in-line with Yang *et al.* [20] and Ratkovich *et al.* [21]. The proprietary software program for this apparatus was used for system control and collection of data. The sludge was tested at a constant room temperature of 23 °C at a MLSS concentration of 6.32 g/L (although data points at MLSS ranges of 5.2 g/L to 7.3 g/L were also analysed). In addition, the shear stress and viscosity were carefully tested in the shear rate range of 10 s^{-1} to 350 s^{-1}. Although full rheology tests were not carried on the activated sludge to ascertain its properties more precisely, the data collected appeared consistent.

TMP Stepping Experiments

A standard TMP stepping procedure was used on the rotating MBR plant as described in detail in Paul and Jones [17]. This same method was also applied to this second rig that employed the square-shaped membrane configuration in its MBR plant. In order to facilitate comparison of operational issues, four TMP steps were carried out for each variation in MLSS concentration for the static square-shaped MBR unit, while for the rotating MBR system only two TMP steps were considered. For the rotating MBR system, TMP steps were carried out at constant TMPs of 15 kPa and 45 kPa. The corresponding initial flow rates for the rotating MBR system were respectively 1.15×10^{-5} m^3/s and 2.45×10^{-5} m^3/s. For the static square-shaped MBR system, the TMP steps up were carried out at constant TMPs of 15 kPa, 30 kPa, 45 kPa and 58 kPa. The corresponding initial flow rates were respectively 1.2×10^{-5} m^3/s, 2.03×10^{-5} m^3/s, 2.5×10^{-5} m^3/s and 3.02×10^{-5} m^3/s. Although data was constantly being logged, for the sake of simplicity and to keep model computation time down to a minimum, only the average data points for every 5 min of filtration time were actually used in the simulation study with the total filtration period being two hours. This meant a total of 25 data points were generated for each individual MLSS concentration.

On unit start-up, for the rotating MBR system, the clean membrane resistance was determined to be 6.26×10^{11} m^{-1}, while the cake water content, τ, was found to be on average 0.456. Conversely on start-up, the square-shaped MBR system had a pristine membrane resistance of 4.55×10^{11} m^{-1}.

RESULTS AND DISCUSSIONS

Rotating MBR Model Validation with Hydrodynamic and Shear Effects

Shear and Viscosity Relationship

A major parameter included in this model formulation was combined shear due to both rotation and aeration as this is a unique feature of this type of MBR configuration. To that end, the viscosity of the MLSS at 6.32 g/L concentration (points at 7.24 g/L were also considered) was measured at different shear rates. The rotational speed of the spindle was 20 RPM and this equated to a shear rate of 26 s^{-1}. Although full rheology tests were not carried on the activated sludge to ascertain its properties more precisely, the data collected appeared consistent. Results indicated that the fluid's viscosity had decreased much faster and at a higher shear rate (by over 32%). That was expected since the calculated radial Reynolds number (R_{eNN}) showed that the flow was laminar during MBR operations (*i.e.*, R_{eNN} of 8.26×10^3 which is less than the 2×10^5 limit). It should be noted that rotating MBR systems can handle shear rates up to 2×10^5 s^{-1}. However, the range of shear rate tests were kept minimal since the pilot MBR in question only operated at very low RPMs. Furthermore, it should be noted that since activated sludge is a shear thinning fluid [7], the rheological measurement range must be kept in the laminar region otherwise the rheometer's output becomes increasingly hard to interpret. The viscometer's output was used to determine parameters *m* and *n* for this activated sludge. These parameters are the coefficients governing the shear rate and viscosity respectively and are related by a logarithmic law. Since the shear equation [17] obeys a power law, it naturally follows that the logarithmic of said equation would produce a polynomial of order one (*i.e.*, a straight line). Consequently, the parameters *m* and *n* were specifically determined by plotting the logarithm of viscosity against the logarithm of the shear rate using a linear curve fitting process in the Matlab and Excel software packages.

Figure 4 shows the viscosity plotted against the shear rate with the solid line being the best fit. The coefficient *m* was found to be 0.0138 whilst *n* gave a value of 0.861. The coefficient of determination for the linear fit process was 0.945, indicating a respectable model fit since the sum of the squared residuals was also minimised. A value of *n* less than one was a clear indication that the fluid had deviated from Newtonian behaviour. Nevertheless, it is worth noting that higher MLSS concentrations are associated with high viscosity that, in turn, strongly hinders mass transfer leading to increases in membrane fouling [20].

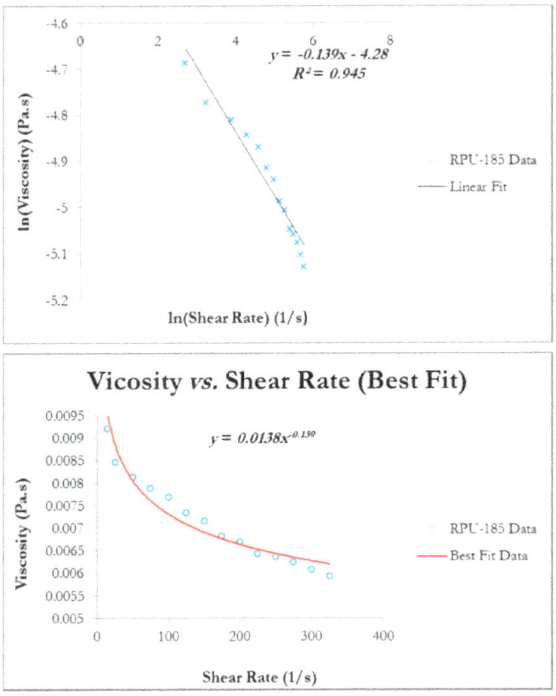

Figure 4. Viscosity plotted against shear rate for rotating MBR system RPU-185.

Rotating MBR Model Validation with Hydrodynamic Regime

In this developed rotating fouling model parameters α, β (σ_a is proportional to β and is therefore less significant) and a combination of $f'.R'$, R_{bo}, g_o and k_{Ab}; each solely and respectively contributed to pore blocking, pore constriction and cake filtration [15].

As described in the earlier study by Paul and Jones [17], the air scouring coefficient, α_v, and the resistance distribution, δ, were obtained via sensitivity analysis. The values found were respectively 0.0292 and 4.6×10^{-4} m^{-1} for α_v and δ. The air scouring flow rate of 3.55×10^{-4} m^3/s was also kept the same for all simulations. Due to varying fluxes, the values of the term k were obtained via sensitivity analysis and are shown in Table 2. These above determined values were kept constant and used in all subsequent simulations to calculate the best fit values for this fouling model. To ensure validity, only the six most important parameters pertaining to all three fouling mechanisms were used for curve fitting during simulations and these *were* $f'.R'$, α, β, R_{bo}/R_m, g_o andk_{Ab}. These best fit simulation values were attained by minimising the sum of squared residuals between the model and the collected experimental data. This was used in conjunction with the "Genetic Algorithm" function in the Matlab software package with an initial population large enough for the data set used to converge to the minimal possible error. The simulations were performed for each TMP at respectively 15 kPa and 45 kPa for MLSS concentrations of 6.32 g/L and 7.24 g/L. The term σ_a was calculated upon obtaining the optimal fitting value of β since the membrane pore size was known.

Table 2. Simulations best fit model parameters for rotating MBR system RPU-185.

Parameters	Unit	15 kPa TMP Step	45 kPa TMP Step
g_o		79.59	24.76
k_{Ab}		0.025	26.57
R_{bo}/R_m		0.206	0.324
$f'R'$	m/kg	0.248×10^9	199.73×10^9
α	m^2/kg	0.118	1.157
β	kg	0.893	0.589
σ_a	kg·m^3	6.816×10^{-17}	4.497×10^{-17}
m	Pa·s^n	0.0138 ± 0.02	0.0138 ± 0.02
n		0.861 ± 0.01	0.861 ± 0.01
k_ω	rad^{-1}	2.812×10^{-5}	5.991×10^{-5}

In summary, the values of shear parameters m and n are within acceptable range as they are in-line with those found in earlier work [17]; however, these values change at a steady rate depending on the range of MLSS concentrations used and since high viscosities are associated with high MLSS, the likelihood of cake filtration (or cake formation) occurring is drastically increased at high TMPs. The values of α_v, δ and k are also adequate since the same conditions found by Paul and Jones [17] were used during experimentation. Moreover, fitting parameters $f'.R'$, α, β, R_{bo}/R_m, $g_o$and k_{Ab} all vary based on the TMP used but stay constant within simulated MLSS concentration range of 6.32 g/L and 7.24 g/L. These best fit values obtained from simulation appear fairly reasonable since they are similar in scope with former work completed by Paul and Jones [17]. Finally, these best fit values also represent an accurate portrayal of experimental data collected and the dominant fouling mechanisms.

Figure 5 displays the TMP step data as normalised volumetric flow rates and the total resistance ratios plotted against the filtration time at a constant TMP of 15 kPa for MLSS concentrations of 6.32 g/L and 7.24 g/L for the rotating membrane system RPU-185; with the dashed lines representing the best fit simulation solution. It should be noted that the total resistance was calculated using Darcy's Law. A sharp decrease in flux of 56% can be observed at roughly both MLSS concentrations. This is in-line with the results produced by the earlier study by Paul and Jones [17] that found the rate of decrease of flux is consistent and steady unless the TMP is dramatically increased. The trend shown by the both sets of curves is sufficient to suggest that fouling occurring during filtration was caused by a combination of all three classical fouling mechanisms. This is further supported by the fact that the resistances' curves displayed a linear best fit. However, as shown in Table 2, it is found upon closer inspection that a relatively small R_{bo}factor and a big cake removal factor, g_o, indicated a weak cake layer formation which is reinforced by a small $f'.R'$ (fractional amount of total foulants multiplied by specific cake layer resistance) and a very small k_{Ab} (blocked pores area constant). Thus, the combination of parameters $f'.R'$, R_{bo}, $g_o$and k_{Ab}, implied that cake filtration was less prevalent which was anticipated since at low TMP of 15 kPa, cake formation is expected to be less dominant [17]. Furthermore, the pore constriction parameter, β, being roughly eight times bigger than the pore blocking parameter, α, ultimately suggested that fouling was dominated by pore constriction (as β >> α).

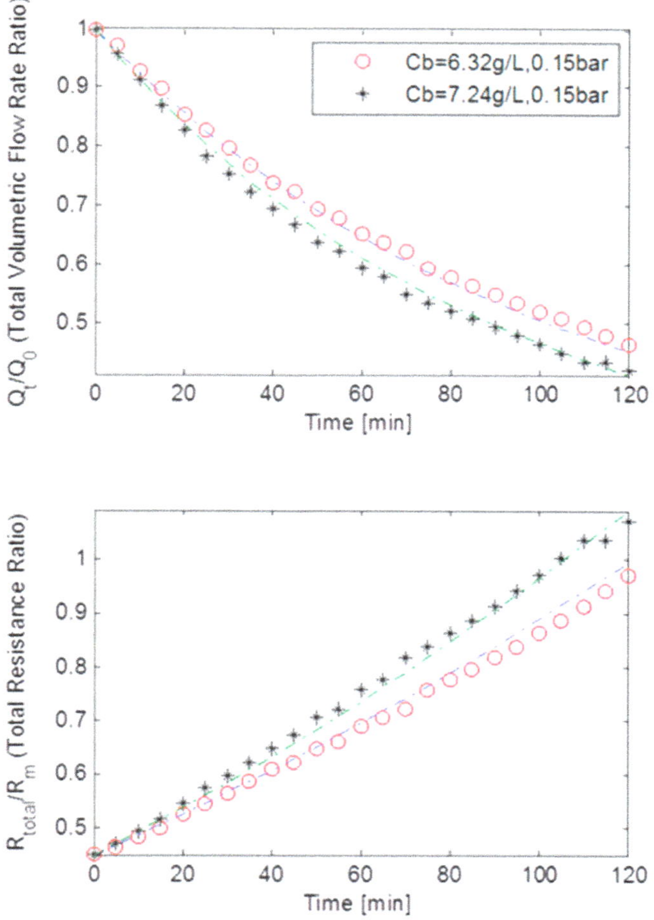

Figure 5. Flux decline and total resistance for TMP step of 15 kPa for the rotating MBR system RPU-185 with best model fits.

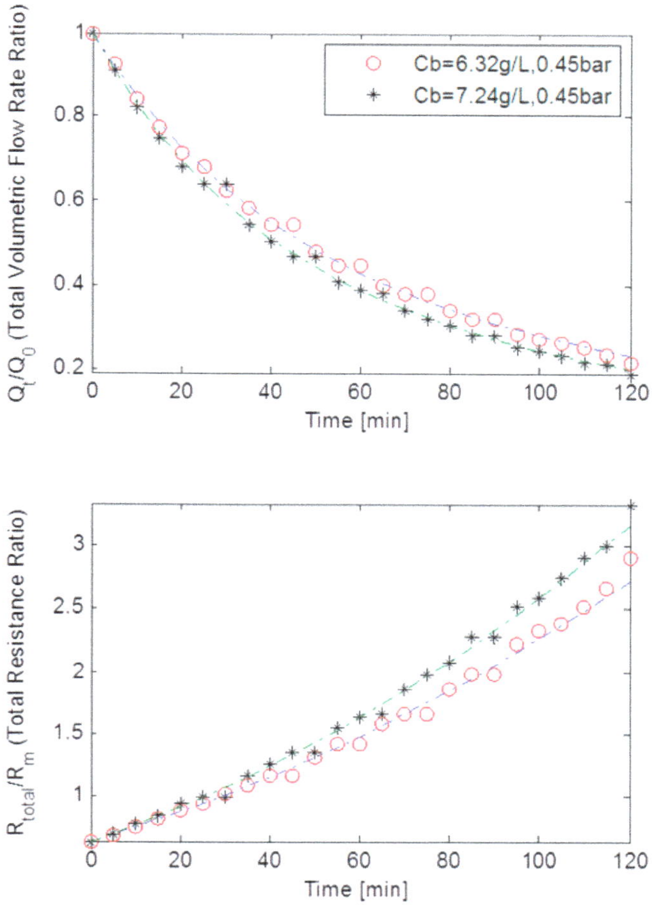

Figure 6. Flux decline and total resistance for TMP step of 45 kPa for the rotating MBR system RPU-185 with best model fits.

At MLSS concentrations of 6.32 g/L and 7.24 g/L and constant TMP of 45 kPa, the starting initial flow rate was 2.45×10^{-5} $m^3 \cdot s^{-1}$. Experimental data showed that at MLSS concentration 6.32 g/L, the flux had diminished drastically and dropped by more than 75%. A similar reduction was detected at MLSS concentration of 7.24 g/L. This meant that not only had the initial flux increased at higher TMPs, but also that the flux decline rate had increased at greater pressure. These findings are in-line with theory since a membrane is likely to foul more quickly when approaching or exceeding critical flux [9,17].

Figure 6 shows the normalised volumetric flow rates and the total resistance ratios plotted against the filtration time at constant TMP of 45 kPa, for MLSS concentrations of 6.32 g/L and 7.24 g/L for the rotating membrane system RPU-185. The massive drop in flux also resulted in the total resistance increasing exponentially. This strongly indicates that the fouling in this case was dominated by both pore blocking and cake filtration although the combined effect of all three fouling mechanisms on the total fouling cannot be precisely extrapolated. Again as presented inTable 2, a relatively bigger R_{bo} and $f'.R'$ when compared to the data from the constant TMP of 15 kPa step, coupled with near equally sized g_o and k_{Ab}, indicated that an adequately strong cake layer was formed. However, with α being roughly twice as big as β, the pore blocking fouling mechanism must have also had a significant impact on fouling. Thus, it is thought that the bulk of the fouling was dominated by both pore blocking and cake filtration.

Static MBR Model Validation with Hydrodynamic and Non-Shear Effects

The flow regimes during filtration processes were all laminar which were well within expectations since calculated R_{eNN} values were much less than 2×10^5.

As mentioned in Section 4.1.2, in this static fouling model, parameter α solely contributed to pore blocking; parameter β wholly contributed to pore constriction and finally a combination of parameters $f'.R'$, R_{bo}, g_o and k_{Ab}, all served to define the cake filtration.

The aeration rates for all the data sets for the static MBR system were similar in scale to that of the rotating MBR system operated under lab scale conditions. Hence, similar constant values for air scouring coefficient, α_v, (*i.e.*, 0.0292), and resistance distribution factor, δ, (*i.e.*, 4.6×10^{-4} m^{-1}) were used during all simulations. Additionally, since the static square-shaped MBR system was not operated under rotation, the shear effects parameters *m*, *n* and *k* were removed and not used during all simulations. As discussed in section 4.1.2 and also as applied to the static fouling model, the key parameters necessary for analysis, curve fitting, and fouling mechanisms determination were $f'.R'$, α, β, R_{bo}/R_m, g_o and k_{Ab}. These output values were determined in Matlab using the "Genetic Algorithm" function in concurrence with the minimised sum of squared residuals between the fouling model and experimental data. Simulations were swiftly performed for each TMP at correspondingly 15 kPa, 30 kPa, 45 kPa and 58 kPa for MLSS concentrations of 6.32 g/L and 7.24 g/L.

Best fit parameters $f'.R'$, α, β, R_{bo}/R_m, g_o and k_{Ab} as presented in Table 3, all appear to be in sensible agreement with prior conducted work [15,17]. A justification for this is that despite variations depending on the TMP used, these values remained constant within simulated MLSS concentrations range of 6.32 g/L and 7.24 g/L. In Table 3, when comparing pore blocking parameter, α, at different data sets, its lowest value of 0.094 is found for the Avanti static square-shaped MBR system for TMP step of 15 kPa. This suggests that the pore blocking's effect on fouling was minimal during filtration. Moreover, this is justified since its corresponding pore constriction parameter, β, is much bigger. Thus, a possible conclusion is that at lower TMPs, pore blocking is less likely to dominate fouling, which is also in-line with Paul and Jones [17] findings. In contrast, the highest value of the pore blocking parameter, α, is 0.907 and is found for the Avanti static square-shaped MBR system for TMP step of 45 kPa. It implies that pore blocking was one of the dominant fouling mechanisms during filtration, which is reasonable since its corresponding pore constriction parameter, β, is much smaller. Therefore an inference is that at very high TMPs, pore blocking is more likely to be one of the dominant fouling mechanisms, which again happens to be in-line with prior work [17].

Table 3. Simulations best fit model parameters including hydrodynamic effects.

Optimised Parameters	$f'R'$ (m/kg)	α (m^2/kg)	β (kg)	σ_a (kg·m^3)	R_{bo}/R_m (−)	g_o (−)	k_{Ab} (−)
Duclos-Orsello *et al.* [15] data	1.81×10^{11}	0.576	2.538	19.4×10^{-17}	0.168	24.58	1.46
Pilot MBR unit at Coors (UK) data	1.76×10^{11}	0.122	0.289	2.21×10^{-17}	0.159	27.24	0.108
Static MBR unit (15 kpa)	0.834×10^{11}	0.094	0.492	3.79×10^{-17}	0.126	63.75	9.36
Static MBR unit (30 kpa)	0.947×10^{11}	0.184	0.305	2.33×10^{-17}	0.692	10.99	2.032
Static MBR unit (45 kpa)	108.8×10^{11}	0.907	0.486	3.78×10^{-17}	0.526	15.36	14.23
Static MBR unit (58 kpa)	32.01×10^{11}	0.221	0.212	1.62×10^{-17}	0.457	29.66	1.26

When comparing pore constriction parameter, β, at various data sets, its lowest value of 0.212 is found for the Avanti static square-shaped MBR system for TMP step of 58 kPa (see Table 3). As such, it can be inferred that pore constriction had a lesser impact on fouling or was equally as influential as other fouling mechanisms. Since its equivalent pore blocking parameter, α, is almost equal in value (though still a bit less), it can be argued that both pore blocking and pore constriction had almost equal importance in the occurrence of fouling. Consequently, a conclusion that can be drawn is that at higher TMPs, pore constriction is less likely to be dominant [17]. On the other hand, the highest value of pore constriction, β, is 2.538 and is found for the Duclos-Orsello *et al.* [15] data set for constant TMP of 14 kPa. It therefore implies that pore constriction was one of the dominant fouling mechanisms during filtration. Additionally, this is justified since its corresponding pore blocking parameter, α, is almost five times smaller. Accordingly, it can be concluded that at lower TMPs, pore constriction is more likely to dominate fouling [17].

As shown in Table 3, when comparing the combination of parameters $f'.R'$ (fractional amount of total foulants multiplied by specific cake layer resistance), R_{bo} (initial resistance of solids deposit),g_o (cake removal factor) and k_{Ab} (blocked pores area constant), which all pertains to cake filtration, its lowest combination is found for the Avanti static square-shaped MBR system for TMP step of 15 kPa. This suggests that cake formation (and by extension cake filtration) was fairly weak and less influential during fouling. Thus, at lower TMPs, cake filtration is expected to be less dominant and prevalent during fouling [17]. Conversely, the highest combination of parameters $f'.R'$, R_{bo}, g_o andk_{Ab}, is found for the Avanti static square-shaped MBR system for TMP step of 45 kPa. It implies that cake filtration was fairly strong and more significant during fouling. As such, it can be deduced that at high TMPs, cake filtration is expected to be one of the dominant fouling mechanisms [17].

Overall, the calculated model parameter values gave very respectable fits when compared to the original data collected for the static square-shaped MBR system, although extreme MLSS concentration values of 8 g/L and 12 g/L respectively for the Duclos-Orsello *et al.* [15] data, and the Coors (UK) data [16], gave rather poor fits. This is largely as expected since the sludge rheological effects which themselves are volatile will dominate all membrane fouling mechanisms in unpredictable ways especially at high MLSS concentrations with associated high viscosities. Additionally for the static square-shaped MBR system it can be seen that at higher TMPs, the simulation fit progressively deteriorated as the flux declined rapidly. Thus, at a TMP step with constant TMP of 45 kPa, the simulation's fit was extremely poor and, by extension, was poor for the 58 kPa case as well. This situation is less prevalent for the rotating MBR system, as both air scouring and rotational shear contribute to the reductions in fouling.

In order to validate the static model, two distinct sets of data were used. The first set was collected from the study carried out by Duclos-Orsello *et al.* [15], and the other was data provided by a pilot MBR plant operated at Coors (UK) [16]. It should be noted that, for the Coors (UK) data, the viscosity (Pa·s) was calculated from the MLSS data provided and by using Equation (19) [20].

$$\mu = 0.0126.(C_{MLSS})^{1.664}.e^{\frac{E}{R_g.(T_{room}+273.15)}} \tag{19}$$

Additionally, four sets of data all obtained via a TMP stepping procedure using the static square-shaped MBR system were used to validate this static fouling model. The data was collected at TMP steps with constant TMPs of 15 kPa, 30 kPa, 45 kPa and 58 kPa respectively. There were obviously no induced shear effects to take into account, which simplified the entire simulation procedure since only the hydrodynamics effects were included with appropriate parameters and coefficients used during simulation runs.

Figure 7 depicts the TMP step data as normalised volumetric flow rates and the total membrane resistance ratios plotted against the filtration time at a constant TMP of 14 kPa for MLSS concentrations varying from 1 g/L to 8 g/L based upon the Duclos-Orsello data; with the solid lines representing the best fit simulation solutions. Again, the total resistance was acquired via Darcy's law. It was immediately noticeable that at the highest MLSS concentration of 8 g/L, the flux decline was much greater than for all lesser concentrations. In fact it was an actual decrease of about 84% making the total membrane resistance climb steeply for this entire filtration event. This indicated that fouling was dominated by pore constriction even though all three types of fouling mechanisms were in operation to a greater or lesser degree. This conclusion appears justifiable since the calculated pore constriction parameter, β, was found to be roughly four times the size of the pore blockage parameter, α (when referring to Table 3). Additionally, a weak deposit layer as depicted by R_{bo}, and a big cake removal factor, g_o, indicated a fairly weak cake layer formation which is reinforced by a smaller *f'.R'* factor and a small blocked pores area constant, k_{Ab}. Thus, the combination of parameters *f'.R'*, R_{bo}, g_o and k_{Ab}, implied that cake filtration was relatively less significant during fouling which was expected for such low TMP operations. These finding seem to concur with the results found in the original study by Duclos-Orsello *et al.* [15]. As with the original study, at bulk concentrations of 1 g/L and 2 g/L the resistances and the volumetric flow rates seemed essentially linear.

Figure 8 shows the normalised volumetric flow rates and the total resistances ratios versus the filtration time at a constant TMP of 18 kPa with initial flow rate of 1.24 × 10^{-5} m^3/s, for normalised MLSS bulk concentrations of 7 g/L, 10 g/L and 12 g/L respectively for the data supplied by the pilot MBR unit operated at Coors (UK) [16]. This MBR system's data set had typically high MLSS values that are usually associated with high overall fluid viscosities. This suggested that the flux was expected to decline rapidly at the highest MLSS concentrations. Indeed it is found that at a MLSS concentration of 12 g/L, the flux had declined almost linearly by 63%. High MLSS concentrations at respectively 7 g/L and 10 g/L exhibited this similar declining behaviour albeit at slightly reduced rates of 59% and 61% respectively. Although slightly below theoretical expectations due to arguably a low TMP regime, the linear increase in resistances was probably caused by the combined effect of all three fouling mechanisms occurring simultaneously with pore constriction being slightly more dominant. As shown in Table 3, and as expected for a low TMP operational regime, an analysis of the combination of parameters *f'.R'*, R_{bo}, g_o and k_{Ab}, indicated that cake formation was moderately weak. Moreover, a much bigger β than α suggested that pore constriction was somewhat more prevalent in fouling.

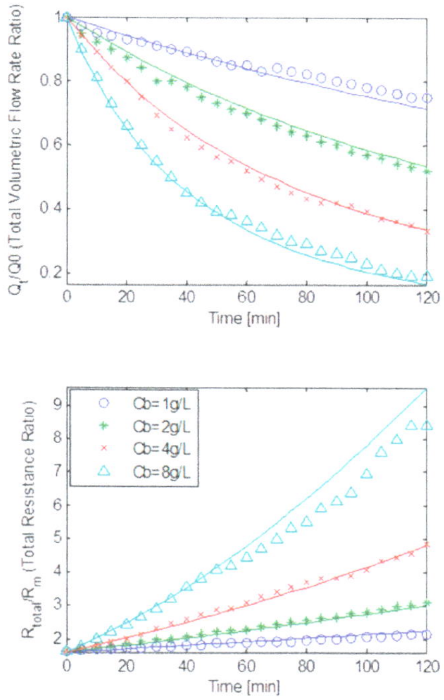

Figure 7. Flux decline and total resistance data obtained from Duclos-Orsello *et al.* [15] with best model fits.

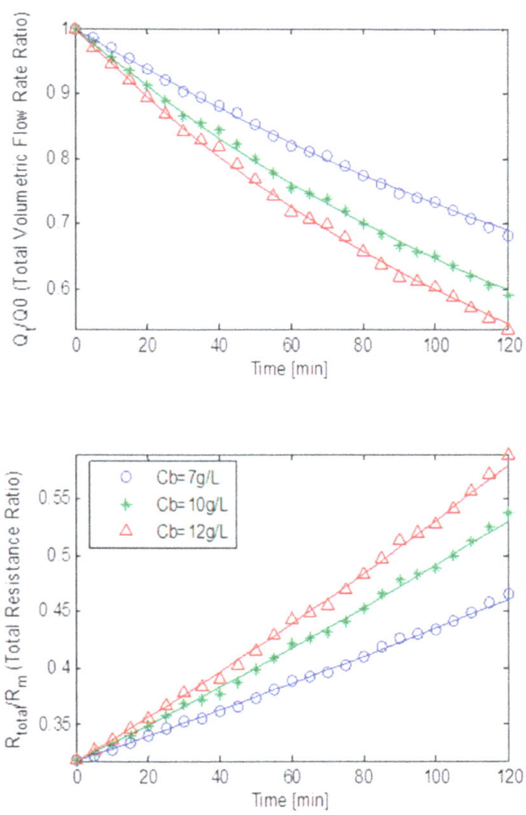

Figure 8. Flux decline and total resistance data obtained from pilot MBR plant located at Coors (UK) with best model fits.

Figure 9 displays the normalised volumetric flow rates and the total resistances ratios plotted against the filtration time at a constant TMP of 15 kPa, for MLSS concentrations of 6.32

g/L and 7.24 g/L for the static square-shaped MBR system. These MLSS concentrations of 6.32 g/L and 7.24 g/L had their fluxes decline at a steady rate and go down approximately by 64%. The resistances displayed an almost linear trend, suggesting fouling was caused by the combination of all three fouling mechanisms. As observed in Table 3, a fairly small R_{bo} factor and a big g_o, indicated the formation of a weak cake layer which was further reinforced by a small $f'.R'$ and k_{Ab}. As a result, the combined parameters $f'.R'$, R_{bo}, g_o and k_{Ab}, suggested that cake filtration was less dominant during fouling which was anticipated since at a low TMP of 15 kPa, cake formation is expected to be less prevalent [17]. Furthermore, β being roughly five times bigger than α, ultimately suggested that fouling was mostly caused by pore constriction.

As can be seen, after two hours of ultra-filtration at MLSS concentrations of 6.32 g/L and 7.24 g/L, and at a constant TMP of 30 kPa, a similar flux decline comparable in size to the 15 kPa TMP case was observed. The rate of decline was steady and this gradual drop was expected since it was in-line with critical flux theory [9].

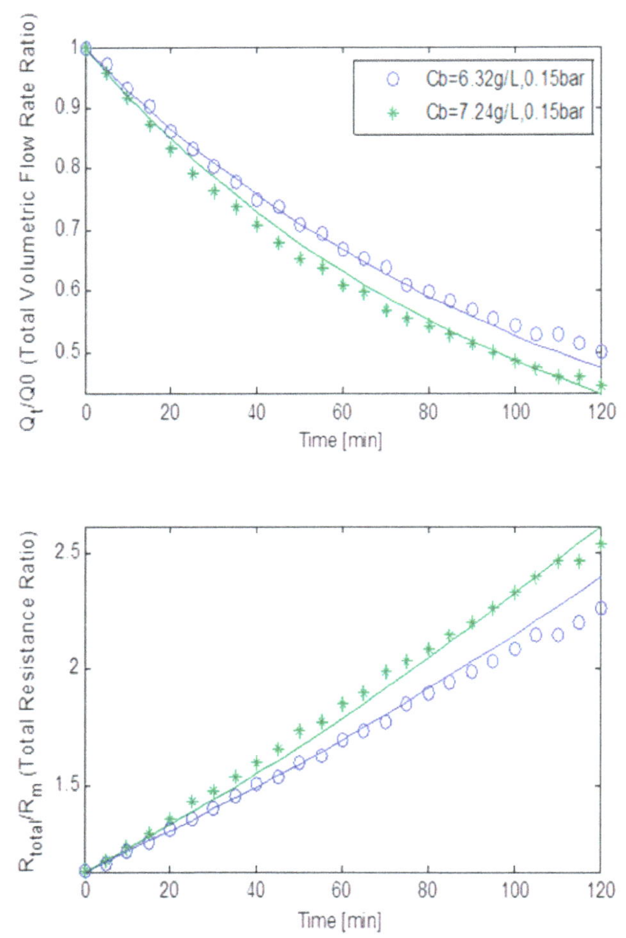

Figure 9. Flux decline and total resistance for TMP step of 15 kPa for the static square-shaped MBR system with best model fits.

Figure 10 portrays the normalised volumetric flow rates and the total resistances ratios plotted against the filtration time at a constant TMP of 30 kPa, for MLSS concentrations of 6.32 g/L and 7.24 g/L for the static square-shaped MBR system. The resistance-time plot, once again, seems to indicate that fouling could be attributed to the combined effect of all three mechanisms. Again as shown in Table 3, a fairly big R_{bo} and a small g_o, indicated a reasonably adequate cake layer formation which was supported by a bigger $f'.R'$ factor and a small k_{Ab} when compared to the data from the constant TMP step of 15 kPa,. Consequently, the combination of parameters $f'.R'$, R_{bo}, $g_o$and k_{Ab}, suggested that cake filtration was adequately prevalent during fouling. In addition α being much smaller than β, all seemingly implied that most of the fouling was dominated by both pore constriction and cake filtration.

Figure 11 represents the effects of the fouling behaviour of the static square-shaped MBR system, using both the normalised volumetric flow rates and total resistance ratios for MLSS concentrations of 6.32 g/L and 7.24 g/L at a constant TMP of 45 kPa. Due to a colossal drop in flux at this high TMP, the total resistance increases at an exponential rate. It is noticeable that the best fit curve at this high TMP is extremely poor (especially after 80 min), but however the curve's trend is of the right scale and in the right direction to allow analysis of the fouling behaviour that is occurring. At first glance, analysis would suggest that fouling may be due to all three fouling mechanisms, but it can be argued that fouling was mainly dominated by cake filtration and pore blocking. A reasonably big R_{bo} and $f'.R'$ coupled with similar sized g_o and k_{Ab}, all strongly indicated that a fairly large cake layer was formed (as seen Table 3). However, with α being roughly twice as big as β, it can be concluded that the pore blocking fouling mechanism was also prevalent during fouling. Hence, the bulk of the fouling was dominated by both cake filtration and pore blocking.

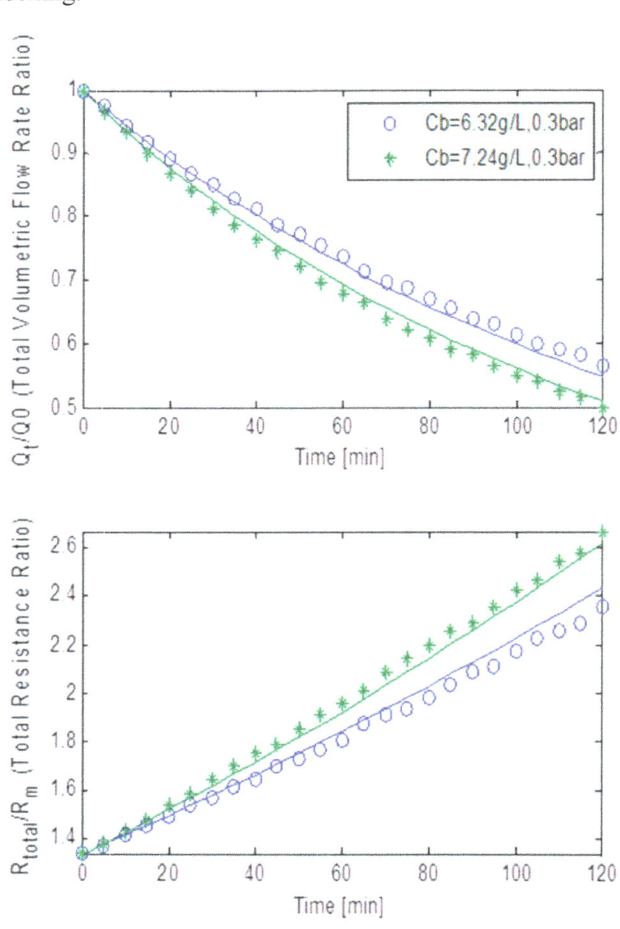

Figure 10. Flux decline and total resistance for TMP step of 30 kPa for the static square-shaped MBR system with best model fits.

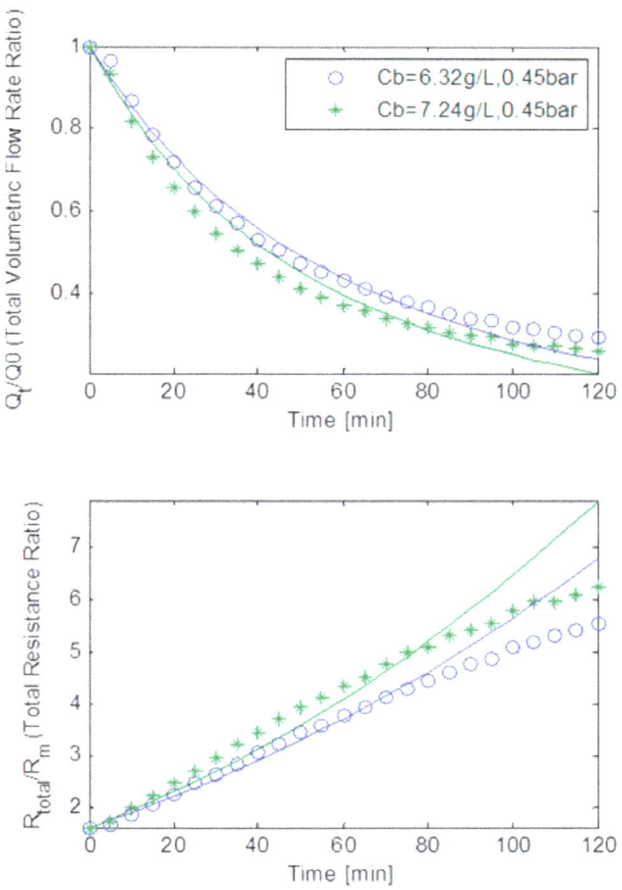

Figure 11. Flux decline and total resistance for TMP step of 45 kPa for the static square-shaped MBR system with best model fits.

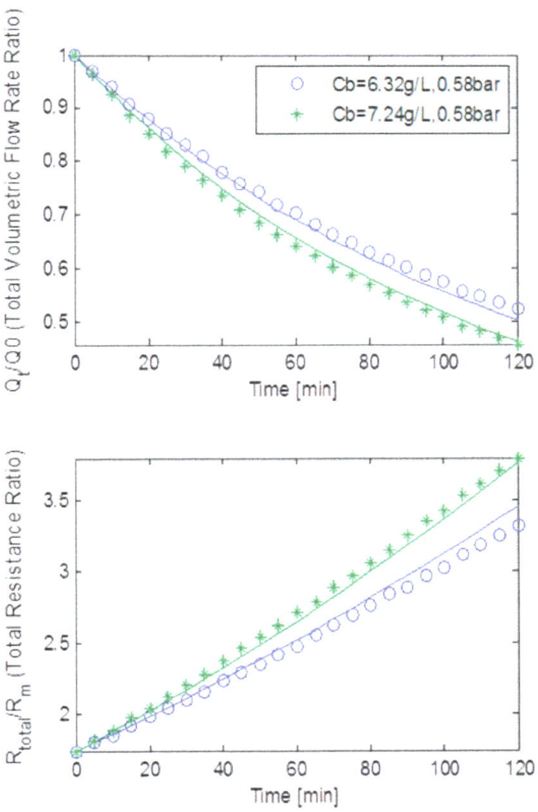

Figure 12. Flux decline and total resistance for TMP step of 58 kPa for the static square-shaped MBR system with best model fits.

Figure 12 shows the fouling behaviour of the static square-shaped MBR system, using both the normalised volumetric flow rates and total resistance ratios for MLSS concentrations of 6.32 g/L and 7.24 g/L at a constant TMP of 58 kPa. The total resistance seemingly increases in a linear fashion with filtration time albeit at a high rate. This is probably because fouling was caused by all three mechanisms happening simultaneously. Parameters α and β being of almost equal size suggests that neither of the two fouling mechanisms was dominant (Table 3). Furthermore, a moderately big R_{bo} and $f'.R'$ when compared to the data from the constant TMP step of 15 kPa, together with an adequate g_o value and a small k_{Ab} value, all indicated that a fairly decent size cake layer was formed. Accordingly, all three fouling mechanisms were of equally great importance during fouling.

Comparison of Simulation Results Generated from Rotating and Static MBR Models

After every simulation done in Matlab, a mean of minimised residuals (*i.e.*, mean fitness) and the best minimised residuals (*i.e.*, best fitness), are automatically calculated. These can in turn be used to accurately determine how well a simulation fit was conducted. Needless to say, the smaller these values, the better analysis can be conducted. Using these facts, a comparison between both models (static and rotating) and their fitness values will be conducted and some conclusions drawn.

When comparing and fitting the data obtained from the rotating MBR system to that obtained from the static square-shaped MBR system, the most prominent difference found for the latter case is the poor fit between the experimental data and the simulation runs. Table 4 summarises the best fitness values after simulations for both systems for the same two TMP steps at constant TMPs of 15 kPa and 45 kPa respectively. As can be seen, the best fitness values for the rotating MBR system are all much smaller than those for the static square-shaped MBR system. In fact the best fitness values for the latter are roughly ten times greater than that for the rotating MBR system. The difference here can be probably attributed to the fact that the shear effects are instrumental in reducing the overall fouling for the rotating MBR system while the static square-shaped MBR system is only aided by standard air scour alone. This concurs with critical flux theory and the area-loss model that explains any rapid TMP jump phenomenon [9], since greater fouling will lead to greater uneven cake build up, leading to loss of clean membrane surface, meaning that locally critical flux could be exceeded, thus promoting the likelihood of rapid declines in fluxes at higher TMPs at unpredictable rates.

Table 4. Statistics with fitness values from simulations as computed in Matlab.

Fitness Statistics	Avanti Rotating MBR		Avanti Static Square-Shape MBR	
Fit Parameters	15 kPa TMP	45 kPa TMP	15 kPa TMP	45 kPa TMP
Best fitness (–)	0.0361	0.0294	0.346	1.293
Mean fitness (–)	0.0849	0.0672	0.968	4.541

In simple terms, this means that for the same constant TMP regime for both systems, the fouling build up expected on the rotational MBR system would be less with a reduced likelihood of local critical flux being exceeded. Thus, data sets obtained under this system would be expected to have flux declines occurring in more predictable and consistent ways, and at reduced levels even at higher TMPs when compared to the static MBR system. Consequently, it would be expected that more consistent and less variable data would give rise to better simulation fits for the rotating MBR system, and conversely produce a much less reasonable and agreeable fit for the static MBR system even though both were manufactured by the same company, using the same pipe manifolds, same membrane materials, same membrane pore sizes, and same spacing between individual membrane sheets.

A logical conclusion from these all these results is that for the rotating MBR system, its additional rotation shear evens out the cake formation on the membrane surface while for the square-shaped MBR system, the cake distribution is highly uneven and less predictable. In physical terms this hypothesis can be confirmed, since it is very evident when carrying out membrane autopsies for both system types, it is clear that huge cake build up occurs on the membrane surface for the static MBR system. Figure 13 and Figure 14 show the actual typical caking patterns observed on both types of membrane sheets when they were individually removed from the bioreactor. It must be particularly noted in Figure 14, the heavy fouling that is occurring on the bottom half of the square membrane sheet where the air scour effect is least prevalent due to the small distance travelled by the coarse air bubbles used. All of these simulation results and actual physical observations, indicate that the rotating MBR system may be a more preferable option if reduced fouling was a key need in the system design.

CONCLUSIONS

Both MBR model types were extensively tested using short term data generated by both pilot units and medium term data from a similar pilot unit operated at Coors (UK) [16]. Reasonable agreement was reached between expected results and simulation outputs, although the rotating MBR system outperformed the static MBR system when both were operated under similar short term TMP stepping regimes, with the former giving much better model fits than the latter.

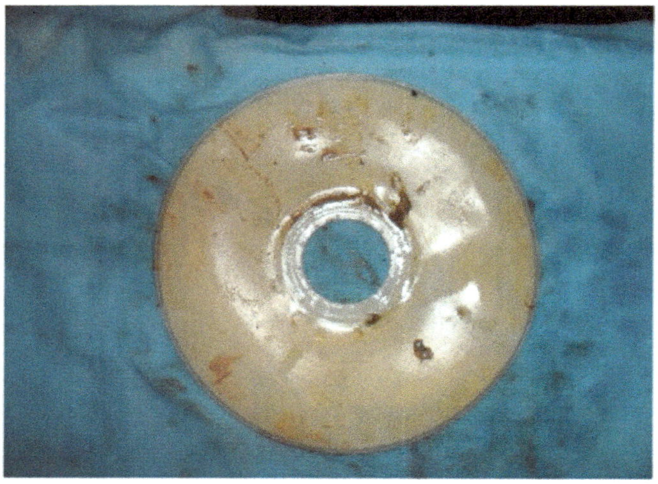

Figure 13. Impact of fouling on the rotating MBR system—even caking is observed.

Figure 14. Impact of fouling on the square-shaped MBR system—uneven caking is observed.

The study concluded that a rotating MBR system could increase flux throughput by a significant amount when compared to a similar static system although there are obvious additional capital and operational cost implications. It is thought that although the slowly rotating spindle induces a very weak crossflow shear, it is still able to even out cake build up across the membrane surface, thus reducing the likelihood of localised critical flux being exceeded and lessening the potential of inducing any subsequent and unpredictable TMP jump phenomenon [9].

Even though the rotating pilot unit exhibited reduced fouling when compared to the static square version, it must be remembered that due to the patented edge sealing method used by Avanti Technology, USA, there are still potential scaling up issues to be addressed for both these MBR configurations.

This critical research work into novel MBR processes for wastewater treatment and recycling adds to the body of research work into the overarching theme of environmental systems engineering. Follow-on work in this area will attempt to confirm and extend these initial findings.

ACKNOWLEDGMENTS

The authors would like to acknowledge Avanti Technology, USA, for their technical contribution and The Royal Society, UK, for providing funding to allow this work to proceed.

AUTHOR CONTRIBUTIONS

The experimental work, data processing, and model development in this study were carried out by Franck Anderson Jones under the direction and supervision of Parneet Paul, and is based upon earlier work originally carried out by Parneet Paul [16]. The first draft of the original manuscript was prepared by Franck Anderson Jones, and later versions were revised and edited extensively by Parneet Paul before publication. This included broadening the scope and nature of this study to be in-line with the work of other researchers in this field of study.

REFERENCES

1. Fenu, A.; Guglielmi, G.; Jimenez, J.; Sperandio, M.;

Saroj, D.; Lesjean, B. Activated sludge model (ASM) based modelling of membrane bioreactor (MBR) processes: A critical review with special regard to MBR specificities. *Water Res.* **2010**, *44*, 4272–4294.

2. Xing, C.H.; Tardieu, E.; Qian, Y.; Wen, X.H. Ultrafiltration membrane bioreactor for urban wastewater reclamation. *J. Membr. Sci.* **2000**, *177*, 73–82.
3. Chang, I.S.; le Clech, P.; Jefferson, B.; Judd, S. Membrane fouling in membrane bioreactors for wastewater treatment. *J. Environ. Eng.* **2002**, *128*, 1018–1029.
4. Zuthi, M.F.R.; Ngo, H.H.; Guo, W.S. Modelling bioprocesses and membrane fouling in membrane bioreactor (MBR): A review towards finding an integrated model framework.*Biores. Technol.* **2012**, *122*, 119–129.
5. Bentzen, T.R.; Ratkovich, N.; Madsen, S.; Jensen, J.C.; Bak, S.N.; Rasmussen, M.R. Analytical and numerical modelling of Newtonian and non-Newtonian liquid in a rotational cross-flow MBR. *Water Sci. Technol.* **2012**, *66*, 2318–2327.
6. Bhattacharjee, C.; Bhattacharya, P.K. Ultrafiltration of black liquor using rotating disk membrane module. *Sep. Purif. Technol.* **2005**, *49*, 281–290.
7. Jaffrin, M.Y.; Ding, L.H.; Akoum, O.; Brou, A. A hydrodynamic comparison between rotating disk and vibratory dynamic filtration systems. *J. Membr. Sci.* **2004**, *242*, 155–167.
8. Jørgensen, M.K.; Malene, T.P.; Morten, L.C. Dependence of shear and concentration on fouling in a membrane bioreactor with rotating membrane discs. *AIChE J.* **2014**, *60*, 706–715.
9. Judd, S. *The MBR Book: Principles and Applications of Membrane Bioreactors*, 1st ed.; Elsevier: Amsterdam, The Netherlands, 2006.
10. Rosenberger, S.; Laabs, C.; Lesjean, B.; Gnirss, R.; Amy, G.; Jekel, M.; Schrotter, J.C. Impact of colloidal and soluble organic material on membrane performance in membrane bioreactors for municipal wastewater treatment. *Water Res.* **2006**, *40*, 710–720.
11. Ahn, Y.T.; Choi, Y.K.; Jeong, H.S.; Shin, S.R. Modeling of extracellular polymeric substances and soluble microbial products production in a submerged MBR at various SRTs. *Water Sci. Technol.* **2006**, *53*, 209–216.
12. Hermia, J. Constant pressure blocking filtration laws-application to power-law non-Newtonian fluids. *Chem. Eng. Res. Des.* **1982**, *60a*, 183–187.
13. Meng, F.; Zhang, H.; Li, Y.; Zhang, X.; Yang, F. Application of fractal permeation model to investigate membrane fouling in membrane bioreactor. *J. Membr. Sci.* **2005**, *262*, 107–116.
14. Liu, R.; Huang, X.; Sun, Y.F.; Qian, Y. Hydrodynamic effect on sludge accumulation over membrane surfaces in a submerged membrane bioreactor. *Process Biochem.* **2003**, *39*, 157–163.
15. Duclos-Orsello, C.; Li, W.; Hob, C.-C. A three mechanism model to describe fouling of microfiltration membranes. *J. Membr. Sci.* **2006**, *280*, 856–866.
16. Paul, P. Development and testing of a fully adaptable membrane bioreactor fouling model for a sidestream configuration system. *Membranes* **2013**, *3*, 24–43.
17. Paul, P.; Jones, F.A. Development of a comprehensive fouling model for a novel rotating membrane bioreactor system. *Water* **2015**, *7*, 377–397.
18. Jaffrin, M.Y. Dynamic shear-enhanced membrane filtration: A review of rotating disks, rotating membranes and vibrating systems. *J. Membr. Sci.* **2008**, *324*, 7–25.
19. Liang, S.; Song, L.; Tao, G.; Kekre, K.A.; Seah, H. A modeling study of fouling development in membrane bioreactors for wastewater treatment. *Water Environ. Res.* **2006**, *78*, 857–863.
20. Yang, F.; Bick, A.; Shandalov, S.; Brenner, A.; Oron, G. Yield stress and rheological characteristics of activated sludge in an airlift membrane bioreactor. *J. Membr. Sci.* **2009**, *334*, 83–90.
21. Ratkovich, N.R.; Horn, W.; Helmus, F.; Rosenberger, S.; Naessens, W.; Nopens, I.; Bentzen, T.R. Activated Sludge Rheology: A critical review on data collection and modelling. *Water Res.* **2013**, *47*, 463–482.

CHAPTER 3

BREEDING OF A NEW WASTEWATER TREATMENT YEAST BY GENETIC ENGINEERING

Miyoshi Kato and Haruyuki Iefuji

Keywords: Wastewater; membrane bioreactor (MBR); rotating membranes; fouling; modelling; cake filtration

INTRODUCTION

Many food factories use wastewater treatment systems that use yeasts (Yoshizawa 1978, 1981, Sato et al. 1986, Moriya et al. 1990, Suzuki et al. 1991, Suzuki et al. 1996). However, some wastewater-containing polysaccharides, such as raw starch and hemicellulose, are difficult to treat because presently used yeasts secrete few enzymes that can digest these polysaccharides. One way to treat these wastewaters is to transform conventional wastewater treatment yeasts with the genes for polysaccharide-digesting enzymes such as raw starch-digesting α-amylase and acid xylanase.

To this end, we isolated *Cryptococcus* sp. S-2 (Iefuji et al. 1994), which secretes several enzymes including raw starch-digesting α-amylase (Iefuji et al. 1996a), acid xylanase (Iefuji et al. 1996b), lipase (Kamini et al. 2000) and polygalacturonase. We then obtained the genes that encode the raw starch-digesting α-amylase and acid xylanase.

H. fabianii J640 is a commonly used wastewater treatment yeast (Saito et al. 1987, Sato et al. 1987, Suzuki et al. 1996). We previously constructed an expression system based on this strain (Kato et al. 1997). A uracil auxotrophic mutant of *H. fabianii* J640, named *H. fabianii* J640 u-1, lacking orotidin-5'-phosphate decarboxylase, was obtained. We constructed a plasmid, pHFura3, that contains the gene encoding orotidine-5'-phosphate decarboxylase of *H. fabianii* J640. In the previous study (Kato et al. 1997), by employing *H. fabianii* J640 u-1 as a host strain and pHFura3 as a vector plasmid, we constructed a transformation system of *H. fabianii* J640. We purified the glucoamylase of *H. fabianii* J640 and cloned its cDNA and genomic DNA (Kato et al. in press). Then, we constructed a new expression vector, pHFGE-1 (Kato et al. in press), which uses pHFura3, and the promoter and terminator regions of the gene encoding glucoamylase from *H. fabianii* J640. We inserted the genes encoding α-amylase and xylanase from *Cryptococcus* sp. S-2 between the promoter and terminator of pHFGE-1. When the pHFGE-1 with one or the other of these foreign genes were transformed into *H. fabianii* J640 u-1, the transformants (named HF-AAMY and HF-XYN, respectively) showed α-amylase and xylanase activities respectively. This showed that pHFGE-1 can derive the expression of foreign genes in *H. fabianii* J640 cells.

In this paper, we investigated the ability of these transformed yeasts, to treat wastewater, and developed a PCR method for monitoring the presence of the foreign gene.

MATERIALS AND METHODS

Strains and Media

Strains *H. fabianii* J640 and *Cryptococcus* sp. S-2 were obtained from the National Research Institute of Brewing culture collection, Japan. A uracil auxotrophic mutant of *H. fabianii* J640, named *H. fabianii* J640 u-1, lacking orotidine-5'-phosphate decarboxylase, was used as a host strain for new expression vector pHFGE-1. *S. cerevisiae*YPH-499 (MATα ura3 lus2 ade2 trp1 his3 leu2) was used as the host for transformation vector pG-1 (Schena et al. 1991). *E. coli* strain HB101 and JM109 were employed as the host of plasmid vector, which were used for DNA manipulation and construction of the gene library.

Yeast cells were grown on YM medium (0.3% yeast extract, 0.3% malt extract, 0.5% peptone and 1% glucose) and

Citation: Kato and Iefuji: Breeding of a new wastewater treatment yeast by genetic engineering. AMB Express 2011 1:7. doi:10.1186/2191-0855-1-7

YPD medium (1% yeast extract, 2% peptone, 2% glucose). Luria-Bertani medium containing ampicillin (100 μg/ml) was used to cultivate *E. coli*. The minimal medium containing 1% glucose and 0.67% yeast nitrogen base (YNB) without amino acids was used to select the yeast transformants. YPM medium was prepared by replacing the glucose of YPD with maltose. The medium used to investigate expression induction, contained 1% yeast extract, 1% casamino acid, and 2% glucose or maltose.

Expression Vector for *H. fabianii* J640

The expression vector pHFGE-1 (Kato et al. in press) (Figure 1A) was used. The cloning site of this vector is a *Bam*HI site between the promoter and terminator from *H. fabianii* J640 glucoamylase DNA. The host cell of this vector is a uracil auxotrophic mutant designated as *H. fabianii* J640 u-1, and it could be transformed by a non-homologous and frequently multicopy integration into the host genomic DNA.

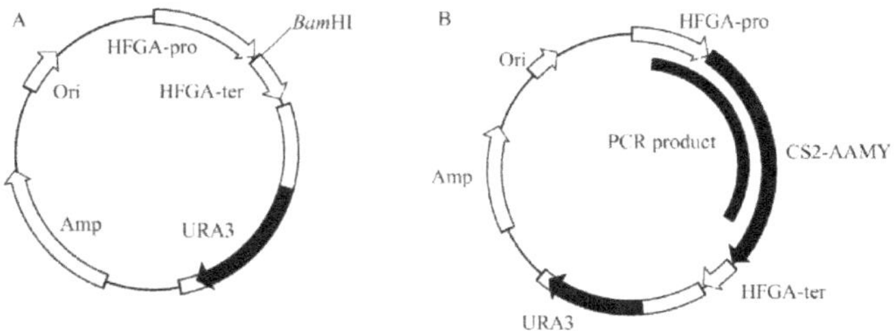

Figure 1 Restriction map. (A) Expression vector pHFGE-1. (B) Position of PCR product in pHFGE-AMY for monitoring (black arc inside circle). CS2-AAMY, α-amylase gene from *Cryptococcus* sp. S-2

Transformation of Yeast

Transformations were carried out by electroporation as described by (Becker et al. 1991). Electroporation was done with a Gene Pulser (Bio-Rad) with settings of 200 V and 25 μF using a 0.2 cm cuvette.

Assay of Xylanase and α-amylase Activity

Xylanase activity was assayed by measuring the amount of reducing sugar liberated from xylan (Iefuji et al. 1996b). One unit of activity was defined as the amount of xylanase needed to liberate 1 μmol of D-xylose per min under the condition just described.

α-Amylase activity was measured with an α-amylase kit (Kikkoman). One unit of α-amylase activity was defined as the amount of enzyme which forms 1 μmol of 2-choloro-4-nitrophenol from 2-choloro-4-nitrophenyl 6^5-azide-6^5-deoxy-β- maltopentaoside under the condition described above.

Preparation of Model Wastewater and Treatment Test

Model wastewater containing soluble starch was made with 1% refined starch (Merck) and 0.25% yeast extract, pH 6.0. The starch was solubilized by autoclaving. Model wastewater containing insoluble starch was made with 0.25% yeast extract, pH6.0, autoclaved and cooled to approximately 55°C. The same amount of starch was sterilized in 70% ethanol. The suspension was centrifuged and decanted. The starch pellet was then added to the autoclaved yeast extract solution.

Yeast cells were incubated at 30°C for 2 days on YM medium. Then 5 × 10^6 cells/ml was inoculated to the model wastewater in an Erlenmeyer flask. Cultures were incubated at 30°C with shaking at 105 rpm and samples were periodically harvested.

Yeast Cells in the Model Wastewater were counted with a Hemocytometer

The model wastewater containing soluble starch was centrifuged at 3000 rpm for 10 min, and chemical oxygen demand (COD) of the supernatant was measured. The decrease in COD of the model wastewater containing soluble starch was used to express the capacity of the yeast to treat the wastewater.

It was not possible to measure COD of the model wastewater containing insoluble starch because of the difficulty in separating the cells and insoluble starch. In this case, degradation of the starch was measured with the iodo-starch reaction (Sato et al. 1987) as follows: 1 ml culture was heated in a micro tube at 100°C for 30 min to solubilize the starch. Yeast cells were then removed by centrifugation. Iodic liquid (0.2 ml; containing 0.0317 g iodine, 0.1 g potassium iodide and 5 ml 3N-HCl in 100 ml water) was added to the supernatant and the optical density was measured at 670 nm. Transmittance at 670 nm was taken as a measure of starch degradation.

Monitoring the Presence of a Foreign Gene in a Transformant

The transformants were cultured in 10 ml YM medium and harvested by centrifugation. DNA was extracted with an Easy-DNA kit (Invitrogen) and used for the PCR template. Unique PCR primers were designed, and the position of the PCR product is shown in Figure 1B. PCR cycling conditions were followed by 25 cycles of 94°C for 1 min, 55°C for 2 min, 72°C for 3 min.

To determine the sensitivity of the PCR, cells were cultured in YM medium, and the cell density was measured. Then a dilution series was made (10^6-10^1 cells/ml). One ml of each dilution was harvested and DNA was extracted with the EASY-DNA kit and used as a PCR template.

RESULTS

Induction of Foreign Gene Expression

Glucoamylase production by *H. fabianii* J640 was induced by maltose and repressed by glucose. Since our constructed expression vector used the promoter and terminator regions of the *H. fabianii* J640 glucoamylase gene, we expected that foreign gene expression in the transformant would also be induced by maltose. As expected, xylanase production by HF-XYN was highest, when maltose was the C source with yeast extract and casamino acid as media components (Table 1).

Table 1 Effect of media components on xylanase activity

Media composition	Xylanase activity (U/ml)
Maltose, YNB w/o amino acids	37
Maltose, Yeast extract, Casamino acid	310
Glucose, YNB w/o amino acids	1
Glucose, Yeast extract, Casamino acid	6

Comparison of Two Vectors

The performance of our expression vector pHFGE-1 was compared with that of the widely used pG-1 transformation vector. *H. fabianii* J640 u-1 (HF-TAMY) cells were transformed with pHFGE-1, and *S. cerevisiae* YPH-499 (SC-TAMY) cells were transformed with pG-1, both of which carried the Taka-amylase gene. The cells were then cultured on YPD and YPM media. Growth on YPD medium was the same for the two cultures (Figure 2A). α-Amylase activity was a little higher in the HF-TAMY cells than in the SC-TAMY (Figure 2B). The α-amylase activity of HF-TAMY cells was highest when maltose was the C source (YPM medium, Figure 2D), indicating that gene expression was induced by maltose.

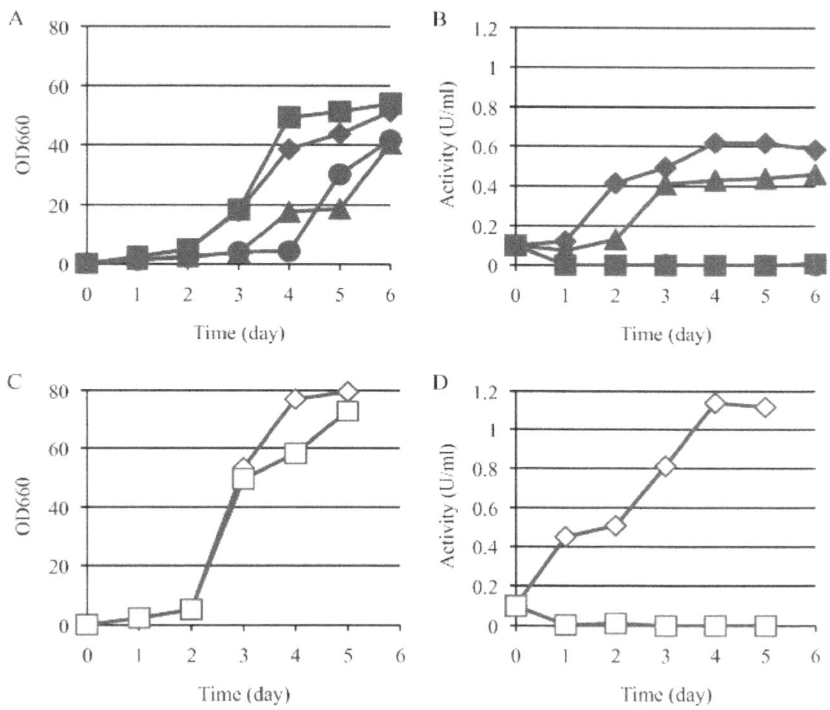

Figure 2 Comparison of two vectors and carbon sources. (A) Growth of transformants on YPD medium. (B) α-Amylase activity of transformant on YPD medium. (C) Growth of transformants on YPM medium. (D) α-Amylase activity of transformant on YPM medium. The strains are: HF-TAMY (diamonds), transformant with pHFGE-1 (not connected to any gene), into *H. fabianii* J640 u-1 (squares), SC-TAMY (triangles), transformant with pG-1 (not connected to any gene), into *S. cerevisiae* YPH-499 (circles).

Treatment of Model Wastewater

HF-AAMY cells and cells of the parent strain *H. fabianii* J640 grew at about the same rate in the model wastewater containing soluble starch or insoluble starch (Figure 3A or 4A). Both the parent strain and HF-AAMY decreased the COD of the model wastewater containing soluble starch to decrease, although the decrease was much faster with the HF-AAMY cells (Figure 3B). The HF-AAMY cells were also much more efficient at degrading the insoluble starch (Figure 4B). These results indicate that HF-AAMY cells have a high capacity to treat wastewater containing starch.

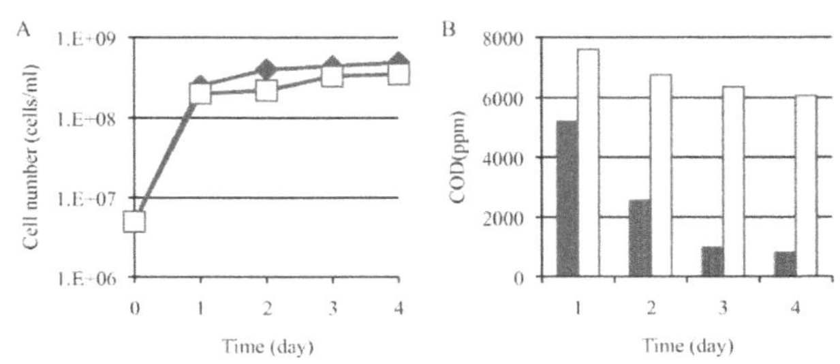

Figure 3 Treatment test of model wastewater containing soluble starch. (A) Growth rate of cells. (B) Decrease of COD. The strains are: HF-AAMY (diamonds, black bars), *H. fabianii* J640 (host strain) (squares, white bars).

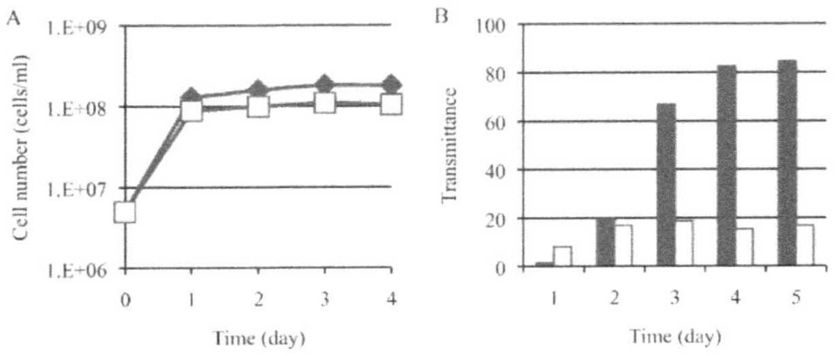

Figure 4 Treatment test of model wastewater containing insoluble starch. (A) Growth rate of cells. (B) Resolution capacity of the insoluble starch. The strains are: HF-AAMY (diamonds, black bars), *H. fabianii* J640 (host strain) (squares, white bars).

Monitoring of Transformant by PCR

Of four strains (*S. cerevisiae* YPH-499, *Cryptococcus* sp. S-2, *H. fabianii* J640 and HF-AAMY (the transformant)), only the transformant showed a PCR product (Figure 5A) corresponding to part of the HFGA promoter and the α-amylase gene (Figure 1B). The detection sensitivity of PCR which uses Taq Plus Long PCR kit (Stratagene) was high, i.e., it could detect only 10^4 cells (Figure 5B). The different intensities of the PCR bands in Figure 5B are presumably the result of the different cell densities in the cultures.

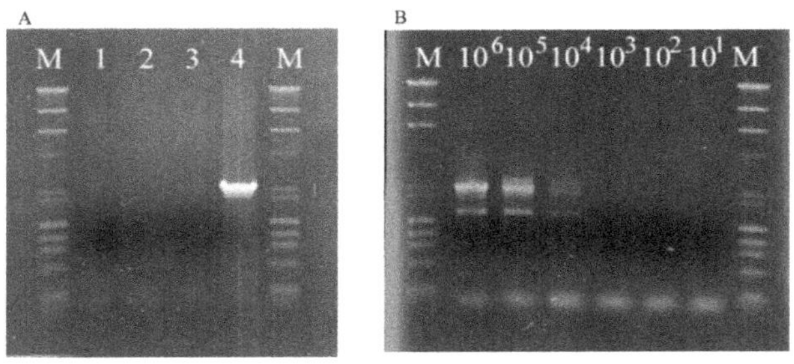

Figure 5 PCR test. (A) Specificity of PCR test for HF-AAMY cells. M, marker The strains are: (1) *S. cerevisiae* YPH-499, (2)*Cryptococcus* sp. S-2, (3) *H. fabianii* J640 (host strain), (4) HF-AAMY (transformant). (B) Sensitivity of PCR test. The number of cells in the reaction mixture are shown at the tops of the lanes.

DISCUSSION

We developed a host vector system for the wastewater treatment yeast, *H. fabianii* J 640, and we created new wastewater treatment yeast transformants (HF-XYN and HF-AAMY). The expression of the foreign gene that was integrated in the transformant was induced by maltose and repressed by glucose. However, the growth rates of the transformants carrying pHFGE-1 and the widely used pG-1 were the same and both transformants strongly expressed the foreign gene, even in medium containing glucose, which was expected to repress expression of the foreign gene. Our host vector system strongly expresses the foreign gene. Because wastewater contains various components, the strong expression of the new strain is an advantage. The HF-AAMY cells were effective in treating the model wastewater.

Because HF-AAMY cells are genetically modified, a sensitive method for monitoring the cells in the environment is needed. Our PCR was shown to satisfy this requirement.

A host vector system was also developed for the methylotrophic yeast *Hansenula polymorpha* (Gellissen et al. 2004,Steinborn et al. 2006). As in these systems, auxotrophic strains (ura-, leu-) were used as the host. The expression cassettes in these systems used the promoters for various genes, including the genes for formate dehydrogenase (FMD), methanol oxidase (MOX), and trehalose-6-phosphate synthase (TPS1). *H. polymorpha* is rapidly becoming the system of choice for heterologous gene expression in yeast. Several production processes for recombinant pharmaceuticals and industrial enzymes have been developed based on gene expression in this strain. Another methylotrophic yeast, *Hansenula ofunaensis*, has also been evaluated for a transformation system (Yamada-Onodera et al. 1999, Yamada-Onodera et al. 2006) but development has not been completed.

A transformation system using *Hansenula anomala*, another wastewater treatment yeast, was developed in the 1990s (Ogata et al. 1992, Ogata et al. 1995). However, none of these studies of wastewater treatment yeasts constructed an expression vector or bred new strains of yeast. With the new transformation system, it should be possible to treat wastewater containing polysaccharides that are presently resistant to degradation.

Our next goal is to use our transformant to treat real wastewater from the food industry. In the future, when genetically engineered yeast is proven to be effective for the treatment of wastewater, a major task will be to prove to the public that the methodology is safe.

REFERENCES

1. Becker DM, Guarente L: High-efficiency transformation of yeast by electroporation. *Methods Enzymol* 1991, 194:182–187.
2. Gellissen G, Hollenberg CP: Encyclopedia of Food Microbiology. *Hansenula* 2004, 976–982.
3. Iefuji H, Iimura Y, Obata T: Isolation and characterization of a yeast *Cryptococcus* sp. S-2 that produces raw starch-digesting **α-amylase,** xylanase, and polygalacturonase. *Biosci Biotechnol Biochem* 1994, 58: 2261–2262. 10.1271/bbb.58.2261
4. Iefuji H, Chino M, Kato M, Iimura Y: Raw-starch-digesting and thermostable **α-amylase** from the yeast *Cryptococcus*sp. S-2: purification, characterization, cloning and sequencing. *Biochem J* 1996, 318: 989–996.
5. Iefuji H, Chino M, Kato M, Iimura Y: Acid xylanase from yeast *Cryptococcus* sp. S-2: purification, characterization, cloning, and sequencing. *Biosci Biotechnol Biochem* 1996, 60: 1331–1338. 10.1271/bbb.60.1331
6. Kamini NR, Fujii T, Kurosu T, Iefuji H: Production, purification and characterization of an extracellular lipase from the yeast, *Cryptococcus* sp. S-2. *Process Biochem* 2000, 36: 317–324. 10.1016/S0032-9592(00)00228-4
7. Kato M, Iefuji H, Miyake K, Iimura Y: Transformation system for a wastewater treatment yeast, *Hansenula fabianii*J640: isolation of the orotidine-5'-phosphate decarboxylase gene (URA3) and uracil auxotrophic mutants. *Appl Microbiol Biotechnol* 1997, 48: 621–625. 10.1007/s002530051105

8. Kato M, Kitajima T, Iefuji H: Isolation and characterization of glucoamylase from a wastewater treatment yeast*Hansenula fabianii* J640, and construction of expression vector. *Appl Microbiol Biotechnol*, in press.

9. Moriya K, Iefuji H, Shimoi H, Sato S, Tadenuma M: Treatment of distillery wastewater discharged from beet molasses-spirits production using yeast. *J Ferment Bioeng* 1990, 69: 138–140. 10.1016/0922-338X(90)90204-A

10. Ogata T, Iimura Y, Obata T, Tamura G: Isolation and nucleotide sequence of *Hansenula anomala* URA3 gene encoding orotidine-5'- phosphate decarboxylase. *J Ferment Bioeng* 1992, 74: 352–357. 10.1016/0922-338X(92)90030-X

11. Ogata T, Okumura Y, Iimura Y, Obata T: Development of an integrative DNA transformation system for the yeast*Hansenula anomala* . *J Ferment Bioeng* 1995, 79: 1–5. 10.1016/0922-338X(95)92734-T

12. Saito K, Nakao T, Shima Y, Shimoi H, Sato S, Tadenuma M: Utilization of waste water discharged from pot-still during spirits-making. *J Brew Soc Jpn* 1987, 82: 444–446.

13. Sato S, Maetani T, Yamamoto N, Hasuo T, Tadeuma M, Yoshizawa K: Selection of flocculent yeast and its application to waste water treatment. *J Brew Soc Jpn* 1986, 81: 621–625.

14. Sato S, Otani M, Shimoi H, Saito K, Tadenuma M: Selection of flocculent mutants of waste water treatment utilization of wild yeasts. *J Brew Soc Jpn* 1987, 82: 515–519.

15. Schena M, Picard D, Yamamoto KR: Vectors for constitutive and inducible gene expression in yeast. *Methods Enzymol* 1991, 194: 389–398.

16. Steinborn G, Böer E, Scholz A, Tag K, Kunze G, Gellissen G: Application of a wide-range yeast vector (CoMed™) system to recombinant protein production in dimorphic *Arxula adeninivorans* , methylotrophic *Hansenula polymorpha* and other yeasts. *Microb Cell Fact* 2006, 5: 33. 10.1186/1475-2859-5-33PubMed Central

17. Suzuki O, Sato S, Iefuji H, Shimoi H, Tadenuma M, Yoshizawa K: Utilization of thermotolerant and flocculent yeast for wastewater treatment. *Hokkokogaku Kaishi* 1991, 69: 83–87.

18. Suzuki O, Kobari O, Iefuji H, Akita O: Breeding of yeast for waste water treatment by protoplast fusion. *J Brew Soc Jpn*1996, 91: 521–526.

19. Yamada-Onodera K, Ono K, Tani Y: Purification and characterization of an enzyme that has dihydroxyacetone-reducing activity from methanol-grown *Hansenula ofunaensis* . *J Biosci Bioeng* 1999, 88: 148–152. 10.1016/S1389-1723(99)80193-5

20. Yamada-Onodera K, Nakajima A, Tani Y: Purification, characterization, and gene cloning of glycerol dehydrogenase from *Hansenula ofunaensis* , and its expression for production of optically active diol. *J Biosci Bioeng* 2006, 102:545–551. 10.1263/jbb.102.545

21. Yoshizawa K: Treatment of waste-water discharged from a Sake' brewery using yeast. *J Ferment Technol* 1978, 56:389–395.

22. Yoshizawa K: Development of the new treating methods of waste water from food industry using yeast. *Nippon Nogeikagaku kaishi* 1981, 55: 705–711.

CHAPTER 4

AN ARTIFICIAL NEURAL NETWORK MODEL FOR WASTEWATER TREATMENT PLANT

Abdullah Erdal Tümer, Serpil Edebali

Keywords: Artificial Neural Network, Modelling, Wastewater Treatment Plant, Performance

INTRODUCTION

In last two decades, Artificial Neural Network (ANN) methods have been applied to various areas of environmental issues such as wastewater treatment. The wastewater treatment process is quite complex. However the developments in intelligent methods make them possible to use in complex systems modelling [1]. It can be used for better prediction of the process performance owing to their high accuracy, adequacy and quite promising applications in engineering [2,3,4]. There are certain key descriptions of variables which can be used to assess the wastewater treatment plant performance. These variables consist of Chemical Oxygen Demand (COD), Biological Oxygen Demand (BOD) and Total Suspended Solids (TSSs). Until now, most of the available studies for modelling Waste Water Treatment Plants (WWTPs) used these variables. The ANN based models find out very satisfactory results. Some of these models based on ANNs are as follows. ANN model was developed for BOD removal process in horizontal subsurface flow constructed wetlands by Akratos et al. [5]. Farouq S et al. in [6] used neural network with single input and multi-input layers and gave comparable predictions of the plant performance criteria. To prediction of biological oxygen demand and suspended solid concentrations based on ANN were presented by Hamed et al. [7]. Total suspended solid (TSS) is an indication of plant performance. A simple prediction models based on neural network for TSS was demonstrated in [8]. Many other ANNs model for wastewater treatment have been proposed either in the past [9-13] or more recently [14- 18]. In this work, an effort has been made to model the wastewater treatment data using the artificial neural networks to predict the performance of Konya Wastewater Treatment Plant. The data set was consist of flow rate, COD, BOD, TSS and ph at 25°C temperature as input layer variables and TSS as output layer variables. The samples collected on a day during for a period of 4 month. The results of the modelling study expressed briefly high correlation coefficient (R) between the measured and predicted output variables reaching up to 0.9.

Citation: Abdullah Erdal TÜMER, Serpil EDEBALİ, An Artificial Neural Network Model for Wastewater Treatment Plant of Konya, DOI: 10.18201/ijisae.65358

MATERIALS AND METHODS

Artificial Neural Network

Artificial Neural Network (ANN) is an information processing system derived from biological nervous systems of brain. The goal of an ANN is to compute between output and input values with some internal calculations [19]. There is a two-stage operation mode of artificial neural networks. One of them is training the other testing stage. Once it must be trained to use an artificial neural network. The training is carried out using some of the inputs and outputs data set. ANN makes a generalization of these data. Artificial neural networks consists of three layers, including inputs, output and hidden layers, and there are many neurons in each layer as shown (Figure.1).

A neuron is consist of five main elements including; inputs, weights, summing function, activation (transfer) function and output. Neurons can have multiple inputs, but it can only have one output. Inputs information comes from outside networks or other neurons. In some cases, neurons can create self-input with feedback. A neuron usually receives many simultaneous inputs. Each input has its own relative weight which gives the input the impact that it needs on the processing element's summation function. Weights are adaptive coefficients within the network that determine the intensity of the input signal as registered by the artificial neuron.

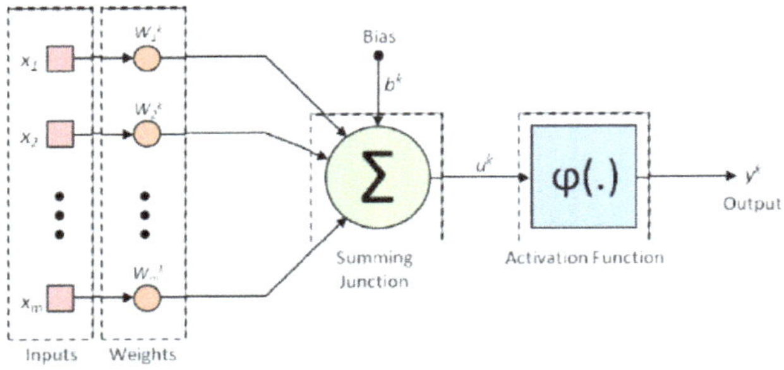

Figure 1. Architecture of ANN

They are a measure of an input's connection strength. These strengths can be modified in response to various training sets and according to a network's specific topology or through its learning rules. The first step in a processing element's operation is to compute the weighted sum of all of the inputs. Summing function operation sums all of the inputs from the processing elements. The formula is as follows:

$$f(x) = \sum_{i=1}^{n}(W_{ix} + a_i) \qquad (1)$$

Table 1. Transfer Function s Used in Algorithms

Transfer Function	***Algorithm***
Tansig	tansig(n) = 2/(1 + exp (-2n)) – 1
Purelin	purelin(n) = n, for all n
logsig	logsig(n) = 1(1 + exp (-n))

Where w is the weight factor and a is the input values.

Transfer function or activation function is a simple function that uses the net value to generate an output. This output is then propagated to the neurons in the next layer.

We used various types of transfer functions as shown table 1.

Also there are many different types of training algorithms. One of the most common training algorithms is Feed Forward Back Propagation algorithm (FFBP) [20].

Konya Wastewater Treatment Plant

Increasing water need in Konya closed basin, which has a large agricultural area, make it necessary to reuse wastewater by refining as a result of drought, major decreases in ground water level. For this purpose, Konya Wastewater Treatment Plant (KWWTP) set into operation in 2010. KAAT is designed for 200.000 m3/day flow and 1.000.000 equivalent population. The Plant is designed according to carbon removal and partial nitrogen removal biological treatment method. Exiting water is discharged by passing through UV disinfection system. KWWTP consists of 4 different units as seen in (Figure. 2).

1. Preliminary Treatment Units:
   - Wastewater inlet structure coarse grid, entrance pump station,
   - Coarse and fine grids
   - Aeration Grid and Grease Removal Unit
   - Preliminary settling pool
2. Biological Treatment Units:
   - Aeration Tanks/ Activated Sludge Reactors (Bardenpho Process)
   - Secondary Sedimantation Tanks
3. Sludge Treatment, Biogas Production and Energy Recovery Units:
   - Sludge Dewatering Tanks
4. Mesophilic Anaerobic Sludge Digesters:
   - Biogas Storage Tanks
   - Heating plant and energy recycling unit.
   - Sludge Dehydratation Plant (Decanter Centrifuge)
5. Disinfection:
   - Open Channel UV disinfection system.

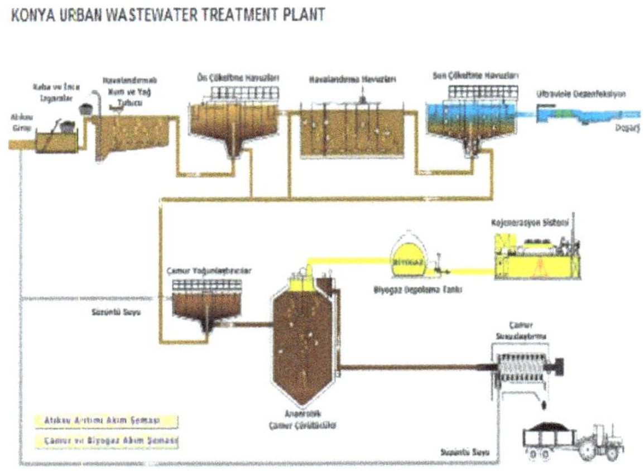

Figure 2. Flow Diagram of the KWWTP

Table 2. The range of data set and their statistic

Parameters	Data Statistic		
	Ranges	Mean ± S.D.	Unit
Input Layer			
Flow rate	129,00 - 214,00	176,00 ± 13,46	m^3 day
Temperature	4,30 - 21,90	14,39 ± 3,35	°C
pH	7,44 - 9,37	8,09 ± 0,43	
TSS	160,00 - 3,47	404,23 ± 335,32	mg/l
BOD	260,00 - 860,00	432,59 ± 110,29	mg/l
COD	382,00 - 1940,0	838,91 ± 238,49	mg/l
Output Layer			
TSS	11,00 - 120,00	27,30 ± 15,49	mg/l

Data Collection

Measurements of the COD, BOD, and TSS were collected 4 month period. This period was satisfactory as it covers all probable seasonal variations in the studied variables. Influent and effluent parameters and their statistic are given table 2. Artificial Neural Network ANN model was developed to modeling a WWTP with the collected data. And in order to obtain convergence within a reasonable number of cycles, the input and output data should be usually normalized.

Normalization method as follows:

$$x' = 0{,}8 * \frac{x_i - x_{min}}{x_{max} - x_{min}} + 0{,}1 \quad (2)$$

where x′ is the normalized value, x is the original data, xmax and xmin, respectively, is the maximum and minimum values, thus the scaled data ranging between 0,1 and 0,9.

Network Properties

The neural network model was created in MATLAB. MATLAB Toolbox opens the Network/Data Manager window, which allows the user to import, create, use, and export neural networks and data [21].

The Network properties are as follows:

- Network inputs: flow rate, pH, COD, BOD, Temperature and TSS.
- Network outputs: TSS.
- Network type: Feed-Forward BackPropagation.
- Training function: TRAINLM.
- Adaption learning function: LEARNGDM.
- Performance function: MSE.
- Number of hidden layers: 1,2,3 used respectively
- Input Layer Transfer function: logsig, purelin and tansig
- Output Layer Transfer function: purelin

The architecture of an ANN model is defined by means of its number of layers and its number of neurons per layer. In this study, we used different numbers of hidden layers and neurons in each hidden layer for find out the best suitable model. Unfortunately, there are no well-developed rules to determine the most suitable values of these parameters for a specific problem. In this case, these values were determined by means of a trial and error method in which the performances of nine different architectures were compared. The architectures considered present the same number of neurons in each hidden layer, range from seven to eleven. These architectures as follows:

with one hidden layer were [3-7-1], [3-8-1], [3-9-1], [3- 10-1] and [3-11-1];

with two hidden layers were [3-7-7-1], [3-8-8-1], [3-9- 9-1], [3-10-10-1], and [3-11-11-1]

with three hidden layers were [3-7-7-7-1], [3-8-8-8-1], [3-9-9-9-1], [3-10-10-10-1], and [3-11-11-11-1].

In order to assess the performance of each architecture the correlation coefficient (R) and mean square error (MSE) were used. The MSE is given by the following expression,

$$MSE = \frac{1}{N} + \sum_{n=1}^{n} \left(q_{(i,pre)} - q_{(i,exp)}\right)^2 \quad (3)$$

where n is the number of data points, is the network prediction, is the experimental response and i is an index of the data [22]. 3.

RESULTS AND DISCUSSION

Artificial Neural Network method, which is previously used in various modelling studies, is used in this study for estimating TSS values of treated wastewater in KWWTP. MATLAB program is used for executing the process. The model uses the trainlm - Levenberg–Marquardt function for training. The MATLAB program randomly divides input variables and target variables into three sets. 60 % of the samples are assigned to the training set, 20% to the validation set, and 20% to the test set. Many different functions such as "logsig", "tansig", and "purelin" were tested as transfer function for hidden layer and "purelin" function was used as the output layer transfer function.

Table 3. 3-2-1 Summary of training and transfer functions used for removal efficiency

Training Functions	Transfer Functions	trainin R	test R	MSE
7	Logsig	0,98	0,81	0,0045913
	Purelin	0,66	0,63	0,0113160
	Tansig	**0,99**	**0,94**	**0,0096608**
8	Logsig	0,99	0,89	0,0053830
	Purelin	0,61	0,50	0,0054988
	Tansig	0,99	0,90	0,0058794
9	Logsig	0,99	0,82	0,0067840
	Purelin	0,57	0,58	0,0041831
	Tansig	0,92	0,86	0,0066576
10	Logsig	0,98	0,87	0,0090310
	Purelin	0,74	0,58	0,0041230
	Tansig	0,99	0,91	0,0039078
11	Logsig	0,99	0,91	0,0054450
	Purelin	0,59	0,65	0,0027967
	Tansig	0,99	0,93	0,0040825

Learngdm (gradient descent with momentum weight and bias learning function) function was used as the adaption learning function and MSE and R was determined as the performance function. The network models repeated with different neurons (7-11), transfer functions and number of ANN configuration mentioned section 2. The correlation coefficient (R) and MSE values belonging to these different ANN configuration's results are presented in Table 3,4,5 respectively in detail.

Models were designed to generate the minimum mean squared error (MSE-it measures the network's performance according to the mean of squared errors) and maximum correlation coefficient (R) values with 7-11 neurons in hidden layers and tansig, logsig and purelin transfer functions.

The suitable training algorithm at different layers, the number of hidden layers, the number of neurons, determination of the transfer and training functions are highly responsive parameters in the design of artificial neural networks. So we tried different ANN architectures.

Table 4. 3-1-1 Summary of training and transfer functions used for removal efficiency

Training Functions	Transfer Functions	trainin R	test R	MSE
7	Logsig	0,94	0,77	0,0027539
	Purelin	0,99	0,87	0,0075834
	Tansig	0,61	0,48	0,0084636
8	Logsig	0,99	0,96	0,0349950
	Purelin	0,96	0,87	0,0059573
	Tansig	0,86	0,52	0,0092192
9	Logsig	0,99	0,95	0,0059296
	Purelin	0,99	0,83	0,0019700
	Tansig	0,71	0,67	0,0094570
10	Logsig	0,98	0,84	0,0063973
	Purelin	0,99	0,85	0,0035657
	Tansig	0,59	0,40	0,0051295
11	***Logsig***	***0,99***	***0,96***	***0,0041654***
	Purelin	0,99	0,90	0,0033590
	Tansig	0,68	0,28	0,0276080
Training	Transfer	trainin	test	MSE

Table 5. 3-3-1 Summary of training and transfer functions used for removal efficiency

Training Functions	Transfer Functions	trainin R	test R	MSE
7	Logsig	0,98	0,93	0,0078904
	Purelin	0,89	0,58	0,0052331
	Tansig	0,93	0,86	0,0044014
8	Logsig	0,99	0,87	0,0075730
	Purelin	0,88	0,58	0,0058282
	Tansig	0,96	0,57	0,0044315
9	**Logsig**	**0,99**	**0,95**	**0,0023310**
	Purelin	0,61	0,38	0,0084474
	Tansig	0,99	0,79	0,0058679
10	Logsig	0,99	0,86	0,0071530
	Purelin	0,79	0,34	0,0049574
	Tansig	0,99	0,83	0,0067905
11	Logsig	0,99	0,79	0,0067391
	Purelin	0,61	0,50	0,0043681
	Tansig	0,81	0,58	0,0443300
Training Functions	Transfer Functions	trainin R	test R	MSE
7	Logsig	0,94	0,77	0,0027539
	Purelin	0,99	0,87	0,0075834
	Tansig	0,61	0,48	0,0084636
8	Logsig	0,99	0,96	0,0349950
	Purelin	0,96	0,87	0,0059573
	Tansig	0,86	0,52	0,0092192
9	Logsig	0,99	0,95	0,0059296

The optimal network size was selected from the one which resulted in maximum R and minimum MSE in training and test data sets and the architecture of the best ANN models for the prediction of the TSS of the KWWTP is shown in Table 3.

According the 3-3-1 model, it was found Rtraining=0,99, Rtest=0,95 and MSE=0,002331. In this case, the network response is satisfactory.

CONCLUSIONS

The results of this study showed maximum correlation coefficient (R) between the measured and predicted output variables, reaching up to 0.99 with 3-3-1 ANN structure. Therefore, the model developed in this work has an satisfactory result and accuracy. As a result, the neural network modeling could effectively predict the performance of KWWTP.

It is concluded that, ANN provides an effective analyzing and diagnosing tool to understand and simulate the plant, and is used as a valuable performance assessment tool for plant operators and decision makers.

ACKNOWLEDGMENT

The authors would like to thank KOSKI, Konya, Turkey, for their help during the study and for providing KWWTP process data.

REFERENCES

1. D. Hanbay, I. Turkoglu, and Y. Demir, "Prediction of wastewater treatment plant performance based on wavelet packet decompositionand neural networks," Expert Systems with Applications, vol. 34(2), pp.1038-1043, 2008.
2. R.S. Govindaraju, "Artificial neural network in hydrology. II:hydrologic application, ASCE task committee application of artificial neural networks in hydrology," Journal of Hydrologic Engineering vol 5, pp. 124–137, 2000.
3. H.R. Maier, and G.C. Dandy, "Neural networks for prediction and forecasting of water resources variables: a review of modeling issues and applications," Water Resources Research vol. 15, pp. 101–124, 2000.
4. T.R. Neelakantan, T.R. Brion, and S. Lingireddy, "Neural network modeling of cryptoposporidium and giardia concentrations in Delware River, USA," Water Science and Technology vol 43, pp. 125–132, 2001.
5. C.S. Akratos, J.N. Papaspyros, and V.A. Tsihrintzis, "An artificial neural network model and design equations for BOD and COD removal prediction in horizontal subsurface flow constructed wetlands," Chemical Engineering Journal, vol. 143(1), pp. 96-110, 2008.
6. F.S. Mjalli, S. Al-Asheh, and H.E. Alfadala, "Use of artificial neural network black-box modeling for the prediction of wastewater treatment plants performance. Journal of Environmental Management," vol. 83(3), pp. 329-338, 2007.
7. M. M. Hamed, M. G. Khalafallah, and E. A. Hassanien, "Prediction of wastewater treatment plant performance using artificial neural Networks," Environmental Modelling and Software, vol. 19, pp. 919–928, 2004.
8. L. Belanche, J. J. Valde´s, J. Comas, I. R. Roda, and M. Poch, "Prediction of the bulking phenomenon in wastewater treatment plants," Artificial Intelligence in Engineering, vol. 14(4), pp. 307–317, 2000.
9. D. J. Choi, and H. Park, "A hybrid artificial neural network as a software sensor for optimal control of a wastewater treatment process," Water research, vol. 35(16), pp. 3959- 3967, 2001.

10. K. P. Oliveira-Esquerre, M. Mori, and R. E. Bruns, "Simulation of an industrial wastewater treatment plant using artificial neural networks and principal components analysis," Brazilian Journal of Chemical Engineering, vol. 19(4), pp. 365-370, 2001.

11. A. G. El-Din, and D. W. Smith, "A neural network model to predict the wastewater inflow incorporating rainfall events," Water research, vol. 36(5), pp. 1115-1126, 2002.

12. S. Geissler, T. Wintgens, T. Melin, K. Vossenkaul, and C. Kullmann, "Modelling approaches for filtration processes with novel submerged capillary modules in membrane bioreactors for wastewater treatment," Desalination, vol. 178(1), pp. 125-134, 2005.

13. A.G. El-Din, D.W. Smith, and M.G. El-Din, "Application of artificial neural networks in wastewater treatment," Journal of Environmental Engineering and Science, vol. 3(S1), pp. S81-S95, 2004.

14. S. Yordanova, R. Petrova, N. Noykova, and P. Tzvetkov, "Neuro-fuzzy modelling in anaerobic wastewater treatment for prediction and control," International Journal of Computing, vol. 5(1), pp. 51-56, 2014.

15. Y. M. Guo, Y. G. Liu, G. M. Zeng, X. J. Hu, W. H. Xu, Y.Q. Liu, and H. J. Huang, "An integrated treatment of domestic wastewater using sequencing batch biofilm reactor combined with vertical flow constructed wetland and its artificial neural network simulation study," Ecological Engineering, vol. 64, pp. 18-26, 2014.

16. M. Bagheri, S. A. Mirbagheri, M. Ehteshami, and Z. Bagheri, "Modeling of a sequencing batch reactor treating municipal wastewater using multi-layer perceptron and radial basis function artificial neural networks," Process Safety and Environmental Protection, vol. 93, pp. 111-123, 2015.

17. P. Kundu, A. Debsarkar, S. Mukherjee, and S. Kumar, "Artificial neural network modelling in biological removal of organic carbon and nitrogen for the treatment of slaughterhouse wastewater in a batch reactor," Environmental technology, vol. 35(10), pp. 1296-1306, 2014.

18. G. B. Gholikandi, S. Jamshidi, and H. Hazrati, H. "Optimization of anaerobic baffled reactor (ABR) using artificial neural network in municipal wastewater treatment," Environmental Engineering and Management Journal, vol. 13(1), pp. 95-104, 2014

19. V.N. Delgrange, N. Cabassud, M. Cabassud, L. DurandBourlier, and J.M. Laine, "Neural networks for prediction of ultrafiltration transmembrane pressure: application to drinking water production," Journal of Membrane Science vol. 150, pp. 111–123, 1998.

20. H. Demuth, M. Beale, and M. Hagan, "Neural Network Toolbox 5: Users Guide" Natick, MA, The MathWorks Inc., 2007.

21. M.S. Nasr, M. A, Moustafa, H. A. Seif, and G. El Kobrosy, "Application of Artificial Neural Network (ANN) for the prediction of EL-AGAMY wastewater treatment plant performance- EGYPT," Alexandria Engineering Journal, vol. 51(1), pp. 37-43, 2012.

22. A.R. Khataee, G. Dehghan, A. Ebadi, M. Zarei, and M. Pourhassan, "Biological treatment of a dye solution by Macroalgae Chara sp.: effect of operational parameters, intermediates identification and artificial neural network modeling," Bioresource Technol, vol. 101(7), pp. 2252-2258, 2010

CHAPTER 5

SLUDGE REDUCTION BY LUMBRICULUS VARIEGATUS IN AHVAS WASTEWATER TREATMENT PLANT

Yalda Basim, Mahdi Farzadkia, Nematollah Jaafarzadeh and Tim Hendrickx

Keywords: Sludge reduction, Aquatic worms, Ahvaz wastewater treatment plant, Lumbriculus variegatus

INTRODUCTION

Biological treatment is the most used technology in wastewater purification. Generation of large amounts of sludge is one the major features undertaken in the some biological wastewater treatment plants [1]. Such excess sludge has to be properly treated prior to final disposal, even though the cost of sludge treatment is extremely high, accounting for up to 60% of the total operating cost in a wastewater treatment plant [2].

At the present time, there are not any principle activities for pollution control of disposal sludge from municipal wastewater treatment plants (WWTPs) in Iran, and therefore it is generally unstable [3]. Reuse or discharging of raw sludge would lead to many hazardous materials which would pollute natural resources such as water, soil and agricultural products [4]. Activated sludge process is one of the most comprehensive and widely-used biological processes in domestic and industrial treatment plants in Iran [5]. This process is generated of organic and inorganic sludge to a great deal, that is considered a disadvantage [6]. Based on the results of the study that was done by Farzadkia in four local municipal wastewater treatment plants in Tehran which worked based on activated sludge process, except one of them, the others disposed untreated sludge [7].

Strict environmental regulations regarding waste sludge in wastewater treatment plants and the high expenditure of treatment, transportation and disposal of sludge have intensively led to the sludge minimization [5]. The important methods for the reduction of excess sludge are: endogenous metabolism, uncoupling metabolism, increase of dissolved oxygen in the reactor, oxic settling-anaerobic, ultrasonic cell disintegration, alkaline heat treatment, and growth of controllable predators. Also oxidation of a part of produced sludge is done by oxidizing materials such as chlorine and ozone [2, 8].

Recently, the use of protozoa and metazoa has been proposed as a biological method for sludge reduction. This method, which is based on relationships present in the food chain and causes the overall reduction of the biomass, has gained more attention due to its low energy consumption and the lack of subsequent pollution [9]. A number of researches have been carried out on the reduction of disposed sludge by aquatic worms. Hendrickx *et al.* compared the performance of a species of aquatic worms called *Lumbriculus varigatus* in a reactor with perforated media containing worms with a usual reactor without any worms. The results of this research showed that the rate of total suspended solids (TSS) reduction in the reactor containing worms, in most of the cases, was three times greater than the reactor without worms [10].

In addition, Huang *et al.* studied the reduction of the sludge produced by activated sludge process by a species of aquatic worms called *Tubifex tubifex* in a reactor [11]. An ecological method regarding the use of four types of microfauna for the reduction of waste sludge has been surveyed by Liang *et al.* The results of this survey showed that the rate of sludge reduction depends on the classification and body size of the microfauna, and the species in the kingdom of *Oligochaeta*[9]. Wei *et al.* found out that worms, according to body size, are the biggest organisms in the sludge treatment cycle, compared with protozoa. They are easier to maintain and due to body size, have enough capability in sludge reduction [12]. Wei and Liu designed a combined reactor for the reduction of sludge using both free-flowing worms and sessile ones. In this experimental

---

Citation: Basim et al.: Sludge reduction by lumbriculus variegatus in Ahvas wastewater treatment plant. Iranian Journal of Environmental Health Sciences & Engineering 2012 9:4. doi:10.1186/1735-2746-9-4

reactor, the TSS of sludge was reduced by 48 percent, which is mainly due to the presence of *Tubificidea*[13]. Rastak studied the possibility of reducing activated sludge in wastewater treatment plants using aquatic worms of *Oligochaeta* at experimental scales, and the environmental factors affecting the performance of worms were surveyed. This research confirmed the use of worms as a protein-supply source in the food for fish and domestic animals [14].

Although the advantages of aquatic worm application for sludge reduction are known, this method has not yet been studied in any scale in Iran. The objectives of this study were identification of appropriate and endemic species of aquatic worm called *Lambriculus variegates*, adaptation of life natural environment of worms to laboratory conditions, determination of sludge reduction by *Lambriculus variegates* worm in Ahvaz Wastewater Treatment Plant (WWTP) and determination of effective DO concentration in worm reactor. Ahvaz WWTP operated based on activated sludge process for wastewater treatment and has two stages of anaerobic digester for sludge stabilization. The first stage of anaerobic digester was damaged and was out of service at the time of this study. Comparison of the microbial quality of disposal sludge in Ahvaz WWTP and USEPA criteria showed that the sludge was not in class A or B conditions. Hence, it should not be disposed to the environment or reused for any purpose [15].

MATERIALS AND METHODS

Provision of the worms from the specified species was a chief concern in this research. Due to the possibility of access to this species in areas with organic and decomposed materials, the probable homes of this species were identified. Samples were taken from the benthic materials of Khuzestan water reservoirs including Karoon River, Maleh Stream, Shadegan Wetland, and Dez river by using grab model of Van Vee [10]. Microscopic pictures of the samples were then taken and identified with "Freshwater Biology" [16]. No worms belonging to the subclass *Lumbiculidae* were observed.

Due to the study carried out on the spring pools of Kermanshah province in the west part of Iran [17], a few samples were taken and observed in this area. Finally, the species of *Lumbriculus variegates* from the *Oligochaeta* subclass and *Lumbiculidae* family were found in the sediments deposited of Jabery and Ravansar springs. To maintain the environment of the worms collected as well as their compatibility with the new environment, the vessels containing worms were aerated, and the temperature of the medium was constantly controlled [18]. In order to take samples of aquatic worm species and to carry out field tours, a collection of outdoor equipment including the Sampler VN, a GPS, electrical conductivity meter, mercury and digital thermometers, kit of DO analyzer, a sieve with one millimeter mesh, and other equipments for transportation of samples were used [14].

The worm reactor consisted of a beaker containing both waste sludge and worms. It contained 1500 mg wet weight of the aquatic worm and 100 mL of waste sludge of Ahvaz WWTP. The blank reactor was investigated under the same conditions but without any worm. The waste sludge was daily provided from the return line of Choneibeh WWTP in the west of Ahvaz city. Experiments were carried out in two parallel runs with two different dissolved oxygen contents. Each run consisted of 8 batch steps and took 1 day. Therefore, each run took 8 days; repetition of each batch step was done 8 times. The concentration of the dissolved oxygen in steps one and two were kept up to 3 and 6 mg/L, respectively [10]. pH, temperature and TSS of the incoming and outgoing sludge were determined every 24 hours at the end of each step.

Every 24 hours, aeration was stopped for a short time and, by using a fine mesh; the worms were separated and prepared for entering to the next batch step [14]. The species of the aquatic worms during the days of the experiment were similar: they were selected among the same initial population under the experiment [10]. All experiments were carried out by using the standard methods [19], and the results were assessed based on ANOVA analysis and T statistical test in order to investigate the average differences at a reliable level of 95%. The Smirnov-Kolmogrove test was used to determine the normal distribution of the findings. The effect of water evaporation in aeration was noticed via adding sample volume about 20% in the start time. Selection of 20% was based on the pretests.

RESULTS

TSS in the incoming and outgoing sludge of the reactors and blanks are indicated in Figure 1 and Figure 2, which show that, during the whole days of the two runs, TSS in outgoing sludge was less than that the incoming sludge. The minimum and maximum amounts of TSS of sludge entry the reactors and blanks were 2456 and 7184 mg/L, respectively. The efficiency of sludge reduction in the reactors, as well as the blanks are shown on Figure 3, which indicates that the efficiency of sludge reduction in the second-phase reactor was 33%, slightly higher than 32% of the first-phase reactor. The rates of sludge reduction in the blank reactors in the two phases were 16% and 12%, respectively. The average sludge reduction by the species *Lumbriculus variegatus* was found to be about 0.45 mg of sludge per one mg of the worm per day. The Smirinove-Kolmogrove test on the suspended solids in the sludge and the dissolved oxygen showed that the data followed a normal distribution. Based on the data obtained from runs (1) and (2) in the reactors, no significant difference was observed between the sludge reductions in two runs ($p<0.05$). These results were confirmed by the sludge reduction in two runs in the blanks ($p<0.05$). In other words, variation of oxygen concentration did not affect the sludge reduction efficiency in these experiments.

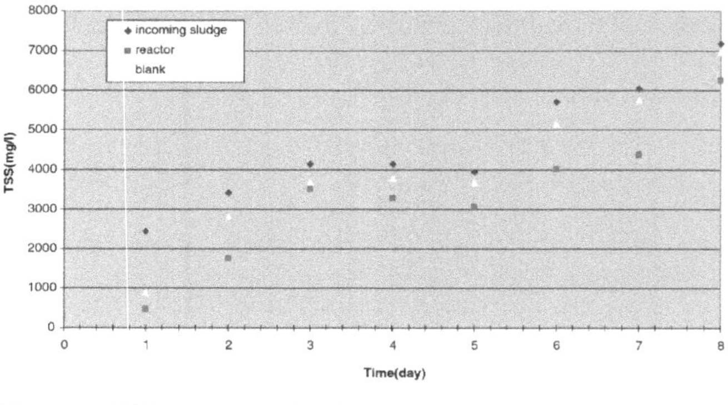

Figure 1 TSS concentration in reactor and blank-Run 1.

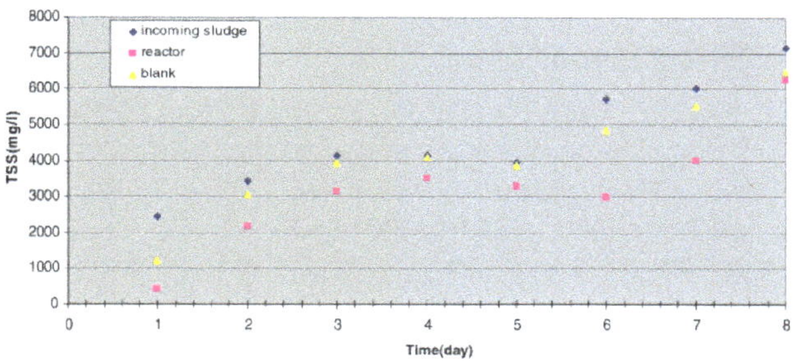

Figure 2 TSS concentration in reactor and blank-Run 2.

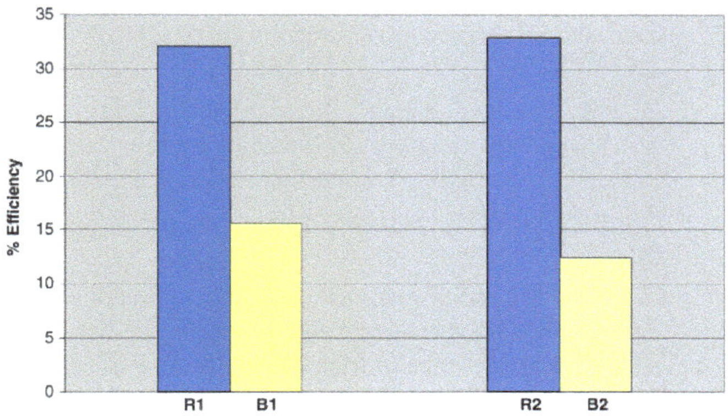

Figure 3 Efficiency of sludge reduction in the reactors (R) and blanks (B) in runs (1) and (2).

However, statistical analyses indicated that there is a meaningful difference between the rates of sludge reduction in the reactors with the blanks in two runs ($p>0.05$). Due to similar medium conditions of the reactors and the blanks, this meaningful difference highly confirmed the effectiveness of the worms in the reactors for sludge reduction.

DISCUSSION

In the present study, the rate of sludge reduction in the first and second runs of the reactors were 32% and 33% and in the runs (1) and (2) of blanks were 16% and 12%, respectively. In the reactor containing the aquatic worms of *Lumbriculus variegatus*, which were experimented by Hendrickx *et al.*, the rate of sludge reduction was almost three times higher than in the blank (without worms) [10]. In the present study, the sludge reduction rate in the worm reactors were 2 to 2.75 times higher than in the blank. This difference may be based on the differences between the methods of experiments in two researches. In another survey in China, sludge reduction was studied by using of a combination of protozoa and metazoa and the rate of reduction (in MLSS) was from 45% to 58%, compared to that of the blank. In this survey, the DO concentration and hydraulic retention time of sludge were considered as 1–4 mg/L and 6–13 hours, respectively [20]. The higher efficiency of sludge reduction in this study may be resulted from the use of a combination of protozoa and metazoa in one reactor.

In another study, carried out in China on the aquatic worm species of *Tubificide*, the average TSS reduction in the sludge was 48% and the DO concentration was $\geq$2 mg/L [13]. A comparison of the 48% reduction of the above-mentioned research with that of the present research (33%) shows that the aquatic worm *Tubificide* has a better performance than the species of *Lumbriculus variegatus*. It may be noted that, Wei and Liu alternately used two reactors as series containing worms.

In another research which was performed in the Netherlands in an aerobic reactor on the activated sludge containing worm species of *Lumbriculus variegates*[14], almost similar result for sludge reduction (up to 30%) compared to our study (33%), was obtained, which is because of similar worm species and treatment method (activated sludge). In China, another survey was carried out by Guo *et al.*,[21] in which, a reactor containing worm species of *Tubificide* was used to reduce sludge and an efficiency of 46.4% was obtained. The SRT (Solid Retention Time), HRT (Hydraulic Retention Time) and DO concentration in this research were 30 days, 15.4 hours, and 0.5-3 mg/L, respectively [21]. Better results in this study may be related to longer SRT (4 times more) and type of the worms (*Tubificide*).

Wei *et al*, showed that the performance of the worm species *Oligochaeta* in the conventional activated sludge (CAS) system could lead to considerable reduction in sludge and improve its settling characteristics [22]. The similar results of sludge reduction was obtained in our study on Ahvaz WWTP with CAS process.

The daily rate of sludge reduction by four species of worms was studied by Liang *et al*[9]. These rates were 0.1 to 0.54 mg sludge/mg worm, based on the order or body size of the worm [9]. The average sludge reduction in our study (0.45 mg sludge/mg worm per day) is almost near to the maximum rate of sludge reduction in the above-mentioned research.

The quantitative results of this research, regarding the sludge reduction by worm species *Lumbriculus variegatus*, were in agreement with Hendrickx's study (36%) [23]. But, no meaningful relationship was found in our research between DO concentration and the rate of sludge reduction. This disagreement may be arising from differences in worm habitats and their living conditions, the type of reactors and operational conditions.

DECLARATIONS

Acknowledgements

The authors hereby appreciate the valuable assistance of Dr I Mobedi, professor, in Tehran University of Medical Sciences (TUMS) and Dr S M B Nabavi for identifying the aquatic worm species. Special thanks go to engineer Raisi and other personal of the Laboratory of Ahvaz wastewater treatment plant.

AUTHORS' CONTRIBUTIONS

YB: conception and design, generation of data, collection of data, assembly of data, analysis of data, interpretation of data, drafting of the manuscript. **MF:** conception and design, interpretation of data, drafting of the manuscript, revision of the manuscript, approval of the manuscript. **NJ:** interpretation of data, drafting of the manuscript, revision of the manuscript, approval of the manuscript. **TH:** conception and design, revision of the manuscript, approval of the manuscript. All authors read and approved the final manuscript.

REFERENCES

1. Mahvi AH: Sequencing batch reactor: A promising technology in wastewater treatment. Iran J Environ health Sci and Eng. 2008, 5: 79-90.
2. Baghapour MA, Jabbari E, Baskaran K: Reducing of excess sludge production in wastewater treatment using combined anaerobic/aerobic submerged biological filters. Iran J Environ Health Sci Eng. 2011, 8: 207-218.
3. Farzadkia M, Mahvi AH: Comparison of extended aeration activated sludge process and activated sludge with lime addition method for biosolids stabilization. Pakistan Journal of Biological Sciences. 2004, 7: 2061-2065.
4. Mantovi P, Baldoni G, DalRe L, Piccinini S, Rossi L: Effects of 15 years sludge application on cropland. In: Sludge disposal; Proceeding book of IWA conference of Sludge management, Moscow. Sec. 2006, 10: 10-16.
5. Jamal A, Norieh N, Farzadkia M: Comparison of aerobic and lime stabilization methods for evaluation of sewage sludge reuse. Journal of Environmental Science and technology. 2011, 4: 182-190.
6. Hosseini Koupaie E, Alavi Moghaddam MR, Hashemi H: Comparison of overall performance between moving-bed and conventional sequencing batch reactor. Iran J Environ Health Sci Eng. 2011, 8: 235-244.
7. Farzadkia M: Investigation of sludge stabilization and reuse in four small treatment plants of Tehran city. Scientific Journal of Hamadan University of Medical Sciences. 2002, 9: 51-55.
8. Takdastan A, Mehrdadi N, Azimi AA, Torabian AN, Bidhendi G: Investigation of intermittent chlorination system in biological excess sludge reduction by sequencing batch reactors. Iran J Environ Health. Sci Eng. 2009, 6: 53-60.
9. Liang P, Huang X, Qian Y, Wei Y, Ding G: Determination and comparison of sludge reduction rates caused by microfaunas' predation. Journal of Bioresource Technology. 2006, 97: 854-861. 10.1016/j.biortech.2005.04.018.
10. Hendrickx TLG, Elissen HJH, Temmink H, Buisman CJN: A new reactor concept for sludge readuction using aquatic worms. Water Res. 2006, 40: 3713-3718. 10.1016/j.watres.2006.08.029.
11. Huang X, Liang P, Qian Y: Excess sludge reduction induced by Tubifex tubifex in a recylced sludge reactor. Journal of Biotechnology. 2007, 127: 443-451. 10.1016/j.jbiotec.2006.07.035.
12. Wei Y, Van H, Borger A, Fan Y, Eiklboom D: Minimization of excess sludge production for biological wastewater treatment. Water Res. 2003, 37: 4453-4467. 10.1016/S0043-1354(03)00441-X.
13. Wei Y, Liu J: Sludge reduction with a novel combined worm-reactor. Journal of Hydrobiologia. 2006, 564: 213-222. 10.1007/s10750-005-1720-1.
14. Rastak CH: Sludge reduction by predatory activity of aquatic Oligechaetes in wastewater treatment plants. Journal of Hydrobiologia. 2006, 564: 197-211. 10.1007/s10750-005-1719-7.
15. Farzadkia M, Jaafarzadeh N, Loveimi Asl L: Optimization of bacteriological quality of biosolids by lime addition. Iran J Environ Health Sci and Eng. 2009, 6: 29-34.
16. Edmondson WT: Fresh-water Biology. 1959, Second Edition
17. Khatami SH, Shayegan SJ: Physico - chemical and biological characteristics of Sarabs (spring pools) in the Kermanshah province of Iran. Iranian Journal of Science & Technology. Transaction B, Engineering. 2006, 30: 621-635.
18. Pierret: Anatomie syntemitique biologie. Grasse. 1957, France: Praite do zoologie, 435-
19. APHA: Standard Methods for the examination of water and wastewater 21st edn. 2005, Washington, D. C.: American Public Health Association
20. Wang Q, Ai H, Li X, Liu H, Xie W: Pilot-scale study of biomass reduction in wastewater treatment. Journal of Water environment research. 2007, 79: 521-527. 10.2175/106143006X123030.
21. Guo X, Liu J, Wei Y, Lin L: Sludge reduction with Tubificidae and the impact on the performance of the wastewater treatment process. Journal of Environmental Sciences. 2007, 19: 257-263. 10.1016/S1001-0742(07)60042-4.
22. Wei Y, Van H, Borger A, Eikelboom D, Fan Y: Comparison performances of membrane bioreactor and conventional activated sludge processes on sludge reduction induced by Oligochaete. J. Environmental Science Technology. 2003, 37: 3171-3180. 10.1021/es026176d.
23. Hendrickx TLG, Temmink H, Elissen HJH, Buisman CJN: The effect of operational conditions on aquatic worms eating waste sludge. Water Res. 2009, 43: 943-950. 10.1016/j.watres.2008.11.034.

CHAPTER 6

COD CAPTURE: ENERGY SELF-SUFFICIENT DOMESTIC WASTEWATER TREATMENT

JunfengWan, JunGu, Qian Zhao & Yu Liu

INTRODUCTION

The conventional activated sludge (CAS) process has been employed for domestic wastewater treatment for more than 100 years, where organic matters are mainly converted to biomass and carbon dioxide. In practice, the CAS process has been extended and modified to handle soluble nutrients (e.g. nitrogen and phosphorus) present in domestic wastewater. One of such examples is the CAS-based various biological nitrogen removal (BNR) processes including the latest short-cut nitrification coupled with anaerobic ammonium oxidation (anammox) process[1,2]. These new development may lead to more sustainable and economically viable biological processes for nutrient removal. However, it should also be noted that the CAS process is receiving increasing pressure due to its low energy efficiency. For example, 35% of the energy efficiency was achieved in the Jurong Water Reclamation Plant of Singapore, whereas only 31% in the Gaobeidian Wastewater Treatment Plants of China[3]. It had been reported that about 3% of the annual electrical energy was consumed for wastewater treatment in USA[4]. Moreover, increasing in-plant energy consumption would be inevitable since discharge standards of effluent are being tightened in more and more countries due to the serious concerns on public health and environmental protection. These imply that energy-intensive treatment processes may possibly be required for meeting such needs. With no doubt, the substantial energy consumption in current wastewater treatment plants (WWTPs) is a critical issue that challenges the global wastewater treatment practice, especially when environmental sustainability is concerned. In WWTPs, a substantial amount of electrical energy that generated essentially from fossil fuels is consumed to drive wastewater treatment. This is inevitably linked to the issue of global climate change. In China, about 114 million metric tons of carbon dioxide is generated and released annually from the production of electrical energy required to drive the WWTPs operation countrywide[5]. Meanwhile, a great amount of greenhouse gases, such as carbon dioxide, methane and nitrogen oxide, is produced during various biological conversions of organic and nitrogenous matters present in domestic wastewater. Obviously, substantial in-plant energy consumption will have considerable impact on the future wastewater industry in consideration of global climate change.

In practice, the chemical oxygen demand (COD) has been commonly employed to determine the total amount of organic matters in domestic wastewater. In the WWTPs with the CAS as core treatment technology, given a growth yield of 0.3–0.5 g dry biomass produced per gram COD removed, 1 kg COD removed will generate 0.3–0.5 kg dry biomass. As a result, a large amount of waste sludge is currently generated during biological treatment of domestic wastewater globally. In order to reduce electric energy input to a WWTP, the present practice is to recover energy from waste sludge through anaerobic digestion. Despite advanced development of anaerobic digestion technology, only about 30–50% of the total COD or volatile solids can be degraded towards biogas production, implying that the overall energy efficiency of anaerobic digestion process is very low due to the fact that microorganisms cannot directly utilize waste sludge, which instead should be hydrolyzed to soluble COD prior to anaerobic digestion. This indeed is the major drawback of the present WWTP settings in which soluble COD is first converted to biomass from which energy is subsequently recovered through anaerobic digestion at low energy efficiency. For

Citation: Wan, J. et al. COD capture: energy self-sufficient domestic wastewater treatment. Sci. Rep. 6, 25054; doi: 10.1038/srep25054 (2016).

example, the overall in-plant energy efficiency of 20–50% has been reported in various countries[3,6,7,8]. Therefore, there is an urgent global need to substantially improve the overall energy efficiency of WWTP by adopting innovative process configuration. Ideally, organic maters present in domestic wastewater should be captured as much as possible for direct anaerobic digestion prior to biological conversion to sludge. This in turn may lead to the paradigm shift of the current WWTP with the benefits of (i) enhanced energy recovery; (ii) reduced energy consumption associated with aeration and (iii) lesser excessive sludge production. In this study, we described that WWTP should be designed according to the concept of A-B process for significantly improving the energy recovery potential, while reducing in-plant energy consumption. Hence, various possible configurations of A-B process with energy analysis are discussed. With the advances in environmental microbiology along with recent development in autotrophic ammonia removal, A-B process appears to be a feasible option towards energy self-sufficiency in domestic wastewater treatment plant.

RESULTS

Energy Situation in Current WWTPs

Nowadays, domestic wastewater cannot be simply considered as "waste", but a valuable resource full of energy, valuable materials and clean water. The current energy consumption in the conventional CAS process is normally in the range of 0.3 to 0.6 kWh per m^3 of wastewater treated with an average of 0.45 kWh/m^3 equivalent to 1620 kJ/m^3,[3,9] whereas a typical COD concentration in domestic wastewater is often around 500 mg/L. Therefore, the energy consumption in the conventional CAS process can be estimated as 1620 kJ m^{-3}/500 g COD m^{-3} = 3.20 kJ/g COD. Theoretically, the potential energy in typical domestic wastewater has been estimated in the range of 14.7–17.8 kJ free energy/g COD, averaged at 16.2 kJ/g COD[10,11] which is nearly 5-fold of electrical energy (i.e. 3.2 kJ/g COD) utilized to drive wastewater treatment in the CAS process. These numbers potentially suggest that WWTPs may be energy self-sufficient so long as 20% of the total energy in domestic wastewater could be completely converted to the electrical energy. Figure 1illustrates the COD mass flow in a typical domestic wastewater treatment plant with the CAS as a core treatment unit. It is obvious that in such a process, the total energy that can potentially be recovered essentially comes from anaerobic digestions of both primary and secondary sludges.

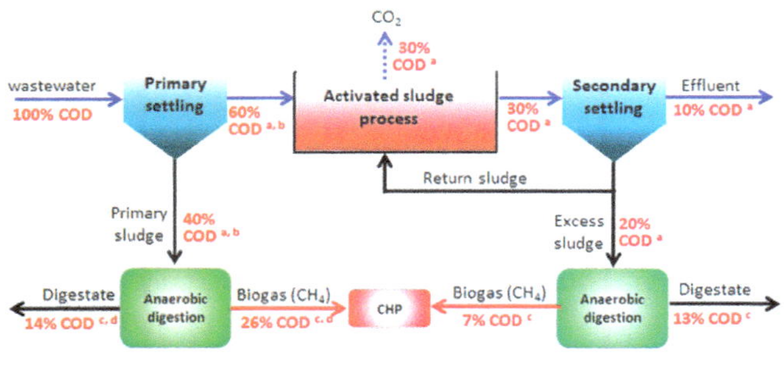

Figure 1: COD mass flow through a conventional domestic wastewater treatment plant. (**a**) data from Metcalf and Eddy12; (**b**) data from Rossle and Pretorius18; (**c**) data from Parkin and Owen47; (**d**) data from Miron *et al.*48.

As can be seen in Fig. 1, 26% and 7% of total COD from primary and secondary sludges could be converted to methane through anaerobic digestion, respectively. As theoretical chemical energy derived from one gram of methane-COD is about 13.9 kJ[11], the total recoverable chemical energy may be calculated as the sum of (13.9 × 0.26) from primary sludge and (13.9 × 0.07) from secondary sludge, i.e. 4.58 kJ/g COD. However, it should be noted that only about 35% of the chemical energy of methane can be converted to electricity through its combustion[9]. Therefore, the maximum recoverable electrical energy from domestic wastewater through anaerobic digestion in WWTPs is about (4.58 × 0.35) = 1.60 kJ/g COD, which is the half of the electrical energy required for removing one gram COD (i.e. 3.20 kJ/g COD). These suggest that the maximum electrical energy recovered from wastewater can only offset at most 50% of total energy needed for the current WWTP operation (Fig. 1), which indeed is consistent with the previous estimate[9]. Zhou *et al.*[3] conducted a series of comprehensive survey on the global energy situation in WWTPs, according to which only about 35% of energy efficiency was realized in the Jurong Water Reclamation Plant of Singapore, whereas only 31% in the Gaobeidian WWTP of China. As illustrated in Fig. 1, anaerobic digestion of primary sludge may contribute to about 78% of the total potential recoverable energy in WWTPs, although only 40% of total COD in domestic wastewater is captured in the primary settling tank. This in turn points to a fact that capture of COD prior to biological oxidation is essential for further improving the plant energy balance. Furthermore, the combined heat and power generation may be considered as an economically viable option for further enhancing the overall conversion efficiency of biogas to recoverable energy in terms of both electricity and heat. In fact, it also appears difficult to achieve energy-neutral or even more challenging for energy-positive operation if the CAS process continues to serve as the core unit of WWTPs. Given such a situation, the current debate on the situation forward is naturally towards the technological options that may lead to the paradigm shift of the WWTP's operation from energy-negative to neutral and ultimately positive.

A-B Process for Domestic Wastewater Treatment

Rational of A-B Process

The central point towards energy recovery is to capture COD from wastewater as much as possible prior to its biological oxidation. As such, the present CAS process should be retrofitted into a two-stage system known as A-B process which is schematically elucidated in Fig. 2A. In A-B process, A-stage is specifically designed to maximize the capture of organic matters from domestic wastewater for direct anaerobic digestion prior to biological oxidation, whereas B-stage is mainly dedicated to handling nutrients. Theoretically, the electrical energy generated should be sufficient to drive the conventional WWTP operation (i.e. 3.2 kJ is required for removing one gram COD) if 65% of total COD in domestic wastewater could be captured in A-stage and converted to methane through anaerobic digestion, i.e. the electrical energy produced from one gram COD = 13.9 × 65% × 35% = 3.2 kJ. As a result of COD capture in A-stage, substantially reduced energy consumption would be expected in B-stage (Fig. 2A). The essence of A-B process is to improve energy recovery through COD capture prior to biological oxidation in A-stage, whereas to reduce energy consumption in B-stage (Fig. 2B). This is the only way possible to realize the energy self-sufficient operation of a WWTP.

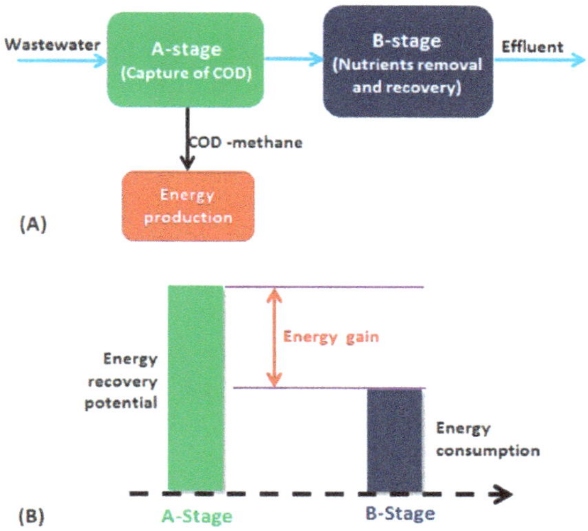

Figure 2: A general configuration of A-B process.

The conceptual A-B process (Fig. 2A) is specifically put forward for maximizing the recovery of the chemical energy from domestic wastewater, instead of transforming it to carbon dioxide during biological oxidation of COD with consummation of substantial amount of electrical energy. However, the feasible options of A-B process towards the process energy self-sufficiency are determined by the choices of A and B stages. In this regard, only three processes with various potentials of the capture of COD from wastewater may serve as A-stage, i.e. (i) chemically enhanced primary treatment (CEPT) process, (ii) high rate activated sludge (HRAS) process and (iii) anaerobic process. It appears from the literature that these three processes would be able to capture at least 60% of total COD in domestic wastewater. Obviously, if they are chosen to serve as A-stage of A-B process, much lesser COD would enter B-stage. According to the denitrification stoichiometry, about 2.86 gram COD should be required for completely denitrifying one gram NO_3^--N[12]. These seem to suggest that the conventional nitrification and denitrification process could not be considered as a choice of B-stage due to the insufficient supply of soluble COD after the COD capture at A-stage. In fact, it appears from the literature that only two processes, i.e. shortcut nitrification-denitrification and partial nitrification combined with anammox processes, may meet the requirements of B-stage. Therefore, Fig. 3 further illustrates various feasible combinations of A-B process.

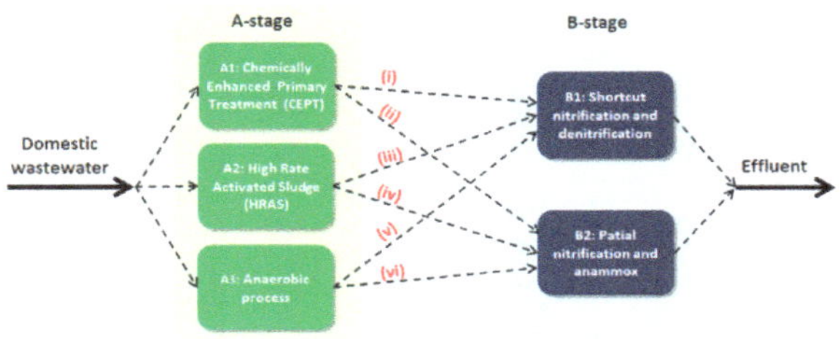

Figure 3: Possible combinations for A-B processes.

A1-B1 or A1-B2 Process

CEPT process has been widely employed to treat domestic wastewater for decades. Up to 80% of TSS and 60% of total COD can be removed through CEPT in full-scale WWTPs[13,14,15,16]. Compared to the conventional primary settling, the quantity of sludge produced by CEPT was found to be increased by nearly 45%, of which 33% was due to enhanced solids capture[13]. Moreover, organic-rich CEPT sludge is more suitable for biogas production through anaerobic digestion[17]. However, it should be noted that CEPT is very poor in removing soluble COD from domestic wastewater, i.e. nearly all soluble COD which is at least 30% of total COD in domestic wastewater may enter into B-stage[18]. In case where CEPT serves as A-stage, soluble COD going into B-stage should be high enough for inhibiting the growth of anammox bacteria against heterotrophic denitrifying bacteria[19,20]. Therefore, it should be more reasonable to consider the shortcut nitrification-denitrification process as B-stage.

In the above discussion, the CEPT-shortcut nitrification-denitrification has been identified as a feasible configuration of A-B process. The COD mass flows in this process configuration are presented in Fig. 4. It appears that 43% of influent COD can be ultimately converted to biomethane, thus the maximum electrical energy potentially recovered from one gram of wastewater COD is calculated as $(13.9 \times 43\% \times 0.35) = 2.09$ kJ/g COD. It had been reported that addition of chemical coagulant would not adversely affect the subsequent anaerobic digestion efficiency of wasted sludge21. It should also be noted that addition of coagulants to CEPT inevitably leads to an increased overall operation cost. According to Diamantis*et al.*[17], the additional operational cost associated with coagulant was about 0.1 €/m^3 (equivalent to 4×10^{-4} €/g COD) when the CEPT process was adapted for the pre-concentration of municipal wastewater. Thus, a detailed cost-benefit analysis should be conducted if CEPT was considered as A-stage in A-B process illustrated in Fig. 4.

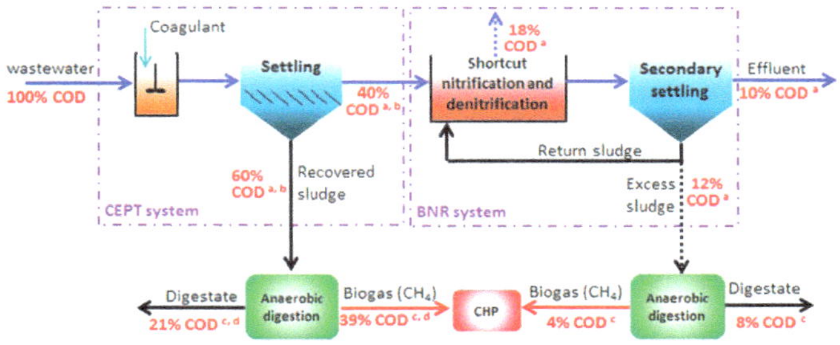

Figure 4 COD flow in A-B process with CEPT as A-stage and short-cut nitrification and denitrification as B-stage; (**a**) data from Metcalf and Eddy12; (**b**) data from Morrissey and Harleman13; (**c**) data from Parkin and Owen47; (**d**) data from Miron *et al.*48.

A2-B1 or A2-B2 Process

HRAS as a variant of the CAS process has attracted great attention due to its high capacity of COD capture at short sludge and hydraulic retention times, e.g. about 55–65% of organic matter could be retained in HRAS process at a SRT of 0.5 day[22], wheras nitrification in HRAS should not be significant due to the shorter sludge retention time less than 3 days. As such, for a domestic wastewater with a typical COD concnetration of 500 mg/L, a substantial amount of

COD at the concnetration level of 175 to 225 mg/L may eventually enter into B-stage, leading to a COD/N ratio of 7 to 9. In general, heterotrophic denitrifiers can easily outcompete anammox bacteria at the soluble COD/N ratio greater than 2[19,23]. Theroretically, 1.71 gram COD is required to denitrify 1 gram NO_2^--N, i.e. COD remained after A-stage is sufficient for removing ammonia through shortcut nitrification-denitrification. It should also be pointed out that the success of the integrated HRAS- anammox process in the Strass WWTP, to a large extent, is due to the selectively retention and return of anammox bacteria harvested from the sidestream deammonification unit by cyclone or screens at B-stage[24]. In addition, the partial nitritation-anammox process could be employed as B-satge if the COD/NH_4^+ -N ratio was lowered to 1.9 in the HRAS process as A-stage[25]. Consequnetly, these seem to suggest that the A2-B2 process is more challenging than the A2-B1 process from the point of view of the system feasibility and stability. Recently, the two HRAS configurations had been investigated for the potential energy recovery from synthetic domestic wastewater[26], of which the HRAS process with the contact and stabilization units showed a much higher COD removal as the stabilization unit is essential for maximizing COD capture[27]. This is due to the fact that the stabilization may ensure a feast-famine cycle, while exerts a selection pressure on microorganisms leading to enhanced COD adsorption and storage in the contact unit[26,28]. As such, the HRAS with the configuration of combined contact and stabilization should be ideal for A-stage. Figure 5 further shows the COD mass flows in HRAS-shortcut nitrification-denitrification process. According to Fig. 5, the recoverable electrical energy can be estimated as $(13.9 \times 47\% \times 0.35) = 2.29$ kJ/g COD which is significantly higher than that achieved in the conventional CAS process.

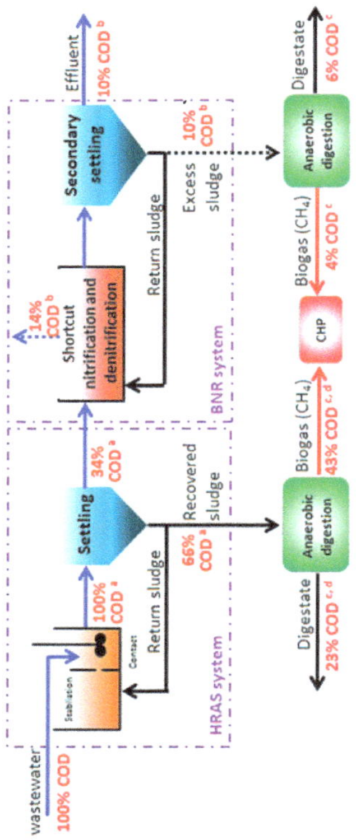

Figure 5 COD flow in A-B process with HRAS as A-stage and short-cut nitrification and denitrification as B-stage; (**a**) data from Meerburg *et al.*26; (**b**) data from Metcalf and Eddy12; (**c**) data from Parkin and Owen47; (**d**) data from Miron *et al.*48.

A3-B1 or A3-B2 Process

Extensive efforts have been dedicated to explore anaerobic process for handling low strength domestic wastewater[29,30,31,32]. Compared with aerobic process (e.g. CAS), it has been well demonstrated that anaerobic process has advantages of no aeration needed, less sludge produced and bio-methane generation which make it more economically viable and energy-efficient. Nowadays, various anaerobic processes are available for treating domestic wastewater, e.g. up-flow anaerobic sludge blanket (UASB), fixed film anaerobic bioreactor, anaerobic baffled reactor, etc. A full scale single-stage UASB system could remove 45–75% of total COD from domestic wastewater at a HRT of 5–19 h, whereas total COD removal of 65–90% was achieved in lab-scale UASB reactor[31,33], whereas 88% of total COD was removed in an anaerobic fluidized bed bioreactor (AFBR)[34]. Despite several full-scale UASB systems have been applied for domestic wastewater treatment, some technical issues still need to be resolved, including relatively long start-up period, sensitivity to seasonal temperature change, slow growth due to low organic strength in domestic wastewater etc[31]. It should be a reasonable consideration that over 80% of total COD in domestic wastewater can generally be removed in anaerobic processes. That is, if anaerobic process is selected to serve as A-stage, remaining COD should not be enough for sustaining the shortcut nitrification-denitrification in B-stage, i.e. the A3-B1 configuration might not be a feasible option in practical application. On the contrary, such low COD concentration achieved is ideal for partial nitrification coupled with anammox at B-stage, thus the A3-B2 configuration appears to be more reasonable in this sense. In fact, the process configurations similar to A3-B2 has been demonstrated to be feasible for the mainstream municipal wastewater treatment, i.e. UASB coupled with moving bed biofilm reactor (MBBR) or integrated fixed film activated sludge (IFAS) for partial nitritation-anammox[35,36]. The anaerobic-partial nitrification-anammox process has been identified in the above discussion as an ideal A-B process which is illustrated in Fig. 6. Highly efficient COD removal in anaerobic process provides a low COD environment for sustaining growth of anammox bacteria in B-stage. For example, 90% of total COD was removed from domestic wastewater in anaerobic baffled reactor, of which 65% of COD was converted methane[37]. According to Fig. 6, if about 65% of COD could be transformed to the COD-methane, the total recoverable electrical energy may be estimated as $13.9 \times (0.65 + 0.08) \times 0.35 = 3.55$ kJ/g COD, which is more than the energy consumed for removing one gram COD (i.e. 3.2 kJ/g COD) in the CAS process.

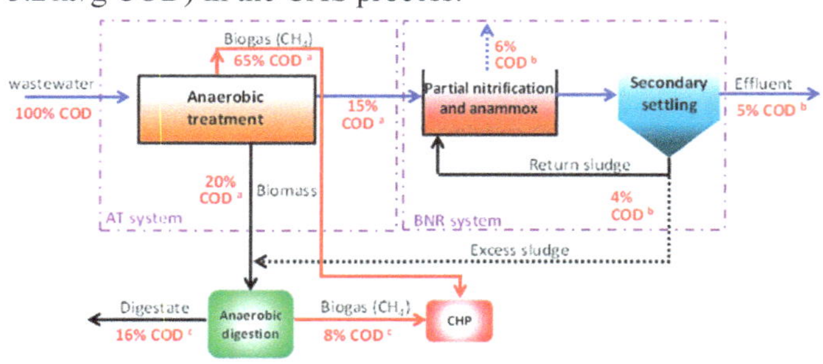

Figure 6 COD mass flow in A-B process with anaerobic treatment as A-stage and partial nitrification and anammox as B-stage; (**a**) data from Gopala Krishna *et al.*37; (**b**) data from Metcalf and Eddy12; (**c**) data from Parkin and Owen47.

DISCUSSION

Nowadays, extensive attention has been paid to the wastewater treatment-energy-greenhouse gas nexus in municipal wastewater treatment plant due to the serious concern on global climate change. Although various anaerobic processes as discussed above may offer perfect option as A-stage, the actual recovery efficiency of biogas from municipal wastewater through anaerobic processes is generally lower than that presented in Fig. 6 due to the dissolution of a substantial amount of methane into treated effluent. According to Yamamoto *et al.*[38], the solubility of methane in water is about 18.6 g/m^3 at 30 °C. This suggests that about 45% of methane produced in an anaerobic system treating municipal wastewater with a soluble COD concentration of 200 mg/L would be present in its dissolved form at 30 °C[39]. In fact, it had been reported that 50% of total methane generated eventually ended up as dissolved methane in an anaerobic MBR treating synthetic wastewater with a COD concentration of 440 mg/L at 15 °C[40]. Similar phenomenon was also observed in an expanded granular sludge blanket reactor treating domestic wastewater[41]. It has been well recognized that methane is at least 25 times more powerful than carbon dioxide in terms of the greenhouse effect. Therefore, the A-B process in Fig. 6 may be challenged by its economic viability and environmental sustainability if recovery methods for dissolved methane are not available for large-scale anaerobic treatment of low-strength domestic wastewater.

It should be pointed out that tremendous amount of sewage sludge is currently generated from WWTPs globally, e.g. about 10 million tons of dry sludge in the European Union, 8 million tons in the United States in 2010. The situation of excessive sludge in China is even more pressing due to increasing number of WWTPs countrywide, i.e. sludge production is expected to increase radically and should have reached more than 11 million tons of dry solids in 2010. The ultimate disposal of excess sludge has been and continues to be one of the most challenging and costly issues. The treatment of the excess sludge may account for 25–65% of the total plant operation cost[42], and landfilling of dewater sludge is no longer an economically and environmentally friendly viable option. The A-B process presented in Fig. 6 indeed offers a feasible solution for ultimately solving sludge problem. In A-stage, the majority of COD in domestic wastewater is directly converted to biogas anaerobically with little sludge production. As a result, the total sludge production in the A-B process illustrated in Fig. 6 should be reduced by at least 60% compared to the conventional CAS process (Fig. 1). Given its pros and cons, a detail cost-benefit analysis of the A-B process (Fig. 6) should be further conducted in a more holistic manner.

Recently, anaerobic membrane bioreactor (AnMBR) has attracted more and more attention for domestic wastewater treatment due to its superior performance, e.g. direct COD capture without production of much excessive sludge, extremely low energy consumption, high-efficiency biogas generation and nearly solids-free permeate[9,34,43]. Different from the A-B processes elucidated above, a new process configuration (Fig. 7) combining AnMBR and nutrient recovery may also be considered as a feasible option for concurrent energy and resources recovery from domestic wastewater. It had been shown that over 92% of COD and nearly 100% of solids removal with the minimum removal of soluble nutrients (e.g. ammonia and phosphate) could be achieved in AnMBR[32,38]. This suggests the possibility of nutrient recovery from the AnMBR effluent. So far, many techniques have been reported for nutrient recovery from wastewater. For example, natural zeolite as an effective adsorbent could adsorb up to 97% of ammonium, while served as seed for phosphate precipitation[44,45]. Moreover, zeolite synthesized from coal fly ash could concurrently remove around 60% of ammonium and 90% of phosphate present in wastewater stream[46]. It should be realized that all the methods reported for recovery of soluble nutrients from various wastewater streams are currently expensive and require strong engineering knowledge and well-trained operators in order to ensure long-term stable operation. In addition to the technology and cost issues, another obstacle in the practice of nutrient recovery from wastewater streams lies in the market acceptance, i.e. potential takers of such products in the market.

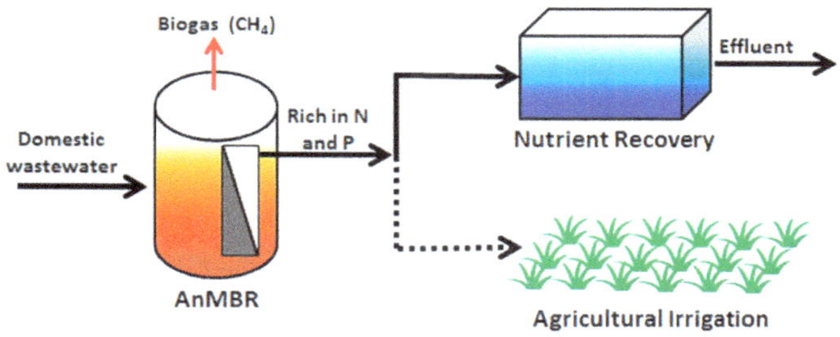

Figure 7: Schematic A-B process with AnMBR as A-stage and nutrient recovery as B-stage.

It has been actively debated if it is necessary to make all water potable. For some countries with increasing population and diverse industries (e.g. China, India, etc), the water policy should allow using different grades of water for different usages. For example, the effluent from AnMBR is nearly pathogen-free, while rich in soluble ammonia and phosphate, thus for the countries with strong agricultural industry, such AnMBR effluent may be explored for direct irrigation[9].

CONCLUSIONS

Although the conventional activated sludge process has been applied with great success for domestic wastewater treatment for more than 100 years, it appears that a substantial reduction in the energy consumption might not be realized through further process optimization. Indeed, as demonstrated in this study, to tackle the present energy challenge as well as tightened effluent discharge standards, we should jump out of the ivory tower of the current treatment philosophy by developing and adopting novel process configurations and emerging technologies. The solutions forward should rely on the ways to improve energy recovery, meanwhile to minimize in-plant energy consumption. In this regard, A-B processes as discussed in this study may offer the most feasible option towards energy self-sufficient municipal wastewater treatment. Making the WWTPs self-sustainable for energy demand is obviously of significance, especially when global climate change is concerned.

METHODS

In environmental engineering, COD is commonly used to measure the amount of organic maters in wastewater[10,11]. In this work we adopt COD in assessing the potential energy consumption and recovery in the proposed A-B processes (Figs 4, 5, 6). A typical COD concentration of 500 mg/L as commonly found in domestic wastewater was used in calculation of the mass and energy flows through the CAS and the proposed A-B processes. The COD mass flow through a specific biological process was calculated based on the reported representative COD removal efficiency in each operation unit of interest[12,13,18,26,37,47,48]. In aerobic or anaerobic process for wastewater treatment, COD is channeled into new biomass, oxidized products (e.g. carbon dioxide and methane), residual COD in the treated effluent. The biomass produced from aerobic or anaerobic degradation of COD is ultimately directed to anaerobic digester for energy recovery. In calculation of the potential energy recovery from the proposed biological processes, it is reasonably thought that energy can only be generated from methane gas produced via anaerobic degradation of directly captured COD or waste sludge as illustrated in Fig. 8. Therefore, the recoverable chemical energy from one gram COD removed can be calculated according to the theoretical chemical energy stored in one gram COD-methane, i.e. 13.9 kJ/gCOD-methane[11]. For example, for an influent COD concentration of 500 mg/L, if 43% of influent COD is ultimately converted to methane, the maximum chemical energy potentially recovered from one gram of wastewater COD can be estimated as (13.9 × 43%) = 5.98 kJ/g COD.

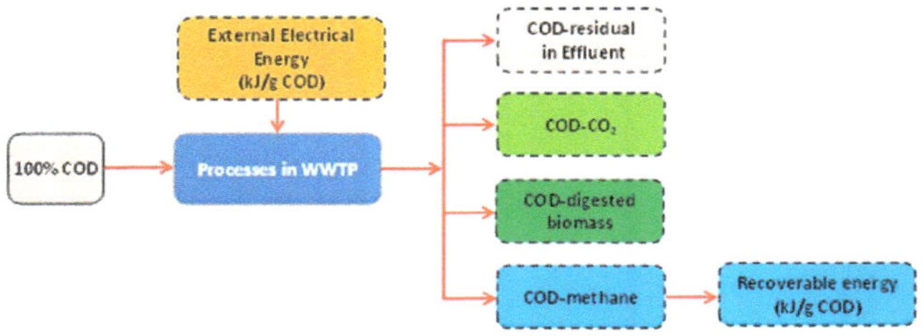

Figure 8: Schematic presentation of COD flux in a biological process.

REFERENCES

1. Mulder, A., Vandegraaf, A. A., Robertson, L. A. & Kuenen, J. G. Anaerobic Ammonium Oxidation Discovered In A Denitrifying Fluidized-Bed Reactor. *FEMS Microbiol. Ecol.* **16,** 177–183 (1995).
2. Peng, Y. & Zhu, G. Biological nitrogen removal with nitrification and denitrification via nitrite pathway. Appl Microbiol Biotechnol. *Appl. Microbiol. Biotechnol.* **73,** 15–26 (2006).
3. Zhou, Y., Zhang, D. Q., Minh Tuyet, L., Puah, A. N. & Ng, W. J. Energy utilization in sewage treatment - a review with comparisons. *J. Water Clim. Change* **4,** 1–10 (2013).
4. US-EPA. *Wastewater Management Fact Sheet, Energy Conservation.* (2006) Available at: http://nepis.epa.gov/Exe/ZyPDF.cgi/ P100IL6T.PDF?Dockey=P100IL6T.pdf. (Accessed: 23th March 2016).
5. Hao, X., Liu, R. & Huang, X. Evaluation of the potential for operating carbon neutral WWTPs in China. *Water Res.* **87,** 424–431 (2015).
6. Jonasson, M. *Energy Benchmark for Wastewater treatment processes-a comparison between Sweden and Austria.* Lund University (2007).
7. Monteith, H., Kalogo, Y. & Louzeiro, N. Achieving stringent effluent limits takes a lot of energy. *Proceedings of the Water Environment Federation* **14,** 4343–4356 (2007).
8. Mizuta, K. & Shimada, M. Benchmarking energy consumption in municipal wastewater treatment plants in Japan. *Water Sci. Technol.* **62,** 2256–2262 (2010).
9. McCarty, P. L., Bae, J. & Kim, J. Domestic Wastewater Treatment as a Net Energy Producer–Can This be Achieved? *Environ. Sci. Technol.* **45,** 7100–7106 (2011).
10. Shizas, I. & Bagley, D. M. Experimental determination of energy content of unknown organics in municipal wastewater streams. J. *Energ. Eng-Asce* **130,** 45–53 (2004).
11. Heidrich, E. S., Curtis, T. P. & Dolfing, J. Determination of the Internal Chemical Energy of Wastewater. *Environ. Sci. Technol.* **45,** 827–832 (2010).
12. Metcalf & Eddy. *Wastewater Engineering: Treatment and Resource Recovery.* (McGraw-Hill international ed., 2014).
13. Morrissey, S. & Harleman, D. F. In *Chemical Water and Wastewater Treatment* II (eds Rudolf Klute & Hermann Hahn) Ch. 26, 401–416, Springer Berlin Heidelberg (1992).
14. Wang, H., Li, F., Keller, A. & Xu, R. Chemically enhanced primary treatment (CEPT) for removal of carbon and nutrients from municipal wastewater treatment plants: a case study of Shanghai. *Water Sci. Technol.* **60** 1803–1809 (2009).
15. Harleman, D. R. F. & Murcott, S. The role of physical-chemical wastewater treatment in the mega-cities of the developing world. *Water Sci. Technol.* **40,** 75–80 (1999).
16. Parker, D. S., Barnard, J., Daigger, G. T., Tekippe, R. J. & Wahlberg, E. J. The future of chemically enhanced primary treatment: evolution not revolution. *WATER* **21,** 49–56 (2001).
17. Diamantis, V. *et al.* Sewage pre-concentration for maximum recovery and reuse at decentralized level. *Water Sci. Technol.* **67,** 1188–1193 (2013).
18. Rossle, W. H. & Pretorius, W. A. A review of characterisation requirements for in-line prefermenters - Paper 1: Wastewater characterisation. *Water Sa* **27,** 405–412 (2001).
19. Jenni, S., Vlaeminck, S. E., Morgenroth, E. & Udert, K. M. Successful application of nitritation/anammox to wastewater with elevated organic carbon to ammonia ratios. *Water Res.* **49,** 316–326 (2014).
20. Xu, G. *et al.* The challenges of mainstream deammonification process for municipal used water treatment. *Appl Microbiol Biotechnol* **99,** 2485–2490 (2015).
21. Zhao, Q.-L., Zhong, H.-Y., Liu, J.-L. & Liu, Y. Integrated coagulation-trickling filter–ultrafiltration processes for domestic wastewater treatment and reclamation. *Water Sci. Technol.* **65,** 1599–1605 (2012).
22. Wett, B., Buchauer, K. & Fimml, C. In *IWA Leading*

Edge Technology Conference, 21-24 Singapore: Asian Water (2007).

23. Lackner, S., Terada, A. & Smets, B. F. Heterotrophic activity compromises autotrophic nitrogen removal in membrane-aerated biofilms: results of a modeling study. *Water Res.* **42,** 1102–1112 (2008).
24. Wett, B. *et al.* Going for mainstream deammonification from bench-to full-scale for maximized resource efficiency. *Water Sci. Technol.* **68,** 283–289 (2013).
25. Lotti, T. *et al.* Anammox growth on pretreated municipal wastewater. *Environ. Sci. Technol.* **48,** 7874–7880 (2014).
26. Meerburg, F. A. *et al.* Toward energy-neutral wastewater treatment: A high-rate contact stabilization process to maximally recover sewage organics. *Bioresour. Technol.* **179,** 373–381 (2015).
27. Huang, J. & Li, L. An innovative approach to maximize primary treatment performance. *Water Sci. Technol.* **42,** 209–222 (2000).
28. Zhao, W., Ting, Y., Chen, J., Xing, C. & Shi, S. Advanced primary treatment of waste water using a bio-flocculation-adsorption sedimentation process. *Acta Biotechnol.* **20,** 53–64 (2000).
29. Mergaert, K., Vanderhaegen, B. & Verstraete, W. Applicability and trends of anaerobic pre-treatment of municipal wastewater. *Water Res.* **26,** 1025–1033 (1992).
30. Von Sperling, M., Freire, V. & de Lemos Chernicharo, C. Performance evaluation of a UASB-activated sludge system treating municipal wastewater. *Water Sci. Technol.* **43,** 323–328 (2001).
31. Seghezzo, L., Zeeman, G., van Lier, J. B., Hamelers, H. V. M. & Lettinga, G. A review: The anaerobic treatment of sewage in UASB and EGSB reactors. *Bioresour. Technol.* **65,** 175–190 (1998).
32. Lettinga, G. *et al.* Anaerobic treatment of domestic sewage and wastewater. *Water Sci. Technol.* **27,** 67–73 (1993).
33. Foresti, E. Anaerobic treatment of domestic sewage: established technologies and perspectives. *Water Sci. Technol.* **45,** 181–186 (2002).
34. Kim, J. *et al.* Anaerobic fluidized bed membrane bioreactor for wastewater treatment. *Environ. Sci. Technol.* **45,** 576–581 (2010).
35. Malovanyy, A., Yang, J., Trela, J. & Plaza, E. Combination of upflow anaerobic sludge blanket (UASB) reactor and partial nitritation/ anammox moving bed biofilm reactor (MBBR) for municipal wastewater treatment. *Bioresour. Technol.* **180,** 144–153 (2015).
36. Malovanyy, A., Trela, J. & Plaza, E. Mainstream wastewater treatment in integrated fixed film activated sludge (IFAS) reactor by partial nitritation/anammox process. *Bioresour. Technol.* **198,** 478–487 (2015).
37. Gopala Krishna, G. V. T., Kumar, P. & Kumar, P. Treatment of low-strength soluble wastewater using an anaerobic baffled reactor (ABR). *J. Environ. Manage.* **90,** 166–176 (2009).
38. Yamamoto, S., Alcauskas, J. B. & Crozier, T. E. Solubility of methane in distilled water and seawater. *J. Chem. Eng. Data* **21,** 78–80 (1976).
39. Liu, Z.-h., Yin, H., Dang, Z. & Liu, Y. Dissolved Methane: A Hurdle for Anaerobic Treatment of Municipal Wastewater. *Environ. Sci. Technol.* **48,** 889–890 (2014).
40. Smith, A. L., Skerlos, S. J. & Raskin, L. Psychrophilic anaerobic membrane bioreactor treatment of domestic wastewater. *Water Res.* **47,** 1655–1665 (2013).
41. Lester, J., Jefferson, B., Eusebi, A. L., McAdam, E. & Cartmell, E. Anaerobic treatment of fortified municipal wastewater in temperate climates. *J. Chem. Technol. Biotechnol.* **88,** 1280–1288 (2013).
42. Zhao, Q. & Kugel, G. Thermophilic/mesophilic digestion of sewage sludge and organic wastes. *J. Environ. Sci. Health, Pt. A* **31,**2211–2231 (1996).
43. Smith, A. L., Stadler, L. B., Love, N. G., Skerlos, S. J. & Raskin, L. Perspectives on anaerobic membrane bioreactor treatment of domestic wastewater: a critical review. *Bioresour. Technol.* **122,** 149–159 (2012).
44. Widiastuti, N., Wu, H., Ang, H. M. & Zhang, D. Removal of ammonium from greywater using natural zeolite. Desalination 277, 15–23 (2011).
45. Karapınar, N. Application of natural zeolite for phosphorus and ammonium removal from aqueous solutions. J. Hazard. Mater. 170, 1186–1191 (2009).
46. Zhang, B.-h. et al. Simultaneous removal of ammonium and phosphate by zeolite synthesized from coal fly ash as influenced by acid treatment. J. Environ. Sci. 19, 540–545 (2007).
47. Parkin, G. F. & Owen, W. F. Fundamentals of anaerobic digestion of wastewater sludges. J. Environ. Eng. 112, 867–920 (1986).
48. Miron, Y., Zeeman, G., van Lier, J. B. & Lettinga, G. The role of sludge retention time in the hydrolysis and acidification of lipids, carbohydrates and proteins during digestion of primary sludge in CSTR systems. Water Res. 34, 1705–1713 (2000).

CHAPTER 7

GREENHOUSE GASES EMISSIONS FROM WASTEWATER TREATMENT PLANTS: MINIMIZATION, TREATMENT, AND PREVENTION

J. L. Campos D. Valenzuela-Heredia A. Pedrouso A. Val del Río M. Belmonte and A. Mosquera-Corral

INTRODUCTION

In the past years most efforts to improve wastewater treatment plants (WWTPs) performance have been focused on obtaining a good effluent quality [1–5]. However nowadays, new challenges are under consideration, oriented to ensure the sustainability of WWTPs in terms of their economic feasibility and environmental impact. Energy consumption and greenhouse gases (GHG) emissions are among the aspects that have become key-factors concerning the overall performance of the WWTPs [6, 7]. Recent studies have identified the WWTPs as potential sources of anthropogenic GHG emissions, contributing to climate change and air pollution [8–10]. WWTPs produce carbon dioxide (CO_2), methane (CH_4), and nitrous oxide (N_2O) during the biological wastewater treatment processes and CO_2 is also emitted during the production of the energy required for the plant operation. The CO_2 released due to the energy demand can be directly reduced by enhancing the energy efficiency of the WWTPs. In this way both the reduction of environmental impacts and the decrease of treatment costs by enhancing the energy savings can be accomplished simultaneously.

With regard to each GHG source the N_2O emitted is generated by nitrification and denitrification processes used to remove nitrogenous compounds from wastewater. Its production occurs mainly in the activated sludge units (90%) while the remaining 10% comes from the grit and sludge storage tanks [11]. N_2O gas is an intermediate of biological processes such as heterotrophic denitrification and nitrification (Figure 1). It is formed during denitrification operated at low pH values and toxic compounds or low dissolved oxygen (DO) concentrations are present in the media [12–14]. Nitrifying bacteria are able to produce N_2O under aerobic or anoxic [15] conditions. In anoxic conditions, both ammonia- and nitrite-oxidizing bacteria are able to produce it, while only ammonia-oxidizing bacteria do it in aerobic conditions. In the latter case, the production is stimulated by the presence of low DO concentrations and presence of nitrite () or organic matter in the liquid media [12, 13]. Nitrous oxide can be produced also from chemical reactions taking place in the presence of hydroxylamine and nitrite [16].

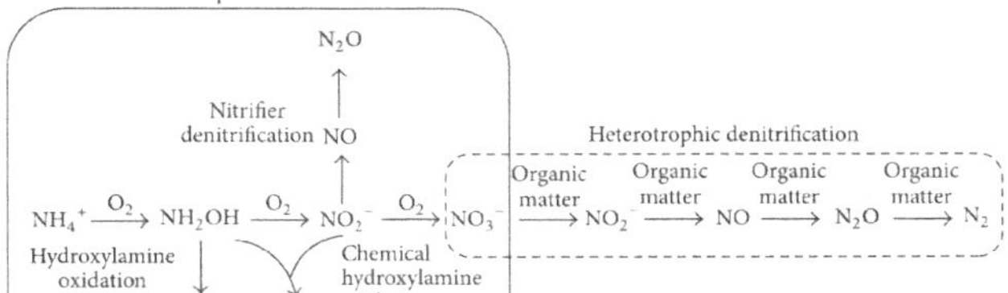

Figure 1: Biological and chemical pathways of N_2O production in the nitrification and denitrification processes.

In practice nitrous oxide is emitted in the WWTP predominantly in the aerobic tank [17]. However, the contribution of the anoxic and aerobic reactors to this production remains still unclear since it can be produced in the anoxic stage and be subsequently stripped to the gas phase in the aerated compartment [18]. Ammonia-oxidizing bacteria have been identified as the main N_2O producers while heterotrophic denitrifying bacteria contribution is only

Citation: J. L. Campos, D. Valenzuela-Heredia, A. Pedrouso, A. Val del Río, M. Belmonte, and A. Mosquera-Corral, "Greenhouse Gases Emissions from Wastewater Treatment Plants: Minimization, Treatment, and Prevention," Journal of Chemistry, vol. 2016, Article ID 3796352, 12 pages, 2016. doi:10.1155/2016/3796352

relevant when nitrite and/or oxygen are present in the anoxic stage [19]. According to Tallec et al. [20], under common operational conditions, the N_2O production occurs mainly via denitrification by nitrifying bacteria. However hydroxylamine oxidation pathway can be the main process responsible for the production of N_2O emissions at high ammonia (NH_4^+) and low nitrite concentrations, when a high metabolic activity of ammonia-oxidizing bacteria is present (at 2 to 3 mg O_2/L) [19].

With regard to CH_4 emissions, Daelman et al. [21] found out that about 1% of the incoming chemical oxygen demand (COD) to the WWTPs was emitted as methane. This amount exceeds the amount of carbon dioxide emission that was avoided by utilizing the produced biogas in anaerobic digestion. The main sources of methane detected by these authors were related to the sludge line units where anaerobic digestion is carried out: the primary sludge thickener, the centrifuge, the exhaust gas of the cogeneration plant, the buffer tank for the digested sludge, and the storage tank for the dewatered sludge. These units contribute to around 72% of methane emissions of the WWTPs while the remaining emissions come from the biological reactors and can be mainly attributed to the CH_4 dissolved in the wastewater which is not totally removed by the biological system. Research works of Yver Kwok et al. [22] and Oshita et al. [23] also showed that most of the methane emissions from WWTPs are closely related to processes involved in the sludge line.

With respect to CO_2 its production is attributed to two main factors: biological treatment process and electricity consumption. In the main stream of the WWTP the organic carbon of wastewater is either incorporated into biomass or oxidized to CO_2. In the sludge line, it is converted mainly to CO_2 and CH_4during anaerobic digestion and, finally, methane is oxidized to CO_2 during biogas combustion.

In recent literature the emissions of GHG from the conventional configurations of WWTPs were determined but the analysis of the possible alternatives to minimize these emissions is generally not done [24–28]. On the other hand, most of the papers studying the application of new processes to remove pollutants from wastewaters are mainly focused on the energy savings [29–32] and only few of them also give an environmental evaluation [33–36]. Therefore, the main objective of this work is to provide an overview of all possible ways to reduce GHG emissions from WWTPs. For that, two possible scenarios are discussed and analyzed: (1) to maintain the present scheme of operation of the WWTP and to modify the operational conditions (minimization) or to implement capture and treatment units for the gaseous streams (treatment) and (2) to change the scheme of operation and to implement new processes which produce lower GHG emissions than the existing ones (prevention).

MINIMIZATION OF GHG EMISSIONS

Perhaps, the most efficient way, in terms of costs, to reduce GHG emissions is to modify the operational conditions of WWTPs units but this is not always possible due to the operational limitations of the installed units. In the following sections, some recommendations about the possible actions to put in practice to operate WWPTs in order to reduce GHG emissions are provided.

N_2O Production

Data obtained from the operation of full-scale WWTPs show a wide range of values for the fraction of nitrogen that is emitted as N_2O (0–14.6% of the nitrogen load) [12]. Such large variation can be related to the different operational conditions imposed in the studied WWTPs. Having this in mind decreasing the amounts of N_2O emitted from activated sludge processes presents a great potential for improvement, by avoiding those operational conditions identified as responsible for its production. Some identified conditions are (i) low dissolved oxygen concentration in the nitrification and the presence of oxygen in denitrification stages, (ii) high nitrite concentrations in both nitrification and denitrification stages, (iii) low COD/N ratio in the denitrification stage, (iv) sudden shifts of pH and dissolved oxygen and ammonia and nitrite concentrations, and (v) transient anoxic and aerobic conditions [12, 13].

Therefore, to minimize N_2O emissions, biological wastewater treatment plants should be operated at high solid retention times (SRT) to maintain low ammonia and nitrite concentrations in the media. Furthermore large bioreactor volumes are recommended to dispose of systems able to buffer loadings and reduce the risk of transient oxygen depletion. N_2O emissions can be also reduced if nitrous oxide stripping by aeration is limited since microorganisms would have more time to consume it [37].

CH_4 Production

CH_4 emissions can be minimized if thickening sludge tanks and sludge disposal tanks are covered to avoid gas leakages and their emissions are captured by hoods which could be burnt with excess biogas in a torch [21]. Besides the methane produced in the plant itself, methane also enters the plant from outside via the influent since it contains CH_4 that has been formed in the sewer. The methane load was estimated as 1% of the influent COD load and is mainly oxidized in the activated sludge tanks (80%), which could be exploited as a means to further decrease methane emissions from wastewater treatment [21].

CO_2 Production

Organic matter oxidation in the biological reactors and combustion of CH_4 are responsible for the direct CO_2 emissions while indirect emissions are attributed to the energy consumption of the WWTP [26]. The SRT applied to the biological reactor is a key operational factor that affects these emissions. The operation of the activated sludge system at high values of SRT promotes endogenous respiration of biomass which increases the amount of COD oxidized to CO_2 and decreases the overall sludge production. This decrease of sludge production implies a decrease of the methane production and, therefore, a decrease of the CO_2emissions associated with its combustion [38]. Both tendencies counteract each other and the addition value of both quantities remains almost constant. Furthermore, the decrease of the SRT also involves an increase of the energy efficiency of the WWTP and, therefore, a decrease of indirect CO_2 emissions. Therefore, CO_2emissions should be minimized by applying the shortest SRT value as possible without negatively affecting the effluent quality. The effect of SRT on the overall CO_2 emissions of a conventional

WWTP can be quantified by performing mass and energy balances according to the methodology described by Campos et al. [39] and using the parameters given in Table 1. The SRT values tested ranged from 10 to 30 d in order to guarantee a stable nitrification. Results showed that an increase of the SRT from 10 to 30 days supposed an increase of 7.6% of the CO_2 emissions (Figure 2).

Table 1: Values assigned to the parameters used to estimate CO_2 emissions.

Parameter	Units	Reference
CO_2 emissions from energy consumption	0.391 kg CO_2/kW·h	
CO_2 emissions from COD oxidation	0.08 kg CO_2/kg COD*	[40]
CO_2 emissions from CH_4 combustion	3.5 kg CO_2/Nm^3 CH_4**	

*Estimated taking into account an elemental composition of $C_{2.43}H_{3.96}O$ for the biodegradable fractions of the COD [41]. **Calculated from stoichiometry and ideal gas law.

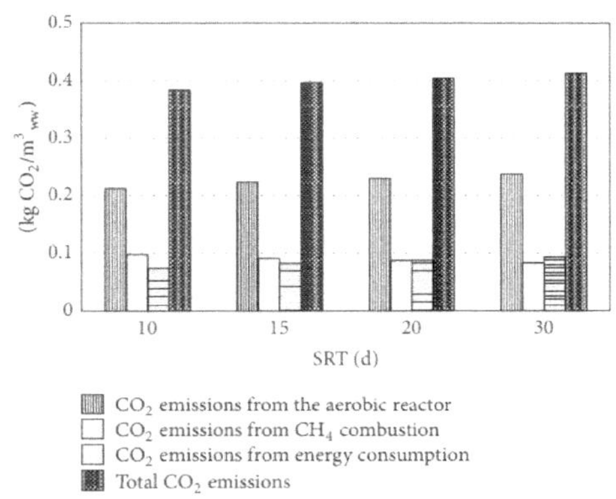

Figure 2: CO_2 emissions estimated for a conventional WWTP operated at different SRT values.

TREATMENT

A second possible option to reduce GHG emissions from WWTPs is to capture and treat them. An important number of technologies are available to destroy or capture N_2O, CH_4, and CO_2 from industrial gaseous streams but there is still a need for the development of efficient low-cost abatement technologies to treat gaseous streams from WWTPs. On the other hand, the capital costs required to cover the different tanks and capture GHG emissions are relatively high [47].

N_2O Removal

Traditional technologies, such as selective catalytic reduction and selective noncatalytic reduction, are currently used to control NO_x emissions from power plants [48–50]. However, both processes require operating at high temperatures or using catalysts, which revert to high installation and maintenance costs. These total costs become prohibitive in large-scale facilities treating air flows containing low-to-moderate concentrations of NO_x [51]. Recently, many different bioprocesses, using nitrifying and denitrifying bacteria or microalgae, have been developed to control NO_x gas emissions. Technologies based on the denitrification process have been successfully used to remove N_2O with efficiencies of 75–99% [51–53]. However, the low aqueous solubility of this greenhouse gas limits the mass transfer rate from the air flow to the liquid phase and, therefore, high hydraulic retention times (HRT) are required to achieve high N_2O removal efficiencies. These long applied HRT result in large bioscrubber (or biofilter) volumes, with the subsequent increase in capital costs [53]. Another alternative is to collect the outlet gaseous stream from the top of the nitrifying unit, containing N_2O, and use it as oxidizer to burn the methane produced in the anaerobic sludge digester [54].

CH_4 Removal

Biological technologies to remove CH_4 from waste gaseous emissions, based on biofilter systems, have been studied since the early 1990s although they are not yet consolidated at industrial scale [55, 56]. Several biological processes are capable of oxidizing methane into CO_2 (1 mol to 1 mol) which allow reducing the total GHG emissions in terms of CO_2 equivalents since the warming factor of CO_2 is lower than that of methane. In aerobic conditions CH_4 is oxidized by methanotrophic bacteria in the presence of oxygen. Another option relies on the application of anaerobic conditions and exploitation of the activity of bacteria and archaea to oxidize CH_4 using sulfate, nitrite, nitrate, Mn^{+4}, or Fe^{+3} as electron acceptors [57]:

$$CH_4 + SO_4^{-2} \longrightarrow HCO_3^- + HS^- + H_2O \quad \Delta G = -16.6\ \mathrm{kJ/mol} \tag{1}$$

$$3CH_4 + 8NO_2^- + 8H^+ \longrightarrow 3CO_2 + 4N_2 + 10H_2O \quad \Delta G = -928\ \mathrm{kJ/mol} \tag{2}$$

$$5CH_4 + 8NO_3^- + 8H^+ \longrightarrow 5CO_2 + 4N_2 + 14H_2O \quad \Delta G = -765\ \mathrm{kJ/mol} \tag{3}$$

$$CH_4 + 4MnO_2 + 7H^+ \longrightarrow HCO_3^- + 4Mn^{+2} + 5H_2O \quad \Delta G = -556\ \mathrm{kJ/mol} \tag{4}$$

$$CH_4 + 8Fe(OH)_3 + 15H^+ \longrightarrow HCO_3^- + 8Fe^{+2} + 21H_2O \quad \Delta G = -270\ \mathrm{kJ/mol} \tag{5}$$

As in the case of the N_2O gas, the low solubility of CH_4 implies the necessity to operate the biofilters at high residence time values (2–30 minutes) [58]. For these reasons, nowadays, the interest has moved to remove the CH_4 directly from the liquid phase, before it is stripped to the atmosphere. To accomplish this removal the anaerobic methane oxidation is coupled to a denitrification process ((2)-(3)) which uses the methane as electron donor. In this case methane and both nitrite and nitrate are removed from wastewater [59]. Furthermore not only is the GHG removed but also the electron donor requirements for the denitrification processes diminish, reducing the costs of potential addition of external carbon source.

After the biological processes the methane remaining in the exhausted gaseous stream can be submitted to a postcombustion process [60].

CO_2 Removal

For the CO_2 gas removal extensive research has been carried out on the study of its capture by chemical or physical sorption and membrane separation processes from power cycles and industrial processes [61, 62]. However, the application of these technologies is generally associated with high capital and operating costs and the generation of waste streams. For these reasons, nowadays the cultivation of microalgae is being considered as an attractive alternative for CO_2 gas sequestration. In WWTPs microalgae can be used for precombustion CO_2 capture, as an economic way for biogas purification [63], or for postcombustion CO_2capture in order to maximize the microalgae production for their use as biofertilizer [64] or as substrate to increase biogas production [65].

PREVENTION

Most of the efforts to improve WWTPs performance are being focused on economic aspects related to energy consumption reduction, minimization of sludge production, and maximization of the amount and quality of biogas generated. To face these topics is important not only in terms of operational costs but also in terms of environmental impacts since it allows reducing direct and indirect GHG emissions [9, 66, 67].

Nowadays, only around 35–45% of the energy contained in the raw wastewater, as organic compounds, is converted into CH_4 during anaerobic digestion of primary and secondary sludge. The remaining part is wasted under aerobic conditions due to the use of conventional nitrification and denitrification processes to remove nitrogen and organic matter simultaneously [68]. An alternative is to apply autotrophic processes to remove nitrogen, such as those based on the combination of the partial nitrification plus Anammox processes or the use of microalgae and even the application of biochemical processes. In this way both organic matter and nitrogen compounds can be removed in separated processes as the former is not required for denitrification but directed to the anaerobic digestion for biogas production. Thus oxygen requirements are minimized while methane production is maximized [69–71].

Application of Partial Nitritation and Anammox Processes to Remove Ammonia

In principle according to the metabolism of Anammox bacteria these are not directly involved in the production of N_2O [72] and, therefore, the application of the Anammox process in the WWTPs instead of the conventional nitrification-denitrification processes is expected to reduce N_2O emissions. However, in practice during the operation of full-scale Anammox reactors treating the reject water from sludge anaerobic digesters N_2O emissions have been detected and accounted for up to 0.6% of the converted nitrogen [46]. This value is much higher than the percentages previously measured in lab-scale Anammox enriched reactors fed with synthetic media of 0.03–0.1% [73, 74]. For this reason the results at full scale can be attributed to the presence of nitrifying bacteria, entering the Anammox reactor in the stream coming from the previous partial nitrification unit [74].

Furthermore Kampschreur et al. [46] measured also N_2O emissions in a partial nitritation full-scale reactor, attributed to denitrification carried out by ammonia-oxidizing bacteria, which corresponded to the 1.7% of the inlet nitrogen load [18]. From the previous results, obtained from full-scale systems, about 2.3% of the nitrogen load can be converted to N_2O in nitritation-Anammox systems. In these conditions a two-stage partial nitritation-Anammox process appears as a nonsuitable alternative to reduce N_2O emissions in WWTPs [66, 67].

Emitted percentages can be reduced down to 0.8–1.2% if a one-reactor nitritation-Anammox system is used [43, 75–77]. At full scale this configuration is the most applied one for the treatment of reject water from the sludge line [78]. For this reason, all the latest studies of the application at the main stream of the Anammox based processes have been carried out in single-stage systems. However, this kind of systems must be operated at low dissolved oxygen concentrations to maintain the balance between ammonia oxidation and Anammox rates and, therefore, the achieved nitrogen removal rates are relatively low [68]. However to operate at low dissolved oxygen levels promotes the development of nitrite-oxidizing bacteria and favours the oxidation of ammonia to nitrate instead of its desired conversion to N_2 [78]. Due to this difficulty in avoiding the activity of the nitrite-oxidizing bacteria a change of concept has occurred and most of the research in course to implement the Anammox process at the main stream is focused on the two-stage reactor configuration. In this way operational strategies to avoid the development of nitrite-oxidizing bacteria, without affecting the Anammox bacteria, can be evaluated [68]. This might imply that the emissions of N_2O would hamper the practical application of the partial nitritation-Anammox process, from the energy-saving and cost-effective point of view. Nevertheless, since the total amount of N_2O emission from the partial nitrification unit is correlated to the nitrite concentration present, N_2O emissions about 0.1% of the inlet nitrogen load are expected considering a nitritation unit operated for the treatment of the main stream (20–25 mg NO_2^--N/L) [44]. Until now data of N_2O emissions from Anammox systems in operation in the main stream conditions are not available in the literature, although if the entrance of nitrifying bacteria inside the Anammox system is minimized the expected emissions would be limited to 0.1% of the inlet nitrogen load and mainly due to the presence of heterotrophic denitrifying bacteria [45, 79]. This means that the partial nitritation-Anammox system treating the main stream would emit in total around 0.2% of the inlet nitrogen load as N_2O.

Taking into account the fact that WWTPs with nitrogen removal carried out by nitrification-denitrification processes have a median emission factor of 0.01 kg N_2O-N/kg $N_{influent}$ [42], meaning that 0.6% of the inlet nitrogen is converted into N_2O, the application of partial nitritation-Anammox processes in both sludge line (20% of the total nitrogen load with a conversion of 0.8% into N_2O) and main stream (80% of the total nitrogen load with a conversion of 0.2% into N_2O) will signify an important decrease of the N_2O emissions.

CANDO Process

Recently, Scherson et al. [54] introduced a new N removal process called CANDO (Coupled Aerobic-anoxic Nitrous Decomposition Operation) which involves three steps: (1) biological conversion of NH_4^+ to NO_2^-; (2) biological or chemical partial anoxic reduction of NO_2^- to N_2O; and (3) N_2O conversion to N_2 with energy recovery.

Then from steps (1) and (2) ammonia is converted to N_2O, which is used in step (3) as a cooxidant for CH_4 combustion or decomposed over a metal oxide catalyst to recover energy. The end product of the reaction is the N_2. The innovation consists of utilizing N_2O as a renewable energy source and reducing the requirements of organic matter which is consumed during denitrification. Combustion of CH_4 with N_2O releases roughly 30% more heat than using O_2 ((6) and (7)) and mitigates the release of N_2O to the atmosphere:

$$CH_4 + 4N_2O \longrightarrow CO_2 + 2H_2O + 4N_2 \qquad \Delta H = 1219\,\text{kJ/mol } CH_4 \tag{6}$$

$$CH_4 + 2O_2 \longrightarrow CO_2 + 2H_2O \qquad \Delta H = 890\,\text{kJ/mol } CH_4 \tag{7}$$

Steps (1) and (3) of the CANDO process have been already applied at full scale while step (2) is still under study [54, 71, 80]. In these research works, two ways of producing nitrous oxide from nitrite are proposed: (1) abiotic reduction by Fe(II) with conversions over 90% and (2) partial heterotrophic denitrification (62% of NO_2^- converted to N_2O).

Application of Microalgae

One of the main operating costs of conventional activated sludge systems, where nitrogen removal takes place, is associated with the large aeration requirements. Alternative systems like those based on microalgae are being considered as potential substitutes. In these systems nitrogen is removed via assimilation for biomass growth without oxygen consumption (8) [81], decreasing energy requirements. Moreover, a low N_2O production is expected (0.005% kg N_2O-N/kg N_{applied}) if microalgae are used to remove nitrogen [82]

$$106CO_2 + 236H_2O + 16NH_4^+ + HPO_4^{-2} + \text{light} \longrightarrow C_{106}H_{181}O_{45}N_{16}P + 118O_2 + 171H_2O + 14H^+ \tag{8}$$

When microalgae are applied for wastewater treatment cultivation the process is generally carried out in open raceway ponds, since the capital costs of these systems are lower than those of photobioreactors [83]. These microalgae ponds occupy large land areas which limits their use to rural areas. Another disadvantage of the microalgae application relies on the poor settling properties of the microalgae, which implies the use of coagulants and flocculants for separation from the treated wastewater [84]. For this reason, a novel approach consisting in the use of algal-bacterial cocultures has received significant attention in recent years as well. In this way the bacterial population would profit from the O_2 produced by algae reducing the aeration requirements of treatment processes and at the same time greenhouse gas emissions are mitigated by the CO_2 consumption during algal photosynthesis (Figure 3). According to (8), microalgae produce 17 kg O_2/kg N_{removed}. Then, for typical urban wastewater, the amount of oxygen produced by microalgae would be higher than the O_2 amount needed to remove organic matter by the activity of the heterotrophic biomass [85]. Therefore, both organic matter and nitrogen could be simultaneously removed in an open raceway pond without oxygen external supply. In addition, challenges associated with the high energy requirements for algal biomass harvesting might be overcome by means of the better settleability properties of the algal-bacterial coculture. Su et al. [86] demonstrated that an algal-bacterial coculture is able not only to achieve high COD and nutrient removal efficiencies but also to settle completely over 20 minutes. They also argued that the shares of algae and sludge inoculated in the pond have an influence on the nutrient removal efficiency and settleability, and they identify the ratio value of 1 : 5 (algae/sludge by weight) as that providing the biomass with the best settleability.

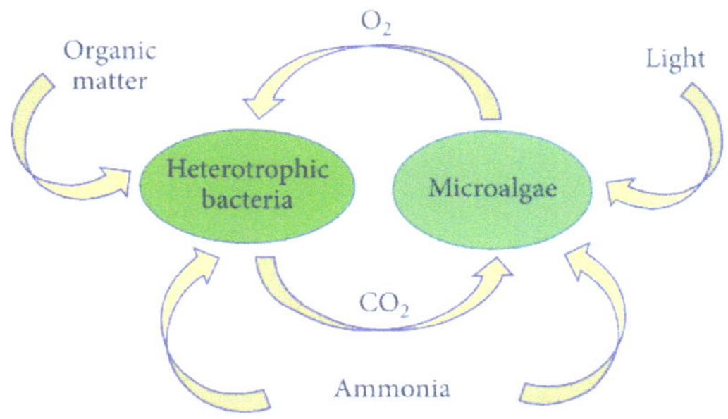

Figure 3: Interactions of a mix culture containing heterotrophic bacteria and microalgae.

CASE STUDIES

In order to quantify the potential reduction of GHG emissions due to the implementation of new processes in WWTPs (prevention strategy), five different configurations were evaluated for comparison purposes.

Case A. A conventional activated sludge system was used as a base case, performing the nitrification-denitrification process to remove both organic matter and nitrogen (Figure 4). The operational conditions of this system were SRT of 15 d, hydraulic retention time (HRT) of 12 h, internal recycle ratio of 3, external recycle ratio of 1, and aerobic volume percentage of 45%. In the primary settler, a particulate COD removal efficiency of 45% was assumed. Primary and secondary sludge are treated in an anaerobic digester (SRT: 30 d) in order to produce biogas, used in a cogeneration unit, and reduce the amount of sludge generated. The following cases take this one as a base case and only the descriptions of modifications applied to this configuration are included.

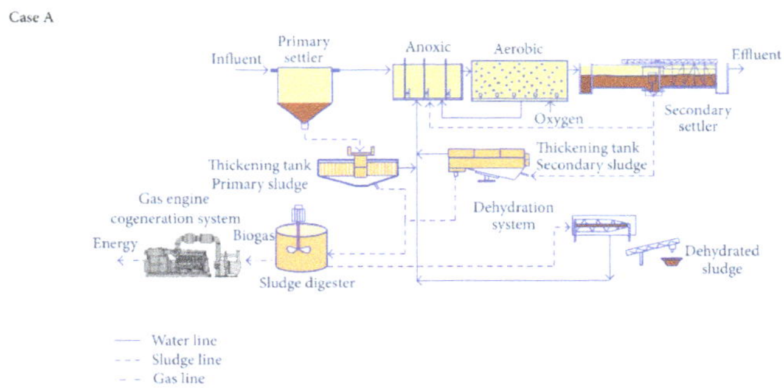

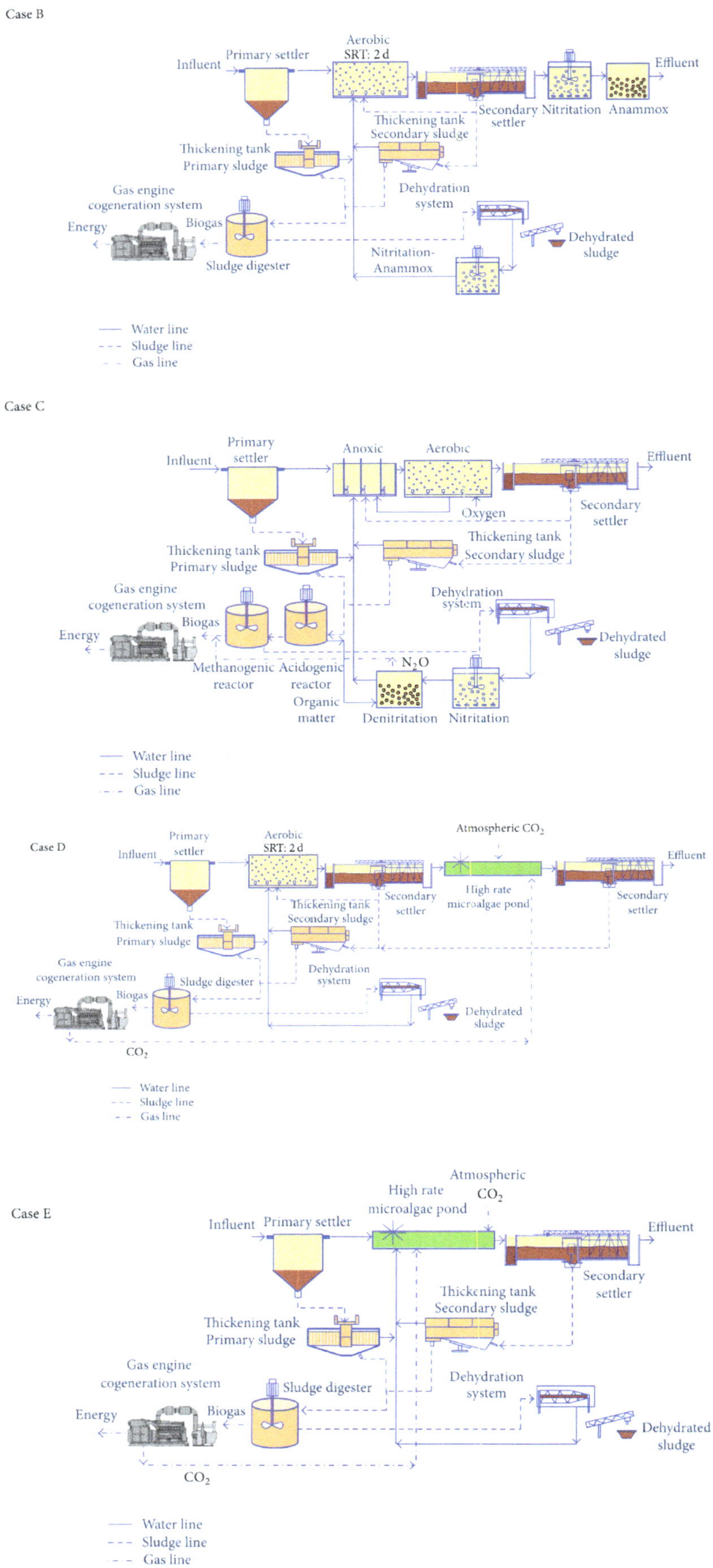

Figure 4: Layout of the five WWTPs configurations evaluated in terms of GHG emissions.

Case B. The activated sludge system was substituted by an aerobic reactor operated at a SRT of 2 d, to remove organic matter and to maximize the sludge generation, in order to increase methane production, followed by a partial nitritation and an Anammox reactor to remove nitrogen. In this case, the primary settler is intended for removing all the particulate COD to promote the anaerobic route of organic matter. Nitrogen is treated from the return sludge stream by a 1-stage partial nitritation-Anammox reactor.

Case C. A CANDO system comprising a partial nitrification and a partial denitrifying reactor is implemented in the sludge line. In this case, organic matter separated in the primary sludge is increased 20% since the nitrogen load applied to the activated sludge system is decreased in the same percentage. In the sludge line, an acidogenic reactor is used to provide organic matter to the partial denitrifying reactor. Produced N_2O is used, instead of O_2, to burn methane.

Case D. The activated sludge system was substituted by an aerobic reactor operated at a SRT of 2 d, to remove organic matter, followed by a high rate microalgae pond, operated at a HRT of 6 days, to remove nitrogen.

Case E. A high rate microalgae pond (SRT and HRT of 6 days) where the microalgae remove the nitrogen and provide the oxygen required for the heterotrophic bacteria to oxidize the organic matter was evaluated in substitution of the activated sludge system.

Mass and energy balances were performed by using Excel spreadsheets according to the methodology described in Campos et al. [39]. Finally, the emissions of CO_2, CH_4, and N_2O were estimated considering the results obtained from the mass balances and the parameters given in Tables 1 and 2. From these values the global GHG emissions of each configuration expressed as kg CO_2/m^3 of wastewater treated were calculated taking into account the greenhouse gas production impact factors of 21 and 310 for CH_4 and N_2O, respectively.

Results obtained from the calculations indicate that systems using microalgae to remove nitrogen are the most suitable systems to reduce GHG emissions during wastewater treatment (Figure 5). This fact is mainly due to the high amount of CO_2 captured by the microalgae together with the contribution of three other factors: (1) the improvement of the WWTP energy efficiency: in Case D, since the application of microalgae to remove nitrogen does not require the presence of organic matter, most of it can be converted into methane while, in Case E, oxygen generated by microalgae allows an important energy saving in terms of aeration; (2) the very low emissions of N_2O observed in the high rate microalgae ponds; and (3) the additional biogas production due to the anaerobic digestion of the generated microalgae. When the partial nitritation and Anammox processes are used to remove ammonia, instead of conventional nitrification and denitrification processes, the WWTP energy efficiency is also improved which also causes a decrease of GHG emissions. However this decrease is considerably lower than that obtained by microalgae systems.

Table 2: Values assigned to the parameters used to estimate GHG emissions for the case studies.

Parameter	Units	Reference
CH_4 emissions from leakages	0.85% of COD treated + 1.3% of CH_4 burnt*	[21]
N_2O emissions from nitrification-denitrification units	0.5% of the nitrogen treated	[42]
N_2O emissions from 1-stage partial nitritation-Anammox reactors (sludge line)	0.8% of the nitrogen treated	[43]
N_2O emissions from PN reactor (main stream)	0.1% of the nitrogen treated	[44]
N_2O emissions from Anammox reactor (main stream)	0.1% of the nitrogen treated	[45]
N_2O emissions from CANDO process	3.4% of the nitrogen treated + 1.3% of N_2O burnt**	[21, 46]

*Taking into account the fact that CH_4 leakage from the cogeneration engine is 15% of the CH_4 emissions. **Taking into account the fact that all the ammonia present in the wastewater is converted into nitrite in the partial nitrification reactor and supposing a leakage of the N_2O from the cogeneration engine similar to that of the CH_4.

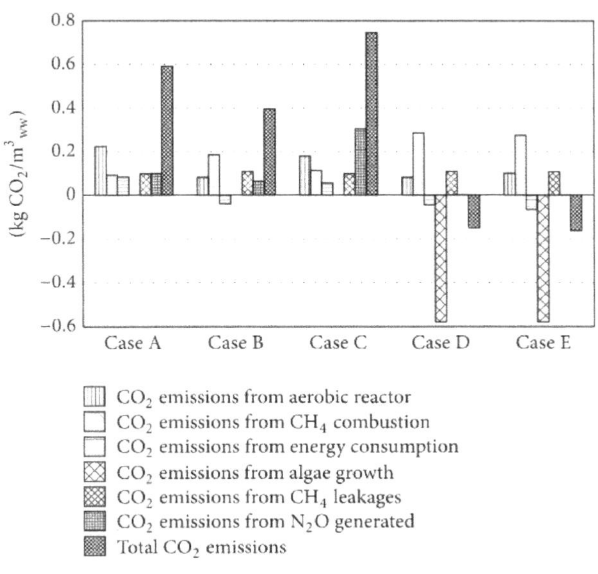

Figure 5: GHG emissions (expressed as kg CO_2, equivalent/m³ of wastewater treated) of different WWTPs configurations.

The WWTP configuration based on the application of the CANDO process (Case C) has associated GHG emissions higher than those of the conventional system. This can be attributed to the increase of N_2O emissions due to the implementation of a partial nitrification reactor in the sludge line and the leakage of nitrous oxide expected in the exhaust gas. On the other hand, in this process organic matter is used to denitrify nitrite into N_2O, decreasing methane production, while N_2O generated can be used to oxidize only around 8% of methane. Those factors limit the energy efficiency improvement achieved by the application of CANDO process.

Nowadays, there are several technologies already implemented at full scale to perform partial nitrification-Anammox processes in the sludge line [87, 88]. However, in spite of the recent advances, their implementation at the main stream is still a challenge due to the strict control of operational conditions needed to maintain the stability of the partial nitrification process [89]. The use of microalgae systems to remove nitrogen from domestic wastewater is a feasible option when enough land is available since this kind of systems would require about ten times the area necessary for activated sludge systems [90]. The CANDO process can be used to improve energy efficiency of WWTPs but can only be applied to the sludge line. Moreover, the need of the development of a reliable technology for its implementation at full scale and its negative environmental impact make it not as attractive as the partial nitrification-Anammox and microalgae systems.

CONCLUSIONS

Minimization. N_2O and CO_2 emissions can be decreased by a good control of the operational conditions of the activated sludge system. CH_4 emissions can be minimized if emissions from the different units of the sludge line are captured by hoods and burnt together with the biogas generated in the sludge anaerobic digester. N_2O emissions will depend mainly on the operational conditions (NO_2^- and O_2 concentrations) of the reactor systems. Treatment. Nowadays most of the technologies available to remove GHG are expensive or even not suitable to be applied to gaseous streams of the WWTPs. Biological systems treatment has low operating costs but their capital costs are high due to their size. The correct selection of the process to be installed in the plant will provide the best results as it is the case of the partial nitritation-Anammox process which is feasible in two units applied in the main stream of the plant but not for the treatment of the sludge line.

Prevention. The configuration of the next generation of WWTPs should maximize the anaerobic pathway for organic matter removal and the use of microalgae, if enough area is available, or partial nitritation-Anammox processes to remove ammonia.

ACKNOWLEDGMENTS

This work was supported by FONDECYT 1150285 (Chile) and Postdoctoral FONDECYT 3140276 (Chile) and by the Spanish Government through FISHPOL (CTQ2014-55021-R) and GRANDSEA (CTM2014-55397-JIN) projects cofunded by FEDER. The authors A. Pedrouso, A. Val del Río, and A. Mosquera-Corral belong to the Galician Competitive Research Group GRC 2013-032, program cofunded by FEDER.

REFERENCES

1. N. Bolong, A. F. Ismail, M. R. Salim, and T. Matsuura, "A review of the effects of emerging contaminants in wastewater and options for their removal," Desalination, vol. 238, no. 1–3, pp. 229–246, 2009. · ·
2. L. Zanetti, N. Frison, E. Nota, M. Tomizioli, D. Bolzonella, and F. Fatone, "Progress in real-time control applied to biological nitrogen removal from wastewater. A short-review," Desalination, vol. 286, pp. 1–7, 2012. · ·
3. W. Luo, F. I. Hai, W. E. Price et al., "High retention membrane bioreactors: challenges and opportunities," Bioresource Technology, vol. 167, pp. 539–546, 2014. · ·
4. A. Santos, W. Ma, and S. J. Judd, "Membrane bioreactors: two decades of research and implementation," Desalination, vol. 273, no. 1, pp. 148–154, 2011. · ·
5. Q. Zhang, J. Hu, and D. J. Lee, "Aerobic granular processes: current research trends," Bioresource Technology, vol. 210, pp. 74–80, 2016. ·
6. W. Mo and Q. Zhang, "Energy-nutrients-water nexus: integrated resource recovery in municipal wastewater treatment plants," Journal of Environmental Management, vol. 127, pp. 255–267, 2013. · ·
7. L. Yerushalmi, O. Ashrafi, and F. Haghighat, "Reductions in greenhouse gas (GHG) generation and energy consumption in wastewater treatment plants," Water Science and Technology, vol. 67, no. 5, pp. 1159–1164, 2013. · ·
8. M. Bani Shahabadi, L. Yerushalmi, and F. Haghighat, "Impact of process design on greenhouse gas (GHG) generation by wastewater treatment plants," Water Research, vol. 43, no. 10, pp. 2679–2687, 2009. · ·
9. T. A. Larsen, "CO_2-neutral wastewater treatment plants or robust, climate-friendly wastewater management? A systems perspective," Water Research, vol. 87, pp. 513–521, 2015. · ·
10. C. Sweetapple, G. Fu, and D. Butler, "Identifying sensitive sources and key control handles for the reduction of greenhouse gas emissions from wastewater treatment," Water Research, vol. 62, pp. 249–259, 2014. · ·
11. P. Czepiel, P. Crill, and R. Harriss, "Nitrous oxide emissions from municipal wastewater treatment," Environmental Science & Technology, vol. 29, no. 9, pp. 2352–2356, 1995. ·
12. M. J. Kampschreur, H. Temmink, R. Kleerebezem, M. S. M. Jetten, and M. C. M. van Loosdrecht, "Nitrous oxide emission during wastewater treatment," Water Research, vol. 43, no. 17, pp. 4093–4103, 2009. · ·
13. Y. Law, L. Ye, Y. Pan, and Z. Yuan, "Nitrous oxide emissions from wastewater treatment processes," Philosophical Transactions of the Royal Society B: Biological Sciences, vol. 367, no. 1593, pp. 1265–1277, 2012. · ·
14. J. Desloover, S. E. Vlaeminck, P. Clauwaert, W. Verstraete, and N. Boon, "Strategies to mitigate N_2O emissions from biological nitrogen removal systems,"

Current Opinion in Biotechnology, vol. 23, no. 3, pp. 474–482, 2012. · ·

15. J. L. Campos, B. Arrojo, J. R. Vázquez-Padín, A. Mosquera-Corral, and R. Méndez, "N_2O production by nitrifying biomass under anoxic and aerobic conditions," Applied Biochemistry and Biotechnology, vol. 152, no. 2, pp. 189–198, 2009. · ·

16. A. Soler-Jofra, B. Stevens, M. Hoekstra et al., "Importance of abiotic hydroxylamine conversion on nitrous oxide emissions during nitritation of reject water," Chemical Engineering Journal, vol. 287, pp. 720–726, 2016. ·

17. J. H. Ahn, S. Kim, H. Park, B. Rahm, K. Pagilla, and K. Chandran, "N_2O emissions from activated sludge processes, 2008-2009: results of a national monitoring survey in the United States,"Environmental Science & Technology, vol. 44, no. 12, pp. 4505–4511, 2010. · ·

18. C. M. Castro-Barros, M. R. J. Daelman, K. E. Mampaey, M. C. M. van Loosdrecht, and E. I. P. Volcke, "Effect of aeration regime on N_2O emission from partial nitritation-anammox in a full-scale granular sludge reactor," Water Research, vol. 68, pp. 793–803, 2015. · ·

19. P. Wunderlin, J. Mohn, A. Joss, L. Emmenegger, and H. Siegrist, "Mechanisms of N_2O production in biological wastewater treatment under nitrifying and denitrifying conditions," Water Research, vol. 46, no. 4, pp. 1027–1037, 2012. · ·

20. G. Tallec, J. Garnier, G. Billen, and M. Gousailles, "Nitrous oxide emissions from secondary activated sludge in nitrifying conditions of urban wastewater treatment plants: effect of oxygenation level,"Water Research, vol. 40, no. 15, pp. 2972–2980, 2006. · ·

21. M. R. J. Daelman, E. M. van Voorthuizen, U. G. J. M. van Dongen, E. I. P. Volcke, and M. C. M. van Loosdrecht, "Methane emission during municipal wastewater treatment," Water Research, vol. 46, no. 11, pp. 3657–3670, 2012. · ·

22. C. E. Yver Kwok, D. Müller, C. Caldow et al., "Methane emission estimates using chamber and tracer release experiments for a municipal waste water treatment plant," Atmospheric Measurement Techniques, vol. 8, no. 7, pp. 2853–2867, 2015. · ·

23. K. Oshita, T. Okumura, M. Takaoka, T. Fujimori, L. Appels, and R. Dewil, "Methane and nitrous oxide emissions following anaerobic digestion of sludge in Japanese sewage treatment facilities," Bioresource Technology, vol. 171, no. 1, pp. 175–181, 2014. · ·

24. D. Gupta and S. K. Singh, "Greenhouse gas emissions from wastewater treatment plants: a case study of Noida," Journal of Water Sustainability, vol. 2, no. 2, pp. 131–139, 2012.

25. H. Yoshida, J. Mønster, and C. Scheutz, "Plant-integrated measurement of greenhouse gas emissions from a municipal wastewater treatment plant," Water Research, vol. 61, pp. 108–118, 2014. · ·

26. D. Kyung, M. Kim, J. Chang, and W. Lee, "Estimation of greenhouse gas emissions from a hybrid wastewater treatment plant," Journal of Cleaner Production, vol. 95, pp. 117–123, 2015. · ·

27. M. Molinos-Senante, F. Hernández-Sancho, M. Mocholí-Arce, and R. Sala-Garrido, "Economic and environmental performance of wastewater treatment plants: potential reductions in greenhouse gases emissions," Resource and Energy Economics, vol. 38, pp. 125–140, 2014. · ·

28. A. Rodriguez-Caballero, I. Aymerich, M. Poch, and M. Pijuan, "Evaluation of process conditions triggering emissions of green-house gases from a biological wastewater treatment system," Science of the Total Environment, vol. 493, pp. 384–391, 2014. · ·

29. H. Bozkurt, M. C. van Loosdrecht, K. V. Gernaey, and G. Sin, "Optimal WWTP process selection for treatment of domestic wastewater—a realistic full-scale retrofitting study," Chemical Engineering Journal, vol. 286, pp. 447–458, 2016. ·

30. A. Mahdy, L. Méndez, M. Ballesteros, and C. González-Fernández, "Algaculture integration in conventional wastewater treatment plants: anaerobic digestion comparison of primary and secondary sludge with microalgae biomass," Bioresource Technology, vol. 184, pp. 236–244, 2015. · ·

31. Y. D. Scherson and C. S. Criddle, "Recovery of freshwater from wastewater: upgrading process configurations to maximize energy recovery and minimize residuals," Environmental Science and Technology, vol. 48, no. 15, pp. 8420–8432, 2014. · ·

32. W. Dai, X. Xu, B. Liu, and F. Yang, "Toward energy-neutral wastewater treatment: a membrane combined process of anaerobic digestion and nitritation-anammox for biogas recovery and nitrogen removal," Chemical Engineering Journal, vol. 279, pp. 725–734, 2015. · ·

33. T. Schaubroeck, H. De Clippeleir, N. Weissenbacher et al., "Environmental sustainability of an energy self-sufficient sewage treatment plant: improvements through DEMON and co-digestion," Water Research, vol. 74, pp. 166–179, 2015. · ·

34. A. B. Bisinella de Faria, M. Spérandio, A. Ahmadi, and L. Tiruta-Barna, "Evaluation of new alternatives in wastewater treatment plants based on dynamic modelling and life cycle assessment (DM-LCA)," Water Research, vol. 84, pp. 99–111, 2015. · ·

35. X. Hao, R. Liu, and X. Huang, "Evaluation of the potential for operating carbon neutral WWTPs in China," Water Research, vol. 87, pp. 424–431, 2015. · ·

36. M. Hauck, F. A. Maalcke-Luesken, M. S. Jetten, and M. A. Huijbregts, "Removing nitrogen from wastewater with side stream anammox: what are the trade-offs between environmental impacts?"Resources, Conservation and Recycling, vol. 107, pp. 212–219, 2016. ·

37. Y. Law, P. Lant, and Z. Yuan, "The effect of pH on N_2O production under aerobic conditions in a partial nitritation system," Water Research, vol. 45, no. 18, pp. 5934–5944, 2011. · ·

38. H. Ge, D. J. Batstone, and J. Keller, "Operating aerobic wastewater treatment at very short sludge ages enables treatment and energy recovery through anaerobic sludge digestion," Water Research, vol. 47, no. 17, pp. 6546–6557, 2013. · ·

39. J. L. Campos, A. Mosquera-Corral, A. Val del Río, et al.,

"Energy and resources recovery in wastewater treatment plants," in Environmental Science & Engineering, vol. 9 of Environmental and Energy Management: Ethics, Laws and Policies, pp. 60–78, Studium Press, 2015.

40. P. Atkins, D. Colbourne, M. Dieryckx et al., "Methologies," in Safeguarding the Ozone Layer and the Global Climate System: Issues Related to Hydrofluorocarbons and Perfluorocarbons, B. Metz, L. Kuijpers, S. Solomon et al., Eds., Cambridge University Press, Cambridge, UK, 2005.

41. I. Takács and P. A. Vanrolleghem, "Elemental balances in activated sludge modelling," in Proceedings of the International Water Association Congress (IWA ‹06), Beijing, China, 2006.

42. J. Foley, D. de Haas, Z. Yuan, and P. Lant, "Nitrous oxide generation in full-scale biological nutrient removal wastewater treatment plants," Water Research, vol. 44, no. 3, pp. 831–844, 2010. · ·

43. S. Wyffels, P. Boeckx, K. Pynaert et al., "Nitrogen removal from sludge reject water by a two-stage oxygen-limited autotrophic nitrification denitrification process," Water Science and Technology, vol. 49, no. 5-6, pp. 57–64, 2004. ·

44. C. M. Castro-Barros, A. Rodríguez-Caballero, E. I. P. Volcke, and M. Pijuan, "Effect of nitrite on the N_2O and NO production on the nitrification of low-strength ammonium wastewater," Chemical Engineering Journal, vol. 287, pp. 269–276, 2016. ·

45. S. Okabe, M. Oshiki, Y. Takahashi, and H. Satoh, "N_2O emission from a partial nitrification-anammox process and identification of a key biological process of N_2O emission from anammox granules,"Water Research, vol. 45, no. 19, pp. 6461–6470, 2011. · ·

46. M. J. Kampschreur, W. R. L. van der Star, H. A. Wielders, J. W. Mulder, M. S. M. Jetten, and M. C. M. van Loosdrecht, "Dynamics of nitric oxide and nitrous oxide emission during full-scale reject water treatment," Water Research, vol. 42, no. 3, pp. 812–826, 2008. · ·

47. M.-S. Chou and W.-H. Cheng, "Gaseous emissions and control in wastewater treatment plants,"Environmental Engineering Science, vol. 22, no. 5, pp. 591–600, 2005. · ·

48. K. Skalska, J. S. Miller, and S. Ledakowicz, "Trends in NOx abatement: a review," Science of the Total Environment, vol. 408, no. 19, pp. 3976–3989, 2010. · ·

49. M. Konsolakis, "Recent advances on nitrous oxide (N_2O) decomposition over non-noble-metal oxide catalysts: catalytic performance, mechanistic considerations, and surface chemistry aspects," ACS Catalysis, vol. 5, no. 11, pp. 6397–6421, 2015.

50. F. Kapteijn, J. Rodriguez-Mirasol, and J. A. Moulijn, "Heterogeneous catalytic decomposition of nitrous oxide," Applied Catalysis B: Environmental, vol. 9, no. 1–4, pp. 25–64, 1996. · ·

51. Y. Jin, M. C. Veiga, and C. Kennes, "Bioprocesses for the removal of nitrogen oxides from polluted air,"Journal of Chemical Technology & Biotechnology, vol. 80, no. 5, pp. 483–494, 2005. · ·

52. O. D. Frutos, I. A. Arvelo, R. Pérez, G. Quijano, and R. Muñoz, "Continuous nitrous oxide abatement in a novel denitrifying off-gas bioscrubber," Applied Microbiology and Biotechnology, vol. 99, no. 8, pp. 3695–3706, 2015. · ·

53. O. D. Frutos, G. Quijano, R. Pérez, and R. Muñoz, "Simultaneous biological nitrous oxide abatement and wastewater treatment in a denitrifying off-gas bioscrubber," Chemical Engineering Journal, vol. 288, pp. 28–37, 2016. ·

54. Y. D. Scherson, G. F. Wells, S.-G. Woo et al., "Nitrogen removal with energy recovery through N_2O decomposition," Energy & Environmental Science, vol. 6, no. 1, pp. 241–248, 2013. · ·

55. J. Nikiema, R. Brzezinski, and M. Heitz, "Elimination of methane generated from landfills by biofiltration: a review," Reviews in Environmental Science and Bio/ Technology, vol. 6, no. 4, pp. 261–284, 2007. ·

56. M. F. M. Abushammala, N. E. A. Basri, D. Irwan, and M. K. Younes, "Methane oxidation in landfill cover soils: a review," Asian Journal of Atmospheric Environment, vol. 8, no. 1, pp. 1–14, 2014. · ·

57. M. Cui, A. Ma, H. Qi, X. Zhuang, and G. Zhuang, "Anaerobic oxidation of methane: an 'active' microbial process," MicrobiologyOpen, vol. 4, no. 1, pp. 1–11, 2015. · ·

58. M. Veillette, M. Girard, P. Viens, R. Brzezinski, and M. Heitz, "Function and limits of biofilters for the removal of methane in exhaust gases from the pig industry," Applied Microbiology and Biotechnology, vol. 94, no. 3, pp. 601–611, 2012. · ·

59. J. Zhu, Q. Wang, M. Yuan et al., "Microbiology and potential applications of aerobic methane oxidation coupled to denitrification (AME-D) process: a review," Water Research, vol. 90, pp. 203–215, 2016. ·

60. M. Kumar, G. Rattan, and R. Prasad, "Catalytic abatement of methane emission from CNG vehicles: an overview," Canadian Chemical Transactions, vol. 3, no. 4, pp. 381–409, 2015.

61. A. A. Olajire, "CO_2 capture and separation technologies for end-of-pipe applications—a review,"Energy, vol. 35, no. 6, pp. 2610–2628, 2010. · ·

62. M. K. Mondal, H. K. Balsora, and P. Varshney, "Progress and trends in CO_2 capture/separation technologies: a review," Energy, vol. 46, no. 1, pp. 431–441, 2012. · ·

63. L. Meier, R. Pérez, L. Azócar, M. Rivas, and D. Jeison, "Photosynthetic CO_2 uptake by microalgae: an attractive tool for biogas upgrading," Biomass and Bioenergy, vol. 73, pp. 102–109, 2015. · ·

64. R. Wang, B. Peng, and K. Huang, "The research progress of CO_2 sequestration by algal bio-fertilizer in China," Journal of CO_2 Utilization, vol. 11, pp. 67–70, 2015. · ·

65. M. D□bowski, M. Zieliński, A. Grala, and M. Dudek, "Algae biomass as an alternative substrate in biogas production technologies—review," Renewable and Sustainable Energy Reviews, vol. 27, pp. 596–604, 2013. · ·

66. H. Gao, Y. D. Scherson, and G. F. Wells, "Towards

energy neutral wastewater treatment: methodology and state of the art," Environmental Sciences: Processes & Impacts, vol. 16, no. 6, pp. 1223–1246, 2014. · ·

67. D. Kim, J. D. Bowen, and E. C. Ozelkan, "Optimization of wastewater treatment plant operation for greenhouse gas mitigation," Journal of Environmental Management, vol. 163, pp. 39–48, 2015. · ·

68. N. Morales, Á. Val del Río, J. R. Vázquez-Padín, R. Méndez, A. Mosquera-Corral, and J. L. Campos, "Integration of the Anammox process to the rejection water and main stream lines of WWTPs,"Chemosphere, vol. 140, pp. 99–105, 2015. · ·

69. H. Siegrist, D. Salzgeber, J. Eugster, and A. Joss, "Anammox brings WWTP closer to energy autarky due to increased biogas production and reduced aeration energy for N-removal," Water Science and Technology, vol. 57, no. 3, pp. 383–388, 2008. · ·

70. R. Khiewwijit, H. Temmink, H. Rijnaarts, and K. J. Keesman, "Energy and nutrient recovery for municipal wastewater treatment: how to design a feasible plant layout?" Environmental Modelling & Software, vol. 68, pp. 156–165, 2015. · ·

71. Y. D. Scherson, S.-G. Woo, and C. S. Criddle, "Production of nitrous oxide from anaerobic digester centrate and its use as a co-oxidant of biogas to enhance energy recovery," Environmental Science and Technology, vol. 48, no. 10, pp. 5612–5619, 2014. · ·

72. B. Kartal, M. M. M. Kuypers, G. Lavik et al., "Anammox bacteria disguised as denitrifiers: nitrate reduction to dinitrogen gas via nitrite and ammonium," Environmental Microbiology, vol. 9, no. 3, pp. 635–642, 2007. · ·

73. M. Strous, J. J. Heijnen, J. G. Kuenen, and M. S. M. Jetten, "The sequencing batch reactor as a powerful tool for the study of slowly growing anaerobic ammonium-oxidizing microorganisms," Applied Microbiology and Biotechnology, vol. 50, no. 5, pp. 589–596, 1998. · ·

74. S. Wyffels, P. Boeckx, K. Pynaert, W. Verstraete, and O. Van Cleemput, "Sustained nitrite accumulation in a membrane-assisted bioreactor (MBR) for the treatment of ammonium-rich wastewater," Journal of Chemical Technology & Biotechnology, vol. 78, no. 4, pp. 412–419, 2003. · ·

75. M. J. Kampschreur, R. Poldermans, R. Kleerebezem et al., "Emission of nitrous oxide and nitric oxide from a full-scale single-stage nitritation-anammox reactor," Water Science and Technology, vol. 60, no. 12, pp. 3211–3217, 2009. · ·

76. J. Yang, J. Trela, E. Plaza, and K. Tjus, "N_2O emissions from a one stage partial nitrification/anammox process in moving bed biofilm reactors," Water Science and Technology, vol. 68, no. 1, pp. 144–152, 2013. · ·

77. S. Lackner, E. M. Gilbert, S. E. Vlaeminck, A. Joss, H. Horn, and M. C. M. van Loosdrecht, "Full-scale partial nitritation/anammox experiences—an application survey," Water Research, vol. 55, pp. 292–303, 2014. · ·

78. B. Ma, S. Wang, S. Cao et al., "Biological nitrogen removal from sewage via anammox: recent advances," Bioresource Technology, vol. 200, pp. 981–990, 2016. ·

79. T. Muangthong-on and C. Wantawin, "Evaluation of N_2O production from anaerobic ammonium oxidation (Anammox) at different influent ammonia to nitrite ratios," Energy Procedia, vol. 9, pp. 7–14, 2011. ·

80. J. Myung, Z. Wang, T. Yuan et al., "Production of nitrous oxide from nitrite in stable type II methanotrophic enrichments," Environmental Science and Technology, vol. 49, no. 18, pp. 10969–10975, 2015. · ·

81. J. B. K. Park and R. J. Craggs, "Nutrient removal in wastewater treatment high rate algal ponds with carbon dioxide addition," Water Science and Technology, vol. 63, no. 8, pp. 1758–1764, 2011. · ·

82. C. Alcántara, R. Muñoz, Z. Norvill, M. Plouviez, and B. Guieysse, "Nitrous oxide emissions from high rate algal ponds treating domestic wastewater," Bioresource Technology, vol. 177, pp. 110–117, 2015. · ·

83. A. I. Barros, A. L. Gonçalves, M. Simões, and J. C. M. Pires, "Harvesting techniques applied to microalgae: a review," Renewable and Sustainable Energy Reviews, vol. 41, pp. 1489–1500, 2015. · ·

84. B. D. Shoener, I. M. Bradley, R. D. Cusick, and J. S. Guest, "Energy positive domestic wastewater treatment: the roles of anaerobic and phototrophic technologies," Environmental Sciences: Processes & Impacts, vol. 16, no. 6, pp. 1204–1222, 2014. · ·

85. N. C. Boelee, H. Temmink, M. Janssen, C. J. N. Buisman, and R. H. Wijffels, "Scenario analysis of nutrient removal from municipal wastewater by microalgal biofilms," Water, vol. 4, no. 2, pp. 460–473, 2012. · ·

86. Y. Su, A. Mennerich, and B. Urban, "Synergistic cooperation between wastewater-born algae and activated sludge for wastewater treatment: influence of algae and sludge inoculation ratios," Bioresource Technology, vol. 105, pp. 67–73, 2012. · ·

87. W. Abma, C. Schultz, J. M. Mulder, et al., "The advance of anammox," Water21, vol. 36, pp. 36–37, 2007.

88. S. Lackner, E. M. Gilbert, S. E. Vlaeminck, A. Joss, H. Horn, and M. C. M. van Loosdrecht, "Full-scale partial nitritation/anammox experiences: an application survey," Water Research, vol. 55, pp. 292–303, 2014. · ·

89. A. Malovanyy, J. Trela, and E. Plaza, "Mainstream wastewater treatment in integrated fixed film activated sludge (IFAS) reactor by partial nitritation/anammox process," Bioresource Technology, vol. 198, pp. 478–487, 2015. · ·

90. R. J. Craggs, S. Heubeck, T. J. Lundquist, and J. R. Benemann, "Algal biofuels from wastewater treatment high rate algal ponds," Water Science and Technology, vol. 63, no. 4, pp. 660–665, 2011.

CHAPTER 8 STABILIZATION PROCESS FOR DOMESTIC WASTEWATER TREATMENT

Nancy Vásquez Sarria, Jenny Rodríguez Victoria, Patricia Torres Lozada And Carlos Madera Parra

Keywords: Activated sludge, contact stabilization, organic matter removal, secondary wastewater treatment

INTRODUCTION

The activated sludge system is a suspended-growth aerobic process which uses the metabolic reactions of microorganisms to convert biodegradable organic wastewater constituents and certain inorganic fractions into new cell mass and byproducts. This process has been widely applied for the treatment of domestic and industrial wastewater, due to the high efficiency obtained, operational flexibility, and possibility of nutrient removal [1-3].

Contact stabilization, as an activated sludge process modification, comprises two aerated reactors (the contact reactor [CR], and the stabilization reactor [SR]) separated by a sedimentation tank [3-7]. The CR receives the influent flow and the biomass in a starved condition so that soluble material is readily adsorbed and absorbed by the starved biomass.

The mixed liquor leaving the CR is settled and the biomass is concentrated. Then, the biomass is sent to SR, where colloidal material removed from the wastewater in the CR is stabilized. Stabilized biomass is returned to the CR by a biomass recycle flow [8]. Figure 1 shows a schematic representation of the contact stabilization process. The main design and operating parameters of activated sludge systems include the hydraulic retention time (HRT), the sludge recycle rate (r), the sludge retention time (SRT), mixed liquor volatile suspended solids (MLVSS), the volumetric organic load (VOL), the food microorganism relation (F/M), the sludge settling properties (sludge volumetric index [SVI]), the characteristics of floc, and the concentration of dissolved oxygen (DO) [6, 9, 10].

Citation: Nancy Vásquez Sarria, Jenny Rodríguez Victoria, Patricia Torres Lozada And Carlos Madera Parra, Stabilization Process for Domestic Wastewater Treatment, ISSN 0012-7353

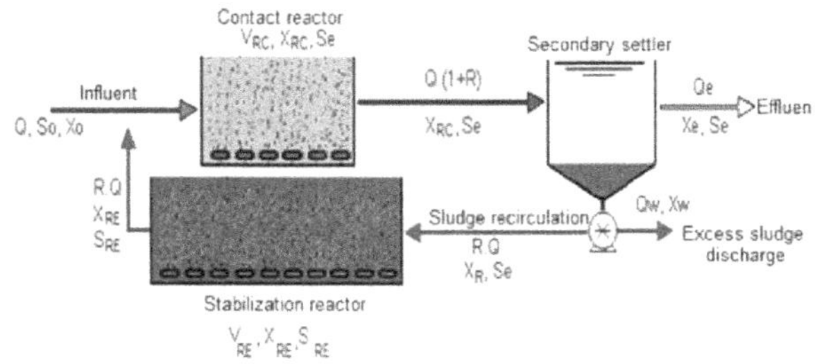

Figure 1. Contact stabilization process

The HRT is one of the main parameters in the activated sludge system as it is implicitly associated with the organic load applied and the reactor volume. The HRT affects the costs of implementation, operation, and maintenance [11]. For CR, the HRT varies between 0.5 and 1.5 h. For SR, the HRT is determined by the recycle rate (flow recirculation) and can vary between 2 and 6 h [9].

The VOL and F/M represent the organic load applied to the system in terms of the reactor volume and the active biomass, respectively [9, 10]. The F/M can create conditions that favor the predominance of filamentous organisms that affect the settling properties of the sludge, causing brown foam in the aeration tank and deterioration in effluent quality [12, 13].

The SRT, used for the design and operation of the system, is the most important parameter in maintaining the MLVSS concentration, as it influences the evolution of the biochemical transformation processes and is related to the rate of growth of microorganisms, because only the microorganisms capable of breeding in this time can survive and enrich the system [6, 14]. According to Liao et al. [15, 16], Liss et al. [17], and Wilen et al. [18], the SRT can affect the floc structure and the settling properties of the sludge. The MLVSS represents the amount of biomass in the system. The typical values for activated sludge systems vary between 500 and 5000 mg.L^{-1} and the recommended concentrations for the contact stabilization process are between 1000 and 3000 mg.L^{-1} for CR, and between 4000 and 10000 mg.L^{-1} for SR [6, 9, 19].

The fraction of biomass in CR is called the sludge distribution fraction (a factor), and values recommended are between 0.1 and 0.3 [4, 8]. The SVI indicates the separation efficiency of the biomass of the mixed liquor. According to Withey et al. [20], the settling properties of sludge formed during the activated sludge process are essential for the clarification of the effluent. Jenkins et al. [12] indicated that high values of SVI are associated with sludge bulking and foam problems that affect the effluent quality. The solids concentration in the secondary settler affect the solids concentration in the recirculation sludge, although if the sludge is concentrated, the recycle rate requirements will be lower in order to guaranteed the MLSSV in the SR, which is also affected by the SRT [6, 9].

The DO concentration is important in the development of processes that occur in the activated sludge systems. The main oxygen requirements are determined by oxidation of organic matter and ammonium through heterotrophic and autotrophic microorganisms respectively [21]. Low levels of DO can affect the sludge settling properties and the metabolic activity of microorganisms, generating an incomplete removal of substrate, which is reflected in the poor effluent quality [22].

Since 1972, studies have been conducted on this process. Between 1975 and 1980 Gujer and Jenkins [5, 7] and Alexander et al. [8] proposed the kinetic models for carbon removal and nitrification, and found that this system can be highly efficient in organic matter removal with efficiencies around 80%. They established that the most relevant operating parameters for this process are the HRT, SRT, *r*, and factor. Sarioğlu et al. [4] applied the contact stabilization process to the domestic wastewater of Istanbul.

The experimental results showed that the treatment efficiency increases when the particulate COD fraction of an influent domestic wastewater increases. Efficiencies observed during the study were around 82%.

Table 1 presents experimental data of WWTP with contact stabilization processes.

Facility	Flow rate (m^3.d)	SRT (d)	F/M (kgBOD. (kgMLSS.d)$^{-1}$)
Akron, USA [23]	454	14	0.072
ECSD, USA [23]	1249	21	0.053
EUVSD, USA [23]	3444	34	0.030
UCSD, USA [13]	57.000	3.5	0.200

More recently, laboratory -scale biological systems studies were carried out by *Universidad del Valle* and Water Company of Cali (EMCALI EICE ESP) [24, 25], to treat domestic wastewater from Cali. Four activated sludge processes were evaluated: extended aeration, adsorption/bioxidation (A/B System), conventional, and contact stabilization. The results showed that the application of the activated sludge systems studied is feasible, since COD, BOD_5, and TSS removal efficiencies were above 80%. In addition, the contact stabilization process allows the highest organic load removal with the smallest area requirements.

This paper presents an analysis of the influence of the HRT and *r* in the performance of a contact stabilization process to remove organic matter and solids (COD, BOD_5, and TSS, respectively) from the domestic wastewater of Cali.

METHODOLOGY

The study was carried out at a pilot-scale wastewater treatment plant in the C-WWTP facilities of Cali.

Pilot Plant

The pilot plant consisted of four sequentially disposed units for preliminary treatment (fine screen and two grit chambers), primary clarification, aeration (for CR and SR), and secondary sedimentation. The wastewater was collected with a submersible pump (Barnes®) located at the entrance of the grit chamber of the C-WWTP. The oxygen was supplied through membrane diffusers and a fine bubble blower system. The recirculation sludge sample was collected by means of a centrifugal pump (Barnes®). Figure 2 shows the pilot plant flowsheet, and Table 2 shows the description of units.

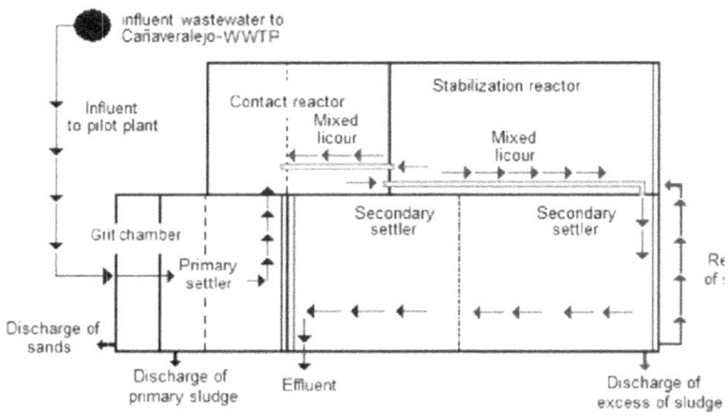

Figure 2. Pilot plant flowsheet

Table 2. Description of units

Unit	Volume (m^3)
Contact reactor (CR)	1.99
Stabilization reactor (SR)	3.91
Secondary settler	5.45

Wastewater

The pilot plant was fed with the raw wastewater which arrived to the C-WWTP that corresponds to Cali domestic wastewater. Table 3 summarizes the plant wastewater influent and primary effluent average quality parameters.

Table 3. Treatment plant wastewater influent and primary effluent average quality parameters

Parameter	Influent (raw wastewater)	Primary effluent
pH (units)	6.1 - 7.7	5.7 - 7.1
T (°C)	23 - 29	23 - 29
Alcalinity (mg.L^{-1})	181 ± 37	166 ± 35
COD (mg.L^{-1})	411 ± 124	331 ± 78
BOD_5 (mg.L^{-1})	275 ± 33	231 ± 67
BOD_5/COD	0.60 ± 0.11	0.64 ± 0.15
TSS (mg.L^{-1})	162 ± 90	83 ± 53
VSS (mg.L^{-1})	121 ± 59	67 ± 38

Operational Conditions

The pilot plant was operated for a period of 183 days. The operation was conducted at different HRT and *r* values. Every HRT variation was called a *phase of operation*, for a total of three phases. In each phase of operation, *r* was varied to provide different operational conditions to assess the performance of the system. Table 4 shows the evaluated conditions in each phase.

Table 4. Evaluated conditions in each phase

Phase	Influent flow (L.s^{-1})	HRT (h) CR	*r* (%)	HRT (h) SR
1	0.33	1.66	70	4.65
			85	3.83
			100	3.26
2	0.50	1.11	55	3.95
			70	3.10
			85	2.56
3	0.66	0.84	40	4.11
			55	2.99

Samples were collected three times a week in the influent and effluent of the system and in the CR, SR, recirculation sludge, and discharge of excess of sludge.

The analyzed parameters were: pH, temperature, alkalinity, COD, BOD_5, TSS, and VSS. All analytical determinations were according to the Standard Methods for the Examination of Water and Wastewater [26] and the structure of activated sludge floc was observed with optical microscopy, according to Vazollér et al. [27] and Jiménez et al. [28].

For the selection of the operational conditions that showed the best performance in terms of COD removal efficiency, a statistical model for a completely randomized design was used. The tests for the data validation were Bartlet for homogeneity of variances and Kolmogorov-Smirnov for normality. In order to correct non-normality and stabilize subgroup variance, a Box-Cox transformation was used [24, 29].

RESULTS AND DISCUSSION

Analysis of Operational Conditions

Behavior of Operation Variables

SRT. Taking into account the results obtained by UNIVALLE-EMCALI [25], the initial value of SRT used in this study was around 10 d. With this condition, the effluent COD concentration was less than 60 mg.L^{-1}; nevertheless, the system showed a high incidence of sludge flotation in the settler, probably caused by the accumulation of the sludge in the secondary settler, whose decomposition processes produced gases that dragged the bottom sludge.

Van Haandel and Marais [1] indicated that for tropical climate conditions, SRT values between 3 and 5 d are enough to ensure the development of predators of bacteria and an effluent with low concentration of organic matter. Considering these aspects, the operation of the system (Phase 2 and 3) was carried out with values of SRT around 6 d. The mean SRT varied between 9 and 12 d, 5 and 8 d, and 5 and 6 d for all three phases, respectively.

The change of SRT to 6 d resulted in a significant reduction of the sludge flotation in the settler unit, without affecting the effluent quality. In all phases the effluent remained with a low COD concentration range (between 20 and 60 mg.L^{-1}), reflecting that the SRT values maintained during the operation of the system allowed for the proper development of biological processes for the removal of organic matter.

Solids in the System. The concentration of solids obtained in the different units showed an effect of increasing *r*, which was attributed to the increased speed of the settled sludge suction and which generated a less concentrated sludge recirculation. Another variable that could affect the stability of the solids concentration was the secondary settler performance, which did not allow a compaction and distribution of sludge proper, generating accumulation and decomposition of this material in this unit. Figure 3 shows the concentration of solids in the system

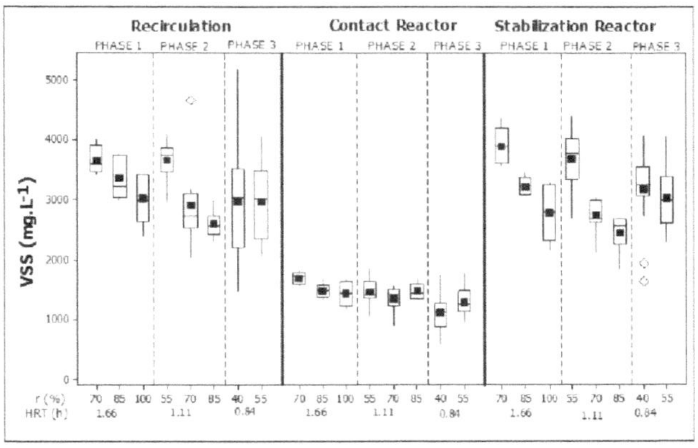

Figure 3. Concentration of solids in the system

An important aspect was that in all operating conditions, the solids concentrations were maintained at values inferior than those typically recommended in the literature, which vary between 1000 and 3000 mg.L^{-1} for CR, and between 4000 and 10000 mg. L^{-1} for SR [9]; however, the system performance in terms of removal of COD and TSS showed that the system can be operated with concentrations of VSS around 1500 mg.L^{-1} for CR, and 3500 mg.L^{-1} for SR.

In activated sludge systems, the biosorption process, the assimilation of organic matter, and a proper operation of aerated reactors depends on an adequate concentration of biomass in the reactor. This condition is improved in the contact stabilization process; because the sludge is separated into two reactors, where the distribution of the sludge is controlled by factor According to Alexander et al. [8], an factor value of 0.1 can be sufficient for ensuring the biosorption of organic matter; however, Sarioğlu et al. [4] recommended an optimal value of 0.30. During the three phases, factor values were between 0.1 and 0.2, which according to the quality of the effluent of the system, were sufficient to ensure good performance.

Organic Load. Influent BOD concentration constant values around 231 ± 67 $mg.L^{-1}$ allowed similar average values of VOL in each phase. Figure 4 presents the behavior of VOL and F/M.

Decreased HRT caused an increase in the VOL and F/M. As a result, values achieved in Phases 2 and 3 exceeded the recommended maximum for this process between 0.2 and 0.6 $kgDBO_5.(kgSSV.d)^{-1}$. Although authors such us Ramalho et al. [30] argue that high values of F/M are related to filamentous organisms, and that this affect the sludge settling properties and quality of the effluent, our results showed that the contact stabilization process can be operated with high VOL without the occurrence of the growth of filamentous microorganisms.

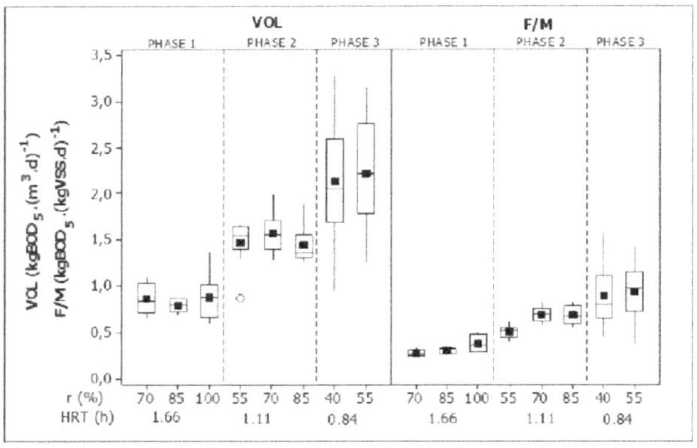

Figure 4. Behavior of the organic load applied

Sludge Quality

Figure 5 shows the behavior of SVI. There is a clear trend of the sludge low settling properties (SVI > 100 $mL.g^{-1}$) associated mainly with the floc structure, which presented a small size, lower floc strength (open structure), and shear sensitivity, due to the low concentration of filaments. The mean floc size varied between 12 to 13 μm for the CR, and 9 to 16 μm for the SR.

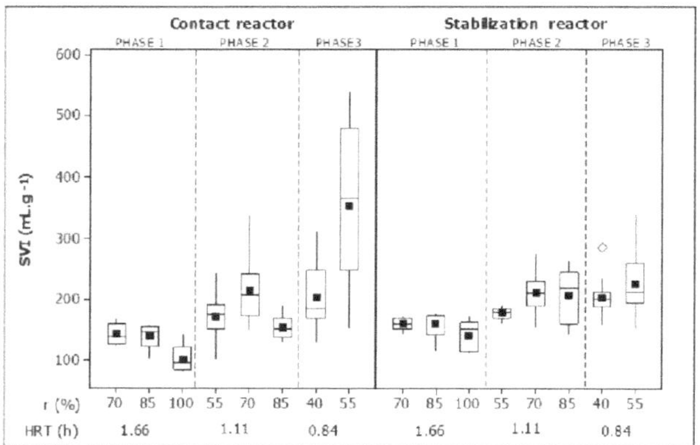

Figure 5. Behavior of sludge volumetric index

According to Li and Ganczarczyk, cited by Wilen et al. [18] this floc size is very small, because typical sizes of flocs usually vary between 20 and 200 μm. Jenkins et al. [12] affirm that the presence of surfactants and toxic substances in wastewater can affect the formation of floc and the bioflocculation process reflected in the predominance of small sizes and dispersed floc.

Another variable that could affect the size of the floc was the turbulence caused by the aeration system. Although DO concentrations were maintained mainly between 1 and 4 $mg.L^{-1}$ during the three phases, is possible that air bubbles could have caused the breakdown of the floc.

The Phase 3 operation was perturbed by electrical supply interruptions every 3 hours during a week. As a result, the system was operated intermittently, affecting the sludge settling properties, which was reflected in the increase in SVI.

System Performance

Effluent Concentration and COD, BOD_5, and TSS Removal

Figure 6 shows effluent quality. Although in all operational conditions the effluent organic matter and solids concentrations were less than 60 $mg.L^{-1}$; the overall behavior of effluent quality showed that the average concentration of COD, BOD_5, and TSS have been affected by changes in *r*, and showed a slight increase with decreasing HRT in the CR.

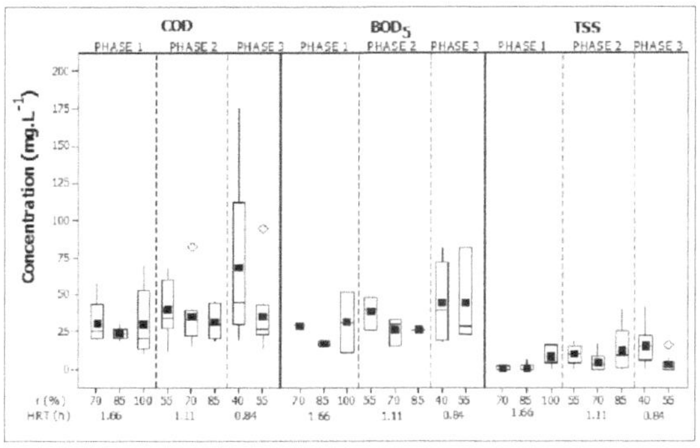

Figure 6. Effluent quality

This condition described was related to the discharge of small flocs in the effluent as a result of the decrease in the sludge settling. However, we considered that this impact was not significant since removal efficiencies during all phases, COD, BOD_5, and TSS, were above 80% (See Figure 7), indicating a good affinity of the biomass with the influent wastewater. For the range of HRT evaluated, the concentrations obtained from MLVSS in CR and SR were appropriate for ensuring an effluent with low COD concentration.

During the evaluation of Phase 3 with 40% r there was an increase in the average influent COD, BOD_5 , and TSS concentration. These increases were attributed to disruptions in the aeration system due to electricity supply interruptions in the C-WWTP, which caused instability in the process and reduced COD, BOD_5, and TSS removal efficiencies up to values around 50%; although most of the data were between 70 and 90%, with average values of 79, 82, and 81% for COD, BOD5 , and TSS, respectively, showing that despite operational drawbacks, the system continued to present good performance in terms of the removal of COD, BOD_5 , and TSS, and that the system can tolerate some external variations without significantly affecting performance.

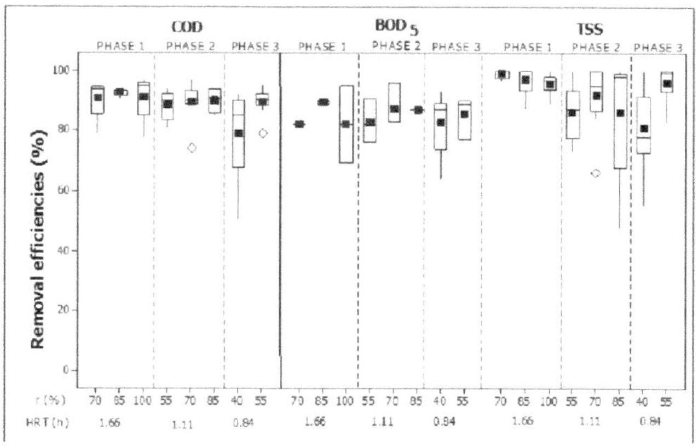

Figure 7. Removal efficiencies of COD, BOD_5, and TSS

Organic Load Removal

Reduction of the HRT and the increase of VOL and F/M caused an increase in organic load removal (see Figure 8). These conditions can be associated with the fact that increasing organic loads in the system allow for better use of the adsorption capacity of the floc, because there is a greater availability of organic matter for the adsorption processes in the CR.

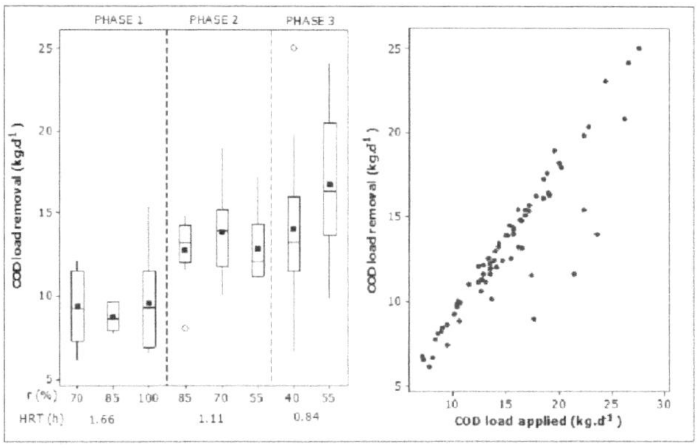

Figure 8. Organic load removal and the relationship between organic load applied and removal

HRT values assessed for Phases 1, 2, and 3, corresponding to 100, 67, and 50 min, respectively (1.66 h, 1.11 h, and 0.84 h) were sufficient to allow for the biosorption and assimilation processes. These results are consistent with studies carried out by Eliosov and Argaman [31] and Guellil et al. [32] who showed that particulate matter is quickly adsorbed by the floc (between 20 and 40 min), although the degradation and absorption is relatively slow.

During the operation of the system with different values of r, HRT in SR varied between 2.56 and 4.65 h, which ensured sludge with a high biosorption capacity reflected in an effluent with low COD, BOD_5, and TSS concentration, and a high organic load removal. This aspect indicated that HRT ranges provided in SR with the different sludge recycle rate applied, were sufficient for allowing for the assimilation of organic matter and renewal of the active surface of the floc, since the hydrolysis, degradation, and assimilation of particulate organic material mainly occur in the SR, which completes the assimilation of soluble material that was not assimilated in the CR.

Selection of the Optimal Operation Conditions

Sludge Recycle Rate. The overall results showed that in the three evaluated phases it is possible to obtain removal efficiencies of COD, BOD_5 , and TSS above 80%. In order to choose a specific operational condition in the system, the average values of COD load removal were compared using analysis of variance (ANOVA) and orthogonal contrast. Table 5 shows the results obtained from the analysis of COD load removed in each r evaluated.

Table 5. Average values of COD load removal

Phase	HRT (h)		*r* (%)	COD load removal ($kg.d^{-1}$)	ANOVA (p-Value)
	CR	SR			
1	1.66	4.65	70	9.36	0.925
		3.83	85	8.74	
		3.26	100	9.58	
2	1.11	3.95	55	12.77	0.625
		3.10	70	13.82	
		2.56	85	12.82	
3	0.84	4.11	40	14.05	0.088
		2.99	55	16.71	

The ANOVA did not show any significant differences between the evaluated r values (p-value > 0.05), indicating that in terms of COD load removal for each phase, it is possible to use any of the r values and obtain similar results for COD load removal. These findings suggest that the variation in the sludge recycle rate do not affect the performance of the system in terms of the COD removal; however these results must be interpreted with caution because the effluent COD and the solids concentration in the system may vary.

For this reason it is important to consider that the sludge recycle rate affects the solids concentration in the SR and the solids are very important for the biosorption processes in the CR [29]; accordingly, the lowest values of r were selected, because for each phase they guaranteed the higher solids concentration in the SR.

Table 6 presents a summary of the best operational condition found for each phase, and average values that must be guaranteed for the main operational variables.

Table 6. Summary of the operational condition found for each phase

Variable	Phase1	Phase2	Phase3
HRT_{CR} (h)	1.66	1.11	0.84
HRT_{SR} (h)	4.65	3.95	4.11
r (%)	70	55	40
SRT (d)	6.0	6.0	6.0
$MLVSS_{CR}$ ($mg.L^{-1}$)	1696	1473	1125
$MLVSS_{SR}$ ($mg.L^{-1}$)	3896	3683	3173
α factor	0.18	0.16	0.14
VOL ($kgBOD_5(m^3.d)^{-1}$)	0.87	1.47	2.13
F/M ($kgBOD_5(kgVSS.d)^{-1}$)	0.27	0.50	0.89

Hydraulic Retention Time. The overall analysis of system performance showed that decreasing the HRT in CR and increasing the organic load applied to the system improves the biosorption process, due to the increased amount of organic matter per unit volume. Table 7 shows the COD load removed in each phase and the results obtained for ANOVA and comparisons to select the HRT that presents de highest removal of COD load.

The ANOVA (p-value < 0.05) showed that there was a significant difference in the removal of COD load in the phases; therefore, partial comparisons were made between phases. The level of p-value for the comparison of the COD load removed between phases 1 and 2 was < 0.001 which shows there was a significant difference in Phases 1 and 2, and Phase 2 presented the major COD load removal.

Table 7. Statistical analysis for the selection of the phase

Pruebas	Value
COD load removal Phase 1 ($kg.d^{-1}$)	9.28 ± 2.30
COD load removal Phase 2 ($kg.d^{-1}$)	13.18 ± 2.31
COD load removal Phase 3 ($kg.d^{-1}$)	15.25 ± 4.44

Transformation	Ln (Y)
ANOVA (p-value)	7.861×10^{-7}
R^2	0.477
Comparison*	Phase1 y Phase2
H_0:	$\sum_{i=1}^{3}\frac{\tau_i}{3} = \sum_{i=4}^{6}\frac{\tau_i}{3}$
H_a:	$\sum_{i=1}^{3}\frac{\tau_i}{3} \neq \sum_{i=4}^{6}\frac{\tau_i}{3}$
Pvalue	<0.001
Comparison*	Phase2 y Phase3
Null hypothesis -H_0:	$\sum_{i=4}^{6}\frac{\tau_i}{3} = \sum_{i=7}^{8}\frac{\tau_i}{2}$
Alternate hypothesis -H_a:	$\sum_{i=4}^{6}\frac{\tau_i}{3} < \sum_{i=7}^{8}\frac{\tau_i}{2}$
Pvalue	0.042
* Compared by orthogonal contrasts	

Likewise, the comparison between Phases 2 and 3 (p-value < 0.05) showed that Phase 3 presented the higher COD load removal, and this confirmed the previous analysis in which we mentioned that an increase in VOL applied to the system caused by the reduction of HRT in CR (increased inflow) favored the removal of the COD load, due to better utilization of the biosorption capacity of the floc.

From these results, the evaluated conditions in phase 3 can be recommended as an operational condition for the contact stabilization process in the secondary treatment of Cali domestic wastewater, for a r of 40%. Table 7 summarizes the operational conditions that must be maintained for such performance.

Table 8. Selected operational conditions for the process

Variable	Value
HRT_{CR} (h)	0.84
HRT_{SR} (h)	4.11
r (%)	40
SRT (d)	6.0
$MLVSS_{CR}$ ($mg.L^{-1}$)	1125
$MLVSS_{SR}$ ($mg.L^{-1}$)	3173
α Factor	0.14
VOL ($kgBOD_5.(m^3.d)^{-1}$)	2.13
F/M ($kgBOD_5.(kgVSS.d)^{-1}$)	0.89

CONCLUSION

Organic load removal by contact stabilization process has a potential application for the secondary treatment of the domestic wastewater of Cali.

For the domestic wastewater of Cali, the operational conditions in the contact stabilization process that allowed us to obtain the best performance in terms of removal efficiencies of COD, BOD_5 , and TSS were an HRT of 0.84 h in CR, and 4.11 h in SR, with an r of 40%. In order to guarantee this performance, average values of SRT of 6 d, VOL of 2.13 $kgBOD_5 .(m^3 .d)^{-1}$, F/M of 0.89 $kgBOD_5 .(kgVSS.d)^{-1}$, and an MLVSS of 1125 $mg.L^{-1}$ in CR, and 3173 $mg.L^{-1}$ in SR must be maintained.

The increase of organic load in the system allows for better use of the biosorption capacity of the floc because there is a greater availability of organic matter for the development of microorganisms in the system.

The hydraulic retention time and the sludge recycle rate have a significant influence on the concentration of VSS in the system, since the increase of the inflow and the sludge recycle rate generate higher suction of sludge, affecting compression, thickening and sludge recirculation concentration, which promotes the recirculation of a diluted sludge and less compacted solids concentration.

ACKNOWLEDGMENTS

The authors wish to acknowledge the institutions that supported this work: EMCALI EICE ESP (Water Company of Cali) and Universidad del Valle.

REFERENCES

1. VAN HAANDEL, A. AND MARAIS, G. O comportamento do sistema de lodo ativado: teoria e aplicações para projetos e operação. Campina Grande, Epgraf, Brasil, 1999.
2. GERNAEY, K. V.; VAN LOOSDRECHT, M. C. M.; HENZE, M.; LIND, M. AND JØRGENSEN, S. B. Activated sludge wastewater treatment plant modelling and simulation: state of the art. Environmental Modelling & Software, 19(9), 763-783, 2004.

3. AL-MUTAIRI, N. Z.; HAMODA, M. F. AND ALGHUSAIN, I. A. Performance-based characterization of a contact stabilization process for slaughterhouse wastewater. Journal of environmental science and health. Part A - Toxic/ Hazardous substances & Environmental engineering, A38(10), 2287-2300, 2003.
4. SARIOĞLU, M.; ORHON, D.; GÖRGÜN, E. AND ARTAN, N. Design procedure for carbon removal in contact stabilization activated sludge process. Water Science and Technology, 48(11 - 12), 285 - 292 2003.
5. GUJER, W. AND JENKINS, D. The contact stabilization activated sludge process--oxygen utilization, sludge production and efficiency. Water Research, 9(5-6), 553-560, 1975.
6. GRADY, C. P. L.; DAIGGER, G. T. AND LIM, H. C. Biological wastewater treatment, Marcel Dekker, 2nd. Edition. USA, 1999.
7. GUJER, W. AND JENKINS, D. A nitrification model for the contact stabilization activated sludge process. Water Research, 9(5-6), 561-566, 1975.
8. ALEXANDER, W. V.; EKAMA, G. A. AND MARAIS, G. R. The activated sludge process part 2. Application of the general kinetic model to the contact stabilization process. Water Research, 14(12), 1737-1747, 1980.
9. OROZCO, A. Bioingeniería de aguas residuales. Teoría y diseño. Asociación Colombiana de Ingeniería-ACODAL, 2005.
10. VON SPERLING, M. Lodos ativados. Departamento de engenharia sanitária e ambiental da Universidade Federal de Minas Gerais, Belo Horizonte - Brasil, 1997.
11. VIERO, A. F. AND SANT'ANNA, G. L. Is hydraulic retention time an essential parameter for MBR performance? Journal of Hazardous Materials, 150(1), 185-186, 2008.
12. JENKINS, D.; RICHARD, M. G. AND DAIGGER, G. T. Manual on the causes and control of activated sludge bulking, foaming, and other solids separation problems. Lewis Publishers, Michigan, 2003.
13. OERTHER, D.; DE LOS REYES, F.; DE LOS REYES, M. AND RASKIN, L. Quantifying filamentous microorganisms in activated sludge before, during, and after an incident of foaming by oligonucleotide probe hybridizations and antibody staining. Water Research, 35(14), 3325-3336, 2001.
14. CLARA, M.; KREUZINGER, N.; STRENN, B.; GANS, O. AND KROISS, H. The solids retention time—a suitable design parameter to evaluate the capacity of wastewater treatment plants to remove micropollutants. Water Research, 39(1), 97-106, 2005.
15. LIAO, B. Q.; ALLEN, D. G.; DROPPO, I. G.; LEPPARD, G. G. AND LISS, S. N. Surface properties of sludge and their role in bioflocculation and settleability. Water Research, 35(2), 339-350, 2001.
16. LIAO, B. Q.; DROPPO, I. G.; LEPPARD, G. G. AND LISS, S. N. Effect of solids retention time on structure and characteristics of sludge flocs in sequencing batch reactors. Water Research, 40(13), 2583-2591, 2006.
17. LISS, S.; LIAO, B.; DROPPO, I.; ALLEN, D. AND LEPPARD, G. Effect of solids retention time on floc structure. Water Science & Technology, 46(1), 431-438, 2002.
18. WILÉN, B. M.; JIN, B. AND LANT, P. Impacts of structural characteristics on activated sludge floc stability. Water Research, 37(15), 3632-3645, 2003.
19. METCALF AND EDDY. Wastewater engineering: treatment and reuse. Metcalf & Eddy. Inc.-McGraw-Hill, New York, 2003.
20. WITHEY, S.; CARTMELL, E.; AVERY, L. M. AND STEPHENSON, T.Bacteriophages - potential for application in wastewater treatment processes. Science of the Total Environment, 339(1-3), 1-18, 2005.
21. ROSSO, D. AND STENSTROM, M. Comparative economic analysis of the impacts of mean cell retention time and denitrification on aeration systems. Water Research, 39(16), 3773-3780, 2005.
22. MADEIROS, U.; CAVALCANTI, P. AND VAN HAANDEL, A. Influência da concentração de oxigênio dissolvido sobre o desempenho de sistemas de lodo ativado. Brasil, 2005.
23. SCHULTZ, J. R.; HEGG, B. A. AND RAKNESS, K. L. Realistic sludge production for activated sludge plants without primary clarifiers. Journal Water Pollution Control Federation, 54(10), 1355-1360, 1982.
24. UNIVALLE-EMCALI. Estudio de evaluación para el tratamiento del efluente de la PTAR Cañaveralejo a través de dos opciones secundarias convencionales. Informe final Fase II. . Universidad del Valle; EMCALI EICE ESP, Santiago de Cali, Colombia, 2007.
25. UNIVALLE-EMCALI. Estudio de evaluación para el tratamiento del efluente de la PTAR Cañaveralejo a través de dos opciones secundarias convencionales. Informe final Fase I. Universidad del Valle; EMCALI EICE ESP, Santiago de Cali, Colombia, 2006.
26. AMERICAN PUBLIC HEALTH ASSOCIATION, AMERICAN WATER WORKS ASSOCIATION, AND WATER ENVIRONMENT FEDERATION. Standard methods for the examination of water and wastewater. Washington DC, USA., 2005.
27. VZAZOLLÉR, R.; GARCÍA, M.; GARCIA, A. AND CONCIÇAO, J. Microbiologia de lodos ativados. CETESB, Sao Paulo, Brasil, 1989.
28. JIMÉNEZ, C.; FERNÁNDEZ, N.; DE LA HORRA, J.; RODRÍGUEZ, E.; ISAC, L.; SALAS, D. AND GÓMEZ, E. Sistema rápido de estimación de los rendimientos de depuración de una EDAR en función de las características macroscópicas del fango activado. Tecnología del Agua, Año 21(216), 40-44, 2001.
29. VÁSQUEZ, S. N. Evaluación a escala piloto del sistema de lodos activados en las modalidades de estabilización por contacto y convencional como alternativa de tratamiento secundario para el agua residual doméstica de la ciudad de Cali. M.Eng. Thesis.. Santiago de Cali, Colombia: Universidad del Valle, 2009.
30. RAMALHO, R. S.; BELTRAN, D. J. AND DE LORA, F. Tratamiento de Aguas residuales. Reverté, 2da. Edición, Barcelona, 1991.

31. ELIOSOV, B. AND ARGAMAN, Y. Hydrolysis of particulate organics in activated sludge systems. Water Research, 29(1), 155-163, 1995.

32. GUELLIL, A.; THOMAS, F.; BLOCK, J. C.; BERSILLON, J. L. AND GINESTET, P. Transfer of organic matter between wastewater and activated sludge flocs. Water Research, 35(1), 143-150, 2001.

CHAPTER 9 SUSTAINABLE WASTEWATER TREATMENT AND RECYCLING IN MEMBRANE MANUFACTURING

Mayamin Razali, Jeong F. Kim, Martin Attfield, Peter M. Budd, Enrico Drioli, Young Moo Lee and Gyorgy Szekely

Keywords: Activated sludge, contact stabilization, organic matter removal, secondary wastewater treatment

INTRODUCTION

Membrane technology has been recognized as the key technology to tackle the global water shortage challenge. Membrane-based technology now accounts for more than 70% of the world desalination capacity[1] and membrane bioreactor technology is now becoming a dominant player in the field of wastewater treatment.[2] In addition, microfiltration and ultrafiltration technologies are now commonly found in a wide range of applications, such as pretreatment of saline water, and in food, textile, and pharmaceutical industries.[3]

Membrane processes are generally considered a green technology, with their unique characteristics such as low energy consumption and convenient compact operation.[4,5] Although membranes certainly have improved the sustainability of chemical processes,[6] it is not widely known that membrane fabrication itself is quite far from being a green process. Membrane fabrication typically employs toxic, dipolar aprotic solvents such as *N*-methyl-2-pyrrolidone (NMP) and *N*,*N*-dimethylformamide (DMF) for phase inversion.[7] These solvents mix with water during polymeric membrane formation and consequently significant quantities of solvent-containing wastewater are generated. Due to the waste-intensive nature of membrane fabrication, it has been shown that downstream processing *via* membrane technology only becomes greener than distillation when more than 100 L m^{-2} of solvent has been processed.[5]

The membrane application market is currently growing at a fast rate of 9% and the overall market size is expected to reach *ca.* $26 billion by 2017.[8] Consequently, the membrane manufacturing industry is expanding rapidly to meet the growing global market demand. There are currently about 4–5 million units of spiral-wound modules (40 m^2 per module) just for the seawater desalination plants, and the number is growing at 10% each year.[9] The amount of wastewater generated to produce 5 million spiral-wound modules can be approximated as 20–100 billion liters, and the generation rate is growing at 10 billion liters per year (see the ESI for calculations). Note that this figure is just for the desalination membrane market (*ca.* $1100 million in 2013)[10] alone. Combined with other fast-growing sectors such as gas separation (*ca.* $150 million in 2002),[11] battery separator (*ca.* $800 million in 2014),[12] pharmaceutical and artificial kidney application (50% of all the polymeric membranes produced),[13] membrane bioreactor, and micro/ultrafiltration market, the figure is likely to be several times higher.

The amount of wastewater generated from membrane manufacturing plants can be estimated to be approximately 100–500 L of wastewater per square meter of membranes produced (see the ESI for calculations). As illustrated in Fig. 1, the solvent contamination in the membrane wastewater always exceeds the minimum allowable level of 100 ppm,[5] requiring appropriate treatment prior to disposal (note the wastewater concentration is a function of the coagulation bath size). The regulations for wastewater threshold levels have been getting tough in the US and particularly in Europe,[14] promoting much interesting research to minimize solvent consumption by employing ionic liquids,[15] supercritical CO_2,[16] and environmentally benign solvents.[17,18] Unfortunately, at the moment, economic feasibility and altered membrane performance are among the factors inhibiting their commercial implementation. On the other hand, direct reuse of the water bath is not possible because the solvent in the coagulation bath affects the membrane morphology during phase inversion. For instance, NMP concentration as low as 3 wt% can decrease

Citation: Mayamin Razali, Jeong F. Kim, Martin Attfield, Peter M. Budd, Enrico Drioli, Young Moo Lee and Gyorgy Szekely, Sustainable wastewater treatment and recycling in membrane manufacturing, DOI: 10.1039/C5GC01937K

the membrane permeability by 25%.[19] Hence, it is vital to maintain solvent concentration in the coagulation bath constant or as low as possible to obtain uniform membranes. This is particularly important for a continuous membrane fabrication process, where the coagulation bath concentration needs to be adequately controlled to produce uniform membranes.

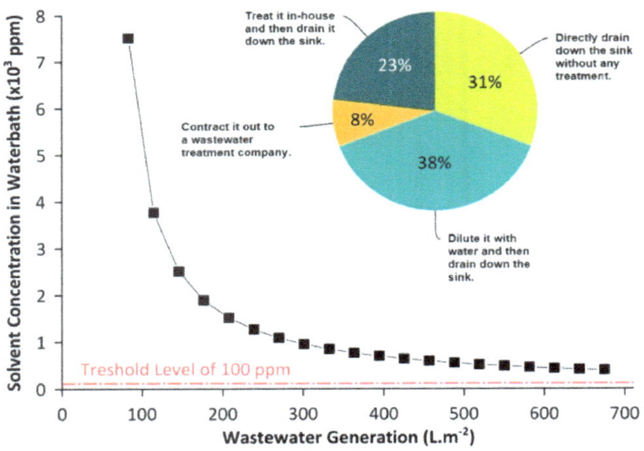

Figure. 1 Calculated solvent concentration in the wastewater during a typical membrane fabrication process. The solvent concentration is above the allowable 100 ppm organic impurity range in wastewater, requiring an appropriate treatment prior to disposal or reuse. The pie chart shows the outcome of our survey how membrane manufacturing companies dispose off the coagulation bath wastewater. Please refer to the ESI for all survey questions and answers.

Our recent survey analysis based on the responses of 13 membrane manufacturing companies has confirmed that membrane wastewater is a serious and urgent challenge to be resolved (see the ESI for full survey details). The survey revealed that the wastewater generated during membrane fabrication (i) contributes to more than 95% of the total waste; (ii) on a scale from 1–10, the average difficulty level of wastewater treatment scored 8.6; and (iii) more than 69% of the companies drained the wastewater without any treatment and only 31% treated the wastewater prior to disposal. Since the wastewater needs special treatment prior to disposal, industries typically dilute the solution using excess water or subcontract it out offsite as shown in Fig. 1. However, it should be emphasized that the wastewater generated from membrane production is relatively clean compared to the wastewater generated from textile or pharmaceutical industries. The current trend in wastewater treatment technology, apart from abating contamination, is to design a sustainable process that does not require high energy consumption and operating costs.[20–22] Hence, the present article examines the practicality and viability of purifying the membrane wastewater *in situ* and recycling it continuously during membrane production using adsorption technology.

Various high performance adsorbents are now available for water purification ranging from state-of-the-art graphene,[23] polymers with intrinsic microporosity (PIM),[24] molecularly imprinted polymers (MIPs),[25] metal organic frameworks (MOFs),[26] to commercial resins,[27] zeolites[28] and charcoal (activated carbon).[27] Thirty-five adsorbents from seven different classes of materials were prepared or purchased and screened for the adsorption of the two most common solvent contaminants[13] in membrane industrial wastewater: NMP and DMF. In search for sustainable wastewater treatment and recycling in membrane manufacturing, the adsorbents' sorption and kinetic behaviors, regenerability, and feasibility for continuous treatment of membrane wastewater in terms of energy savings and waste minimization were critically assessed.

The membrane manufacturing industry, leading universities, research funding agencies and governments are all in agreement that water treatment is one of the grand engineering challenges of today. Only when the wastewater challenge is resolved should the membrane technology be considered green and sustainable.

RESULTS AND DISCUSSION

In Search for the Ideal Adsorbent

The three most important criteria for adsorbent selection are high capacity, fast kinetics, and easy regenerability. Among the 35 adsorbents tested from 7 different classes of materials, Fig. 2summarizes the adsorption capacity for the best performing 6 adsorbents. See Fig. S5 in the ESI for the full screening results. The screening data revealed that acidic adsorbents generally exhibited higher binding capacity than neutral or basic adsorbents from the same class, likely due to the basic nature of the DMF and NMP molecules (pK_a = 16 and 24, respectively). Particularly, the imprinted polymers featuring the most acidic functional monomers, namely methacrylic acid (MIP6), 3-vinylphenylboronic acid (MIP7), and itaconic acid (MIP8), showed the best performance within the MIP class. Similarly, the acidic PHO adsorbent from the charcoal class showed superior performance compared to neutral SAC6 adsorbent from the same class.

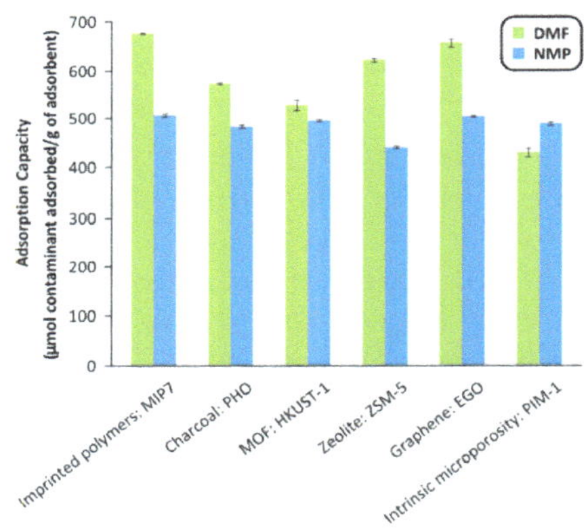

Figure. 2 Adsorption capacity of the best adsorbent from the different classes of materials investigated: imprinted polymers, charcoals, metal–organic frameworks, zeolites, graphene-based materials and polymers of intrinsic microporosity. The adsorbent mass to system volume ratio was fixed at 20 g L^{-1}. See Fig. S5 in the ESI for the full screening results. MIP, HKUST-1, ZSM-5, EGO and PIM stand for molecularly imprinted polymer, Hong Kong University of Science, Zeolite Socony Mobil, exfoliated Graphene Oxide and Polymers of intrinsic microporosity. PHO is part of the internal nomenclature of Eurocarb®.

Although both DMF and NMP are dipolar aprotic solvents with similar polarity indexes (6.4 and 6.7, respectively), the majority of adsorbents showed higher adsorption capacity towards DMF than NMP, likely due to the smaller size and flexible conformation of DMF. The resin-class adsorbents showed relatively poor performance hence were not investigated further.

The best adsorbent from each class was further characterized for its full adsorption isotherm and kinetic behaviors. Each material shows different adsorption behavior as a function of the mass–volume ratio as depicted in Fig. 3(a and b). Analyzing such adsorption behavior can elucidate its adsorption mechanisms by fitting to different isotherm models.[29] See Table S1 in the ESI for full isotherm data. Several trends can be deduced from the data. PHO (charcoal class) adsorption data are the best described by the two-parameter Freundlich model, indicating the surface of charcoal is highly heterogeneous and stronger binding sites are occupied first followed by weaker binding sites.[30] On the other hand, solute-specific adsorbents such as MIP7 (MIP class) are better characterized by the three-parameter models such as Redlich–Peterson isotherm, indicating more homogeneous and specific binding sites as expected.[31] The imprinted polymers were prepared in the presence of DMF and NMP as a template which formed target specific binding sites in the polymer framework. It is notoriously difficult to imprint small and low functionality templates and examples are scarce.[32,33] Graphene is a fascinating new member of carbon materials with honeycomb and one-atom-thick structures which has excited interest in the field adsorption due to its exceptional properties. EGO (graphene-based class) adsorption capacities were among the highest for NMP (503 μmol g^{-1}) and DMF (657 μmol g^{-1}), and are the best fitted by the Langmuir model, indicating the formation of a single adsorption layer on identical and equivalent binding sites.

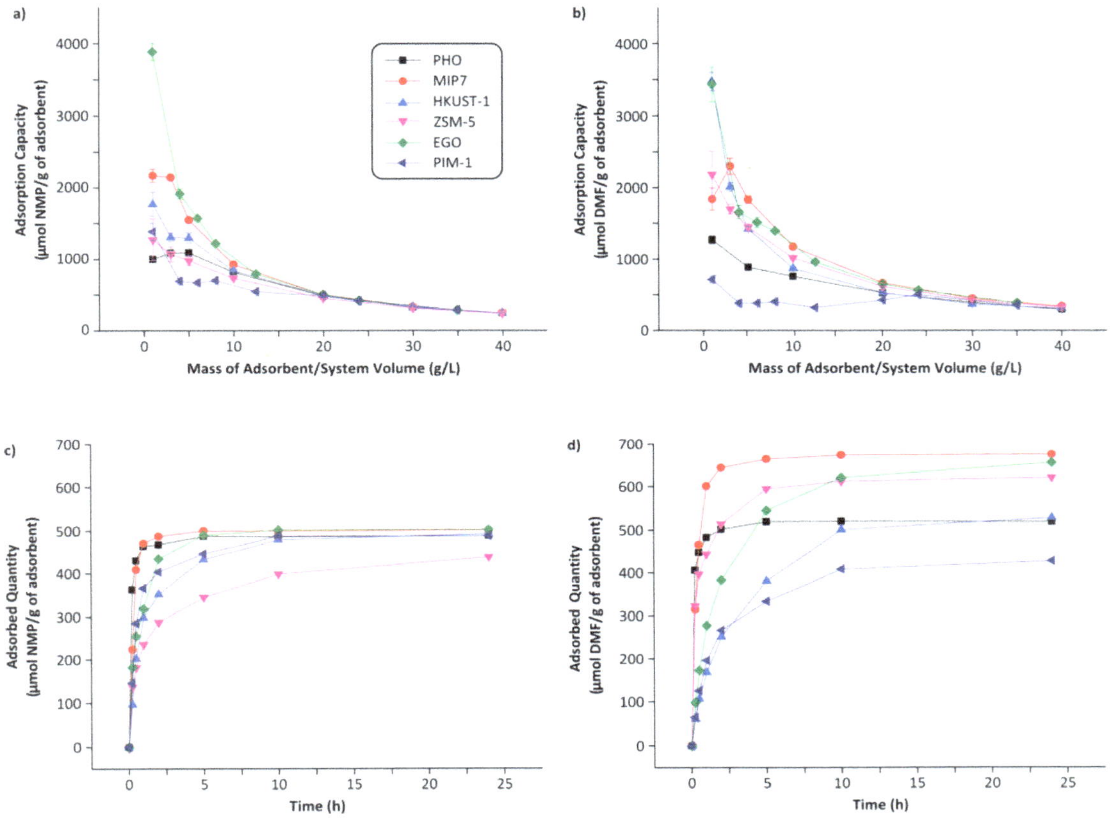

Figure. 3 Comparison of adsorbent isotherm (a, b) and kinetic behavior (c, d), for the best adsorbent from each class, on NMP (a, c) and DMF (b, d), respectively. The shapes of the curves can elucidate the isotherm and kinetic mechanisms. The kinetic data were obtained at a mass–volume ratio of 20 g L^{-1}.

Fig. 4 shows that the adsorption capacities for DMF are generally higher than those of NMP. On the other hand, the adsorption rates are generally higher for NMP than those of DMF. The heterogeneous adsorption behavior (Freundlich isotherm) of PHO (charcoal class) is clearly visible as the sorption profiles for DMF and NMP are quite similar – nonselective adsorption occupying the highest binding sites first. Although PHO showed excellent kinetics for both adsorbents, DMF adsorption capacity was one of the lowest among the tested adsorbents. Fig. 4 summarizes that MIP7, EGO, and ZSM-5 showed the highest adsorption capacities for DMF, whereas the NMP adsorption capacities were relatively similar among the tested adsorbents.

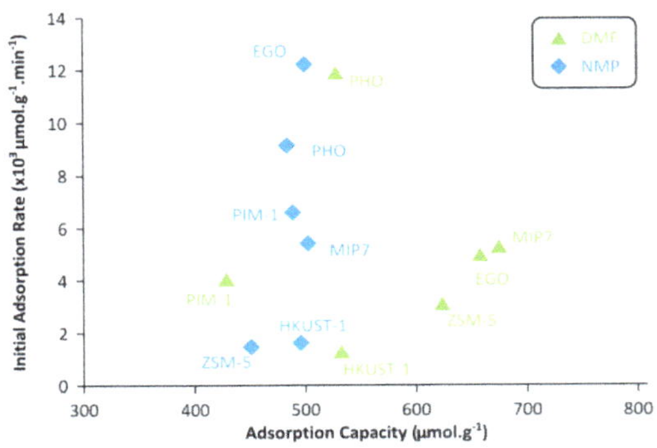

Figure. 4 Adsorption capacity *versus* initial adsorption rate.

Apart from the adsorption capacity and rate it is essential to evaluate the regenerability of adsorbents. Used adsorbents are typically discarded as solid waste posing an environmental issue, and the current trend is to maximize the adsorbent regenerability to improve the process sustainability.[34] Hence, the regeneration efficiency as well as operation lifetime are now two key factors in designing an effective adsorption process. Adsorbents can be regenerated using several different ways such as thermal swing, vacuum swing, pressure swing, steam treatment, and/or chemical treatment methods. The economics of the regeneration method could be a deciding factor when choosing the right adsorbent. For instance, although charcoal is an excellent adsorbent material with low cost, high surface area, and high binding capacity, charcoal can only be regenerated with significant energy input, and typically results in large volumes of solid waste contaminated with toxic pollutants.[14]

To characterize the regenerability of the adsorbents in this work, three different regeneration methods (solvent washing, vacuum swing, and thermal treatment) were evaluated for each adsorbent, summarized in Table 1. Please note thermal regeneration experiments were not performed in the lab and the data were deduced from the literature. It was reported that thermal regeneration of charcoal typically results in 5–15% reduction of the adsorption capacity due to attrition.[34] In addition, on-site thermal regeneration is not cost-effective for adsorbent usage less than 910 kg per day, rendering the thermal regeneration option unsuitable for the membrane fabrication industry. Each adsorbent requires different treatments for regeneration, as summarized in Table 1. For instance, MIP7 can be fully regenerated using 4 ml of MeOH : H_2O (10 : 90) mixture per gram of adsorbent, whereas PIM-1 requires more than 10 ml of pure MeOH per gram of adsorbent (see Fig. S6 in the ESI for details). The HKUST-1 (MOF class) adsorbent was also best regenerated using solvent wash with even lower efficiency. On the other hand, EGO (graphene class), ZSM-5 (zeolite class), and PIM-1 adsorbents could not be regenerated using solvent wash but were effectively regenerated under vacuum.

Table 1 Compatibility of adsorbents with different regeneration methods

	Regeneration method efficiency		
Adsorbent	Washing	Vacuum	Thermal
MIP7	High	High	Low
PHO	Low	Low	Moderate
EGO	Low	High	Low
PIM-1	Moderate	High	Low
ZSM-5	High	High	High
HKUST-1	Moderate	Low	Low

Fig. 5 summarizes the regenerability of each adsorbent up to 10 cycles. It can be seen that MIP7, EGO, PIM-1, and ZSM-5 adsorbents were fully regenerated 10 times without any performance degradation. On the other hand, HKUST-1 adsorption performance slowly decayed with subsequent usage, losing approximately 15% by the 10th cycle. This is likely due to an irreversible decomposition of the HKUST-1 framework due to the partial displacement of organic linkers from the copper centers by either water molecules or the solvents.[35] Nijem *et al.* demonstrated a photo-reduction based modification of the metal center which mitigated such decomposition caused by water.[36] Although it is out of the scope of the present work such approach could increase the potential of HKUST-1 in wastewater treatment.

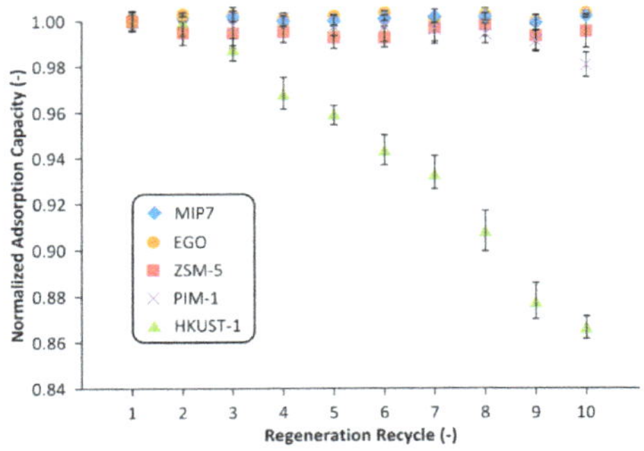

Figure. 5 Regeneration of adsorbents saturated with NMP. MIP7 and PIM-1 were regenerated using solvent washing, ZSM-5, EGO and HKUST-1 were regenerated under vacuum.

Among the tested adsorbents, MIP7 and EGO showed the best performance. The adsorbent regenerability and cost are related, as an expensive adsorbent with high regenerability may be more cost effective than a low-cost adsorbent with low regenerability. To demonstrate the impact of membrane wastewater treatment, MIP7 was tested for a continuous water treatment process. However, as will be shown in the subsequent section, adopting an adsorptive wastewater treatment unit using any of the adsorbents can significantly improve the sustainability of the membrane fabrication process.

Continuous Wastewater Purification and Recycling

In order to demonstrate the feasibility of a continuous adsorptive wastewater treatment process, membrane wastewater was continuously treated using two parallel columns in an alternating sequence as illustrated in Fig. 6(a). Each adsorption column was wet-packed with 50 g of MIP7 adsorbent. Fig. 6(b) shows that a breakthrough of NMP occurred approximately at every 2.5 L wastewater processed, after which the column had to be regenerated (see Fig. S7 in the ESI for the breakthrough curve). Fig. 6(c) illustrates the regeneration profile: 2.4 mL of 1 : 9 v/v methanol–water washing solution per gram of adsorbent was used to fully regenerate the column.

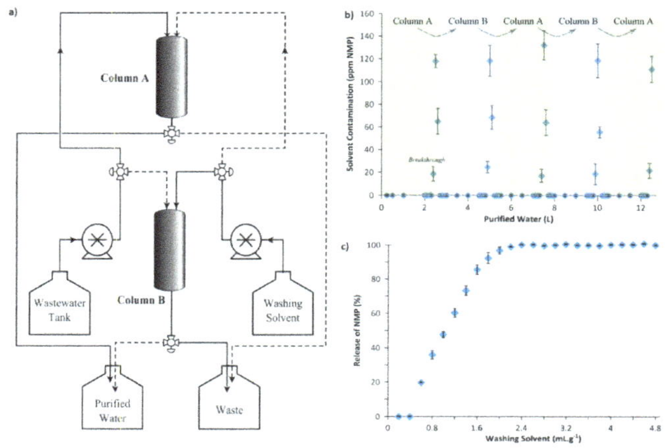

Figure. 6 (a) Continuous adsorption unit process diagram. Two columns were used simultaneously to purify contami-

nated wastewater while regenerating saturated adsorbent. (b) Continuous purification of 12.5 L wastewater from membrane casting. Each column purifies water up to 2.5 L then routed for regeneration (alternate between two columns). (c) Column regeneration profile: 2.4 mL methanol : water (10 : 90) per gram of adsorbent was chosen to fully regenerate the column.

Hence, after processing 2.5 L wastewater using Column A, the feed was routed to Column B while Column A was simultaneously regenerated using the washing solvent. Five cycles of adsorption–regeneration were carried out in a continuous process to purify 12.5 L water to reduce the NMP concentration from 887 ± 90 ppm to only 7.6 ± 0.8 ppm (99.1 ± 0.2% removal). Note that a more stringent breakthrough volume of 2.3 L could achieve 100% NMP removal. Nevertheless, 7.6 ppm NMP contamination is already one magnitude lower than the 100 ppm requirement for the safe disposal of wastewater.

To validate the recyclability of the purified wastewater for membrane fabrication, three different batches of membranes were prepared using three different coagulation baths: (i) fresh water without any NMP contamination, (ii) used wastewater contaminated with 887 ± 90 ppm NMP, and (iii) purified wastewater containing 7.6 ± 0.8 ppm NMP. The performances of the resulting membranes were compared for acetone flux and solute rejection (polystyrene oligomer) as shown in Fig. 7.

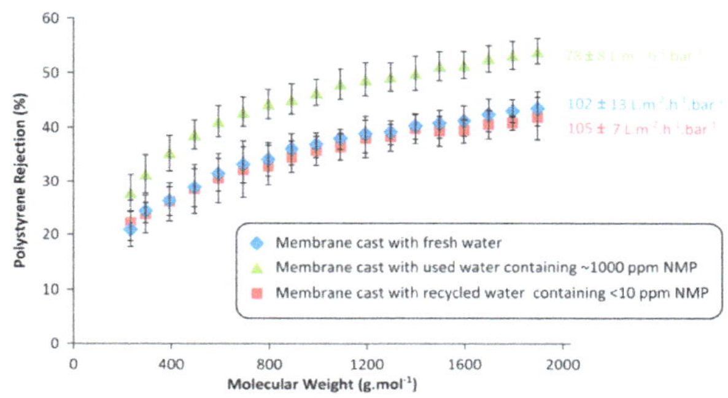

Fig. 7 Flux and rejection data for the three different batches of membranes. Membranes prepared with NMP-containing wastewater exhibit lower flux and tighter rejection profile, whereas membranes prepared with treated and recycled wastewater showed identical performance compared to the one prepared with fresh water, validating the recyclability of the purified water.

In line with the expectation,[19] the membranes fabricated in a NMP-contaminated coagulation bath exhibit 23% lower permeance and 5–10% higher solute rejection compared to the membranes cast in fresh water. This phenomenon is due to the fact that solvent contamination in the coagulation bath affects the membrane formation kinetics by slowing the dope demixing process, resulting in tighter membranes with lower flux. In fact there are two concentration-dependent mechanisms in action when the coagulation bath contains the dope solvent. First, when the solvent concentration in the coagulation bath is high, it dilutes the surface polymer concentration and creates pores.[37,38] On the other hand, lower dope solvent concentration in the coagulation bath slows the diffusion rate of the solvent outflow, ultimately tightening the membranes by delaying the demixing process.[19,39] In this work, 1000 ppm solvent concentration is approximately 0.1 vol% in the bath, which is not high enough to dilute the surface polymer concentration to create pores. Instead, the presence of the solvent in the coagulation bath has rather a stronger effect on the demixing rate as illustrated in Fig. 7.

Furthermore, the membranes prepared using the purified water showed almost identical performance to the membranes prepared using fresh water. This result undoubtedly validates the recyclability of the membrane wastewater, and the feasibility of the proposed continuous adsorptive purification unit. The results obtained reiterate the importance of maintaining a constant coagulation bath composition during a continuous membrane fabrication process in order to obtain uniform membranes. Implementing an adsorption unit operation to recycle membrane wastewater is a simple and effective solution not only to maintain constant membrane performance, but also to significantly improve the sustainability of the membrane fabrication process.

Sustainability Assessment of the Wastewater Treatment Process

The sustainability of the wastewater treatment process was analyzed by employing green metrics in order to assess its realistic impact (see the ESI for detailed calculations). Fig. 8(a) illustrates the required adsorbent amount (MIP7) to reduce the solvent concentration below the threshold value. It can be seen that approximately 1.5 to 1.8 kg of the adsorbent is required per m^2 of membrane fabricated. For this calculation, adsorbent regeneration was not taken into account. If the adsorbents were to be regenerated for 10 cycles, approximately 0.15 to 0.18 kg of the adsorbent would be required per m^2 of membrane fabricated. Note that with more wastewater generated per membrane area (higher L m^{-2}), the longer is the purification time required using the same amount of adsorbent. Hence, to maintain uniform coagulation bath concentration throughout the membrane production process, it is necessary to size the adsorption equipment appropriately considering the membrane production scale and production rate.

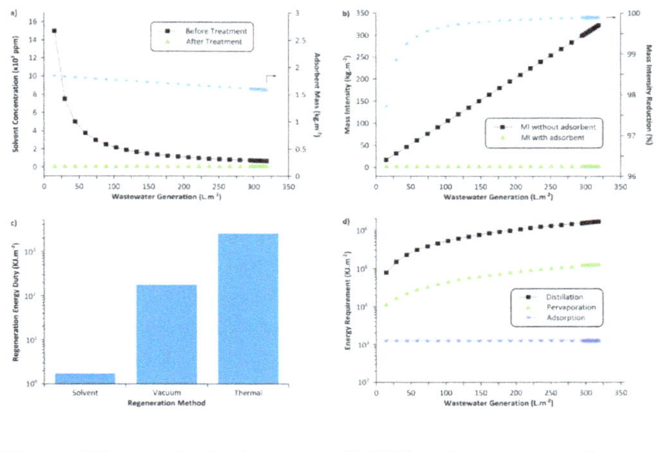

Figure. 8 Required adsorbent mass (MIP7) without regeneration, and solvent concentrations in the wastewater before and after purification (a); mass intensity with and without an adsorption unit: over 99% mass intensity reduction is achieved using adsorption (b); energy duty calculation comparing three different regeneration methods (c); energy duty calculation comparing the three different wastewater treatment methods: the adsorption unit is the simplest method requiring the least amount of energy (d).

Fig. 8(b) shows the process mass intensity of membrane fabrication with and without an adsorption unit. The mass intensity is one of the commonly used metrics to evaluate the sustainability of processes, defined as follows:

$$\text{Mass intensity (MI)} = \frac{\text{Mass of all materials (kg)}}{\text{Mass of product (m}^2\text{ of membranes)}} \quad (1)$$

Since the coagulation bath water is the major contributor to the waste generated during membrane fabrication (see survey question 3 in the ESI), the implementation of the developed adsorptive unit reduces the mass intensity by >99.9% and subsequently transforms the membrane processes to a sustainable technology. On the other hand, approximately 1.67 kg of the MIP7 adsorbent is required to purify 221 kg of wastewater (contaminated with 1000 ppm solvent). Considering the solvent required to fabricate the MIP particles and the washing solvent for adsorbent regeneration, approximately 31 kg of wastewater can be purified per kg of total waste generated (see detailed calculations in the ESI). Given the robust nature of imprinted polymers due to a high crosslinking degree, they can be regenerated about 50 times,[40] improving the ratio up to 54 kg kg^{-1}.

Fig. 8(c) shows the required energy duty for the three different adsorbent regeneration methods discussed in this work: solvent wash, vacuum swing, and thermal treatment. The solvent washing method consumes the least amount of energy and the thermal treatment consumes the most. However, it should be noted that the solvent washing method generates more solvent waste compared to the other two methods. Nevertheless, the solvent volume fraction is negligible relative to the overall process mass intensity and still drastic MI reduction can be achieved. In addition, the solvent washing method is the most convenient and simple method to continuously regenerate the adsorption column, as it simply requires a pump to flow in the washing solvent. Considering the ease of operation and the effectiveness of the method, it is the preferred method to treat membrane wastewater.

Fig. 8(d) compares the required energy duty for the three different wastewater treatment methods: distillation, pervaporation, and adsorption. The most widespread potential competitors of adsorption in water purification (with low organic impurities) are distillation and pervaporation. Distillation consumes the most energy as both DMF and NMP have considerably higher boiling points (T_b = 153 °C and 204 °C, respectively) than water. On the other hand, pervaporation is an emerging technology with a promising potential that can selectively discriminate solvent over water, as commonly used in the bioethanol upgrading plants.[41] In theory, a suitable pervaporation membrane can selectively permeate NMP over water to purify the wastewater. Other membrane technologies like organic solvent nanofiltration can also be employed to recover organic solvents.[42] It can be seen that the energy duty is in between that of adsorption and distillation. The calculation assumed equi-volume permeate concentration. Distillation or pervaporation units would become competitive for a very large membrane production scale, as these technologies can handle large volumes of wastewater effectively.

In summary, considering the ease of operation, feasibility, and energy requirement, it can be concluded that installing an adsorption unit with the solvent-washing regeneration method is the preferred way to recycle wastewater generated by membrane production industries.

CONCLUSIONS

Membrane technology will only become green and sustainable when the wastewater generated during the membrane fabrication process is minimized. In this work, we demonstrated and validated for the first time a practical and effective way to continuously recycle the membrane wastewater, reducing the waste generation by 99%. Seven different classes of adsorbent materials—charcoal, MIPs, zeolites, MOFs, graphene, PIMs, and resins—have been thoroughly screened using membrane fabrication wastewater contaminated with DMF and NMP. The adsorption isotherm and kinetic profile of each adsorbent have been fully elucidated and characterized. Most of the tested adsorbents showed feasible performance to treat the membrane wastewater. Particularly, one of the MIP adsorbents functionalized with phenylboronic acid (MIP7) showed superior performance capable of adsorbing 677 ± 2 and 502 ± 5 mmol DMF and NMP per kg adsorbent, respectively. Regenerability of each adsorbent has been demonstrated up to 10 times without any performance degradation. The proposed continuous adsorptive wastewater treatment is an effective solution to mitigate solvent contamination down to <10 ppm level, allowing safe disposal. In addition, the in-house recyclability of the purified wastewater was validated by crosschecking membrane performances, which revealed the importance of water purity during a continuous membrane fabrication process. The membranes fabricated in a NMP-contaminated coagulation bath exhibited 23% lower permeance and 5–10% higher solute rejection compared to the membranes cast in fresh or recycled water. The process mass intensity of the membrane fabrication process can be reduced by 99.9% with the continuous adsorption unit. Our conservative estimation predicts that the wastewater generation can be reduced by at least 50 billion liters per year. In short, the proposed adsorption process can truly make the membrane processes greener and more sustainable starting from its very fabrication step.

EXPERIMENTAL

General

Reagents (reagent grade) and solvents (analytical grade) were purchased from Sigma-Aldrich and Fisher Scientific, respectively. Charcoal, zeolites, graphene and resins were obtained from Eurocarb, VWR, Sigma-Aldrich and Dow, respectively. MIP, PIM, and MOF were prepared in-house (procedures described below). Acetonitrile porogen was dried over 4 Å molecular sieves. Millipore Type II water was used for all experiments. Polyimide (P84) powder was purchased from HP Polymer GmbH. 5,5′,6,6′-Tetrahydroxy-3,3,3′,3′-tetramethyl-1,1′-spirobisindane (TTSBI, 98%, Synthon Chemicals) was dissolved in methanol, re-precipitated from dichloromethane and dried under vacuum at room temperature before use. Tetrafluoroterephthalonitrile (TFTPN, 98%, Chemos) was used as received. Anhydrous K_2CO_3 (99%, Fisher) was ground to a fine powder and dried in a vacuum oven at 110 °C overnight before use. Anhydrous dimethylacetamide (DMAc), toluene and methanol were purchased from Sigma Aldrich and used as received. An Agilent 6890N GC system equipped with a Variant Factor 4 VF-5 ms (28 m × 0.25 mm DI

× 1 μm DF) column and an Agilent 5973 MS detector was used for NMP and DMF analysis.

Preparation of Imprinted Polymers

MIP microspheres were prepared by a suspension polymerization method according to our reported procedure.[43] Briefly, in a typical MIP fabrication procedure the 3-vinylphenylboronic acid functional monomer (1 mmol), either NMP or DMF template (1 mmol), EDMA cross-linker (20 mmol), AIBN initiator (0.1 wt%), perfluoro polymeric surfactant (PFPS) emulsifier (100 mg), perfluoro methylcyclohexane (PMC) dispersing phase (80 mL) and acetonitrile porogen (15 mL) were stirred at 300 rpm. The imprinted polymers were obtained by polymerization involving irradiation of the stirred mixture with UV light for 6 hours at a wavelength of 365 nm at room temperature under an inert atmosphere. The resulting beads were filtered and the remaining template and unreacted molecules were extracted by sequential washing with methanol. The MIPs were dried under reduced pressure for 24 h at room temperature. Please refer to the ESI for detailed synthesis and characterization of the imprinted polymers.

Preparation of Polymers of Intrinsic Microporosity

PIM-1 was prepared by the method of Du *et al.*[44] To a dry 100 mL three-neck round-bottom flask, equipped with a Dean-Stark trap, mechanical stirrer and nitrogen inlet, 5,5′,6,6′-tetrahydroxy-3,3,3′,3′-tetramethyl-1,1′-spirobisindane (TTSBI) (3.404 g, 0.01 mol), tetrafluoroterephthalonitrile (TFTPN) (2.001 g, 0.01 mol), anhydrous potassium carbonate (4.146 g, 0.03 mol), dimethylacetamide DMAc (20 mL), and toluene (10 mL) were added under an atmosphere of nitrogen gas. The monomers were allowed to dissolve before the reaction mixture was refluxed at 160 °C for 20 min. The viscous solution was then poured into methanol to precipitate the product. The sample was refluxed overnight in deionized water, washed with acetone (150 mL) and dried in a vacuum oven at 110 °C overnight. GPC: M_n = 48 800, M_w = 128 000, M_p = 87 700, M_w/M_n = 2.6. ^{1}H NMR(400 MHz, $CDCl_3$): 6.74 (br, s, 2H), 6.35 (br, s, 2H), 2.25 (br, s, 2H), 2.08 (br, s, 2H), 1.29 (br, s, 6H), 1.24 (br, s, 6H). BET surface area: 705 m^2 g^{-1}. Elemental analysis calculated for $C_{29}H_{20}N_2O_4$ (wt%): C, 75.65 H, 4.35 N, 6.09. Found C, 73.29 H, 4.15 N, 5.74.

Preparation of a HKUST-1 Metal–Organic Framework

HKUST-1 was synthesised by immersing two Cu foil electrodes (~16 cm^2), approximately 2 cm apart, into a 1 : 1 ethanol : deionised water solution containing a 48 mM 1,3,5-benzenetricarboxylic acid linker and 64 mM methyltributylammonium methyl sulphate supporting electrolyte.[45] The solution was maintained at 55 °C and deaerated with N_2 (g) whilst a PGSTAT302N potentiostat (Metrohm Autolab B.V., The Netherlands) was used to apply a fixed potential difference of 2.5 V between the two Cu foil electrodes for one hour. This generated HKUST-1 in solution was collected by centrifugation. The HKUST-1 was then re-suspended in methanol, stirred and collected again by centrifugation before being left to dry. Please refer to the ESI for MOF preparation protocols and characterisation.

Kinetic and Adsorption Isotherms

In a typical procedure 10 mL of 1000 ppm NMP or DMF contaminated water was loaded on 10, 30, 50, 100, 200, 300 and 400 mg adsorbents, the container was sealed, and placed in an incubator shaker at 25 °C and 300 rpm for 24 hours. Samples from the supernatants were taken in order to quantify NMP and DMF. Samples were taken at 0.25, 0.5, 1, 2, 5, 12 and 24 hours to obtain kinetic data.

Adsorbent Recovery

Washing and drying methods were evaluated for adsorbent recovery. In a typical washing protocol 200 mg adsorbent saturated with NMP was placed in an SPE cartridge and washed with 2.25 mL of a washing solvent. 150 μL fractions were collected with a flow of 10 mL min^{-1}. The washing solvents were either methanol, water, 50 vol% or 10 vol% methanol in water. In a typical drying or vacuum regeneration procedure 200 mg adsorbent saturated with NMP was placed under 0.01 mbar vacuum at 25 °C for approximately 12 hours. The weight of the adsorbent was measured at given intervals until no more change was observed. Note that using the Cox chart it can be estimated that the boiling points of NMP and DMF are −5 °C and −20 °C, respectively.[46]

Membrane Fabrication

Polymer membranes were produced *via* phase inversion based on the protocol by Campbell *et al.*[47] Dope solutions were formed by dissolving 25 wt% of polyimide P84 in NMP followed by casting on to polypropylene non-woven sheets using a casting knife set to a thickness of 250 μm at a temperature of 20 °C and a humidity of 35–45%. Membranes were cast using an automatic film applicator Elcometer 4340. The polymer membranes were then precipitated from solution *via* immersion into a water bath (coagulation tank). The membranes were then placed in isopropyl alcohol (IPA) to remove water from the polymer matrix followed by crosslinking with 30 g L^{-1} solutions of hexane-1,6-diamine in IPA at 50 °C for 24 hours. The crosslinked membranes were subsequently washed with IPA to remove the excess crosslinking agent. To avoid pore collapse, the membranes were conditioned with a PEG400 : IPA (50 : 50 v/v) solution for 4 hours and then dried in air.

Membrane Performance Test

All filtration experiments were carried out at 10 bar using a cross-flow filtration system with an effective membrane area of 52 cm^2. The permeance of each membrane was calculated by dividing the acetone flux by the applied pressure. The model system for the solute rejection experiments comprised a mixture of 1 g L^{-1} PS580 and PS1300 polystyrene markers as well as 0.1 g L^{-1} of methyl styrene dimer solution in acetone. The rejection (R) of markers was obtained by measuring the concentration of each polystyrene oligomer in the permeate and the feed, and calculating the ratio of the molecules retained by the membrane.

Continuous Water Treatment

Schematic representation of the continuous adsorption system is shown in Fig. 6A. The system consists of two high pressure pumps (Gilson 305), a feed tank containing the wastewater from the coagulation bath; a wash tank containing methanol : water (10 : 90 v/v) for column regeneration; and a vessel for the purified water and the waste solution; and two adsorption columns wet-packed with 50 g of MIP7. High pressure pumps were used to feed the wastewater and the washing solution at a speed of 25 mL min^{-1} to the columns in an alternating manner. After the collection of 2.5 L purified water the columns were regenerated using 120 mL washing solution. In total 12.5 L wastewater was purified continuously.

ACKNOWLEDGEMENTS

We want to thank Elisabeth Davenport, Dr Jozsef Kupai, Luca Szabo and Christos Didaskalou, all from the School of Chemical Engineering & Analytical Science, Manchester for their technical help. The authors thank Peter Halasz from the School of Mechanical, Aerospace and Civil Engineering for his artistic help. We would also like to acknowledge the help of Dr Wayne Harrison from the School of Chemistry, Manchester with the preparation of PIM-1, Stephen Worrall, Raghidah Wagia and Samaila Jovial from the School of Chemistry, Manchester, for provision of the MOF samples and Eurocarb Ltd for providing activated carbon samples. The work described in this paper was partially supported by EPSRC under grant number EP/K016946/1. The authors would also like to thank financial support from the Nano Material Technology Development Program (2012M3A7B4949745), the National Research Foundation of the Korean Ministry of Science, ICT and Future Planning.

REFERENCES

1. C. Fritzmann, J. Löwenberg, T. Wintgens and T. Melin, *Desalination*, 2007, **216**(1–3), 1–76
2. N. S. A. Mutamim, Z. Z. Noor, M. A. A. Hassan and G. Olsson, *Desalination*, 2012, **305**, 1–11
3. *Handbook of membrane separations: chemical, pharmaceutical, food, and biotechnological applications*, ed, A. K. Pabby, S. S. H. Rizvi and A. M. S. Requena, CRC press, 2015
4. C. Jiménez-González, P. Poechlauer, Q. B. Broxterman, B. S. Yang, D. am Ende, J. Baird, C. Bertsch, R. E. Hannah, P. Dell'Orco, H. Noorman, S. Yee, R. Reintjens, A. Wells, V. Massonneau and J. Manley, *Org. Process Res. Dev.*, 2011, **15**, 900–911
5. G. Szekely, M. F. Jimenez-Solomon, P. Marchetti, J. F. Kim and A. G. Livingston, *Green Chem.*, 2014, **16**, 4440–4473 **RSC**.
6. P. Marchetti, M. F. Jimenez-Solomon, G. Szekely and A. G. Livingston, *Chem. Rev.*, 2014, **114**(21), 10735–10806
7. G. R. Guillen, Y. Pan, M. Li and E. M. V. Hoek, *Ind. Eng. Chem. Res.*, 2011, **50**, 3798–3817.
8. World Membrane Separation Technologies to 2017 – Demand and Sales Forecasts, Market Share, Market Size, Market Leaders, The Freedonia Group, 2015 (http://www.freedoniagroup.com/World-Membrane-Separation-Technologies.html).
9. T. Fane, TCE, July 2013.
10. Global Water Intelligence Report, Volume 49, Issue 18, 2013.
11. R. W. Baker, *Ind. Eng. Chem. Res.*, 2002, **41**, 1393–1411
12. Yano Research Institute, Lithium ion battery material market perspective – four key materials (C54104000), 2012.
13. *Membrane Technology and Applications*, ed. R. W. Baker, Wiley, 2012.
14. T. Schiestel, *Membr. Technol.*, 2015, **2015**(6), 1–7.
15. D. Y. Xing, S. Y. Chan and T. S. Chung, *Chem. Eng. Sci.*, 2013, **87**, 194–203
16. M. Temtem, D. Pompeu, T. Barroso, J. Fernandes, P. C. Simões, T. Casimiro, A. M. Botelho do Rego and A. Aguiar-Ricardo, *Green Chem.*, 2009, **11**, 638–645.
17. N. T. Hassankiadeh, Z. Cui, J. H. Kim, D. W. Shin, S. Y. Lee, A. Sanguineti, V. Arcella, Y. M. Lee and E. Drioli, *J. Membr. Sci.*, 2015, **479**, 204–212.
18. A. Figoli, T. Marino, S. Simone, E. Di Nicolo, X.-M. Li, T. He, S. Tornaghi and E. Drioli, *Green Chem.*, 2014, **16**, 4034–4059.
19. A. K. Ghosh, B. H. Jeong, X. F. Huang and E. M. V. Hoek, *J. Membr. Sci.*, 2008, **311**(1–2), 34–45.
20. W. W. Li, H. Q. Yu and Z. He, *Energy Environ. Sci.*, 2014, **7**, 911–924 CAS.
21. I. Carra, J. A. S. Pérez, S. Malato, O. Autin, B. Jefferson and P. Jarvis, *Chem. Eng. J.*, 2015, **264**, 690–696.
22. L. Fabrizi, B. Jefferson, S. Parsons, A. Wetherill and P. Jarvis, *Environ. Sci. Technol.*, 2010, **44**, 6443–6449.
23. J. Wang and B. Chen, *Chem. Eng. J.*, 2015, **281**, 379–388.
24. N. B. McKeown and P. M. Budd, *Macromolecules*, 2010, **43**(12), 5176.
25. A. Murray and B. Ormeci, *Environ. Sci. Pollut. Res. Int.*, 2012, **19**(9), 3820–3830.
26. B. V. de Voorde, B. Bueken, J. Denayer and D. De Vos, *Chem. Soc. Rev.*, 2014, **43**, 5766–5788.
27. *Handbook of Water and Wastewater Treatment Technology*, ed. N. P. Cheremisinoff, CRC Press, 1994 Search PubMed.
28. S. Wang and Y. Peng, *Chem. Eng. J.*, 2010, **156**, 11–24 CAS PubMed.
29. K. Y. Foo and B. H. Hameed, *Chem. Eng. J.*, 2010, **156**, 2–10 CAS PubMed.
30. F. Haghseresht and G. Lu, *Energy Fuels*, 1998, **12**, 1100–1107 CAS.
31. F. Gimbert, N. Morin-Crini, F. Renault, P. M. Badot and G. Crini, *J. Hazard. Mater.*, 2008, **157**, 34–46 CAS PubMed.
32. Y. Luk, C. J. Allender and T. Wirth, *Tetrahedron Lett.*, 2010, **51**, 5883–5885.
33. G. Szekely, E. Fritz, J. Bandarra, W. Heggie and B. Sellergren, *J. Chromatogr., A*, 2012, **1240**, 52–58.

34. I. K. Shah, P. Pre and B. J. Alappat, *Chem. Sci. Trans.*, 2013, **2**(4), 1078–1088.

35. N. Al-Janabi, P. Hill, L. Torrente-Murciano, A. Garforth, P. Gorgojo, F. Siperstein and X. Fan, *Chem. Eng. J.*, 2015, **281**, 669–677.

36. N. Nijem, H. Bluhm, M. L. Ng, M. Kunz, S. R. Leone and M. K. Gilles, *Chem. Commun.*, 2014, **50**, 10144–10147.

37. S. Bonyadi and T. S. Chung, *J. Membr. Sci.*, 2009, **331**(1–2), 66–74

38. T. He, M. H. V. Mulder and M. Wessling, *J. Appl. Polym. Sci.*, 2003, **87**(13), 2151–2157.

39. *Basic Principles of Membrane Technology*, ed. M. Mulder, Springer, 1996.

40. B. Sellergren, *Molecularly Imprinted Polymers: Man-made Mimics of Antibodies and their Applications in Analytical Chemistry*, Elsevier, Amsterdam, 1st edn, 2001.

41. B. Van der Bruggen and P. Luis, *Curr. Opin. Chem. Eng.*, 2014, **4**, 47–53.

42. J. F. Kim, G. Szekely, M. Schaepertoens, I. B. Valtcheva, M. F. Jimenez-Solomon and A. G. Livingston, *ACS Sustainable Chem. Eng.*, 2014,**2**(10), 2371–2379.

43. J. Kupai, E. Rojik, P. Huszthy and G. Szekely, *ACS Appl. Mater. Interfaces*, 2015, **7**, 9516–9525.

44. N. Du, G. P. Robertson, J. Song, I. Pinnau, S. Thomas and M. D. Guiver, *Macromolecules*, 2008, **41**, 9656–9662.

45. S. S. Y. Chiu, S. M. F. Lo, J. P. H. Charmant, A. G. Orpen and I. D. Williams, *Science*, 1999, **283**, 1148–1150 .

46. E. R. Cox, *Ind. Eng. Chem.*, 1923, **15**(6), 592–593.

47. J. Campbell, G. Szekely, R. P. Davies, D. C. Braddock and A. G. Livingston, *J. Mater. Chem. A*, 2014, **2**(24), 9260–9271

CHAPTER 10 DESIGN OF SUPPORTING MEDIA FOR GROWTH OF BIOFILM IN TREATING INDUSTRIAL WASTEWATER

Hassimi Abu Hasan, Siti Rozaimah Sheikh Abdullah, Nor Ashikin Shafinaz Binti Whayab, Fathiyyah Mohd Zainudin and Muhammad Farhan Md Yusoff

Keywords: Biofilm Media, Biological Wastewater Treatment, Industrial Wastewater, Sequencing Batch Reactor (SBR).

INTRODUCTION

In general, water pollution occurs when there is an excess concentration compared to the original concentration caused by human activity and an adverse effect on the environment or important elements in the environment itself, such as health, living resources, ecosystems, and so on. There are many factors that can contribute to air pollution problems such as disposal of sewage from municipal, industrial, farm animals, farm-based activities and more. Waste resulting from the diversity of these activities lead to water pollution problems that exist in Malaysia. A great selection of granular media can play an important role in maintaining the number of active biomass in many and varied population to produce a surface area per unit volume for biofilm growth. Media and polyester polyvinylchloride float due to the lightweight and sink media such as sand, ceramics and pebbles are heavier and sink to the bottom bioreactor. Floating media have the advantage in getting rid of COD and suspended solids more effectively. In addition, a floating media also have resistance to temperature drop.

The study focused on the design of a supporting media for biofilm attachment for wastewater treatment industries. Media that is designed to function as bacterial adhesion sites that have a surface area per unit volume which is great for maintaining active biofilm and diverse microbial population. Among the commercial media in the market is bioball shaped plastic media with an average diameter of 3-5 cm and a height of 3 cm, which can continue to be used as a medium of waste water treatment. In addition, the media from scouring pad made of nylon is also commercialized as supporters for biofilm growth. Scouring pads typically sheet- diced to the size of 1cm x 1cm x 1cm before it can be used. These media are also known as synthetic media.

Therefore, design of a great supporter media is an important element in the operation of waste water treatment in order to achieve the required standards. In addition, the new design is different compared to designs available in the market in order to be tested and compare the level of efficiency in wastewater treatment processes. The design was to create different variations making the best choice for media supporters' wastewater treatment.

Thus, a comparison can be made between the design of media support and compared with biofilm media in the market in terms of the effectiveness of the water treatment process. Thus, in this scientific project, with design of media with different justifications is designed to test the performance of wastewater treatment. Media is designed to be evaluated in terms of the ability to adhere biofilm growth and effectiveness in wastewater treatment performance will be compared with the performance of existing media support in the market.

The specific objective of this study was to design a supporting media for growth of biofilm in the treatment of industrial wastewater and to measure the effectiveness of supporting media in treating industrial waste water.

EXPERIMENTAL PROCEDURE

AutoCad 2012 software is used in the design of supporting media. The software developed by the company AutoDesk

Citation: Hassimi Abu Hasan, Siti Rozaimah Sheikh Abdullah, Nor Ashikin Shafinaz Binti Whayab, Fathiyyah Mohd Zainudin and Muhammad Farhan Md Yusoff, Design of Supporting Media for Growth of Biofilm in Treating Industrial Wastewater, DOI: 10.17485/ijst/2016/v9i21/95249.

Inc. based in the United States. Media design drawn in the form of Three-Dimensional (3D) with length and thickness specifications. Thus, after fabrication process is complete, the design biofilm media is then being tested in treatment of industrial wastewater where in this experiment, wastewater from pulp and paper mills industry is used. Figure 1 shows the schematic diagram for sequencing batch reactor setup. The proposed experimental set-up for Sequencing Batch Reactor containing design biofilm media can be made as shown in Figure 2. The SBR is operated within 24 hours HRT for each sequential phase of 30 minutes for the feeding phase, 21.5 hours for reaction phase, 1 hour for sedimentation phase and 1 hour for removal phase.

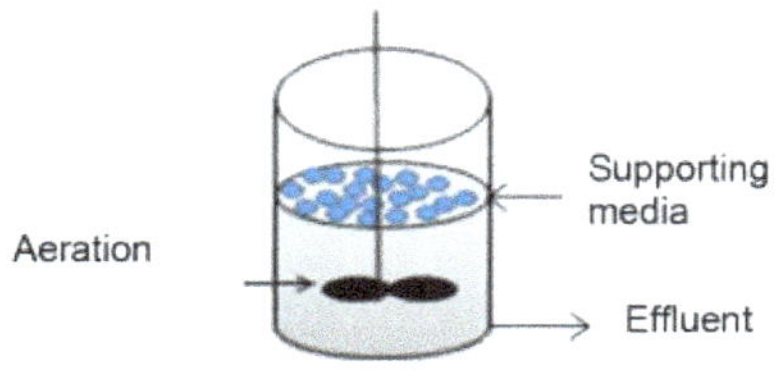

Figure 1. Schematic diagram of sequencing batch reactor.

Figure 2. Experimental setup of sequencing batch reactor.

The Sequencing Batch Reactor (SBR) process uses floating plastic carriers (media) within the aeration tank to increase the amount of microorganisms available to treat the wastewater compared to conventional secondary treatment. The microorganisms consume organic material. The media provides increased surface area for the biological microorganisms to attach to and grow in the aeration tanks. The increased surface area reduces the footprint of the tanks required to treat the wastewater.

The biofilm media (Figure 3.) will be continuously agitated by bubbles from the aeration system that adds oxygen at the bottom of the first compartment of the aeration tank. The microorganisms consume organic material. The bottom portion of reactor will contain stones that act as filter. The waste water will be filtered through stone bed to some extent. After filtering, the effluent sample is then undergone water quality analysis. The parameters tested are removal of Chemical Oxygen Demand (COD), removal or ammonia-nitrogen, the profile of DO and pH, biomass analysis, the colony formation unit and observation using scanning electron microscope.

RESULTS AND DISCUSSION

COD Removal

In this study, the average COD removal efficiency scores obtained in the reactor contains hexafilter media was 67.6% (Figure 4.) while for the commercial media was 63.2% (Figure 5). It can be seen that relatively high COD removal efficiency for the media hexafilter when compared to commercial media. However, the average COD removal has not reached a high percentage because it may be affected by the accumulation of biomass on the surface of a supporting media. However, the COD removal efficiency of experiments in this study was lower when compared to previous studies1. Lo has listed the percentage of COD removal of 75% through activated sludge treatment while recorded2 the COD removal up to 89% using anaerobic SBR treatment. In addition, biological water treatment lagoon recorded using the percentage of COD removal of 70% [3].

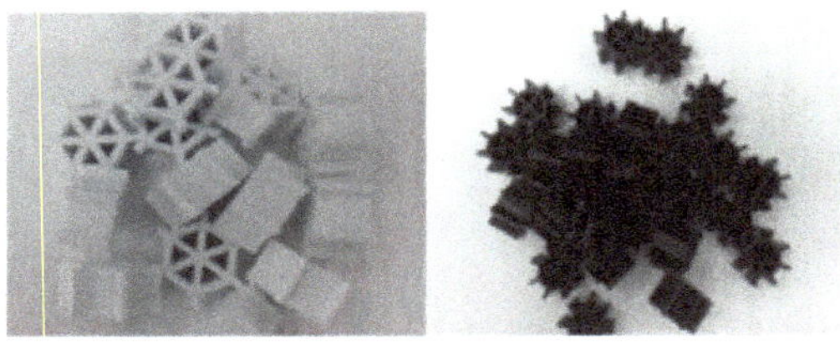

Figure 3. Biofilm media.

pH and DO Profile

Figure 6 and Figure 7 show DO and pH profiles for both compared supporting media. In general, the pH optimum for bacterial growth is between 6.5 and 7.5. pH is an important condition for the occurrence of nitration process and denitrification in wastewater treatment process.

Therefore, it is important to maintain optimum pH. Based on the trend graph, the pH of the two reactors are in the optimum pH range for the activity of bacteria. In addition, aeration must always be carried out in the reactor to maintain the DO concentration to a high level for rapid microbial activity. Conversely, in the event of a reduction in DO can affect the rate of COD removal efficiency. Through previous studies in4 say that the neutral pH of 7 is the optimum pH conditions in the treatment using rotating biological contactor. While treatment with trickling filter is recorded at pH 6.5-8.5 filter as optimal pH conditions5.

While for DO analysis, the Lo says that the optimal DO range is between 20-50 mg/L in water treatment using activated sludge. In5 also says DO optimum is between 25-70 mg/L using rotating biological contactor treatment.

Removal of Ammonia Nitrogen

Monitoring of ammonia-nitrogen removal efficiency was performed and compared between the two media.

The average ammonia-nitrogen removal efficiency for hexafilter media was 40.6% while the commercial media noted the removal of a high percentage of 43.6%. The low increase in ammonia - nitrogen removal trend can be attributed to activities of less active biofilm layer on the surface of a supporting media. Figure 8 and Figure 9 show the profile of ammonia-nitrogen removal during wastewater treatment performance by both supporting media. In addition, the decrease in the concentration of organic matter and nitrogen can also contribute to ammonia removal efficiency and low oxygen reaction by microorganisms. The result of ammonia-nitrogen removal efficiency was lower in this study compared with previous studies as it depend upon the availability of the bacteria present, oxygen and source of inorganic carbon6

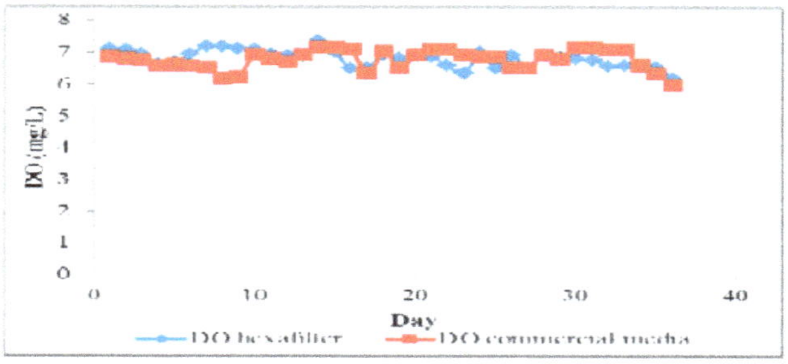

Figure 6. DO profile.

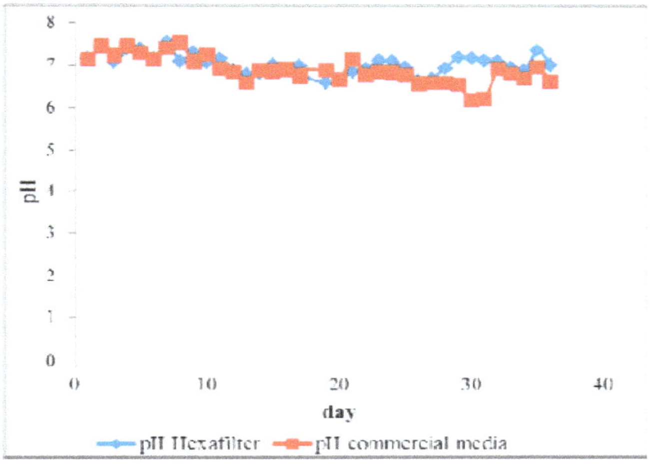

Figure 7. pH profile.

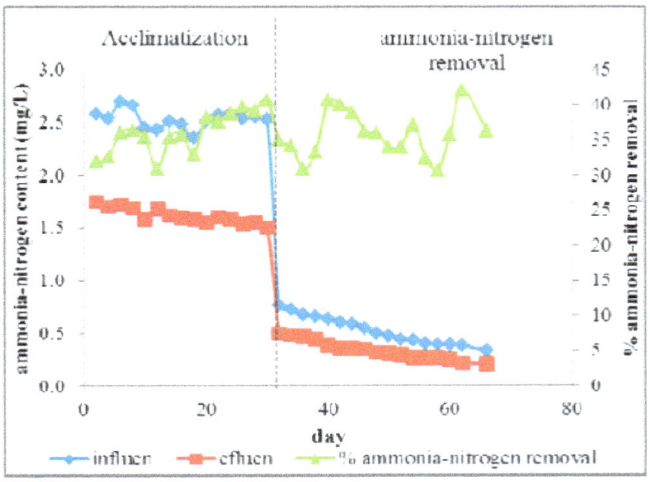

Figure 8. Ammonia-nitrogen removal for hexafilter.

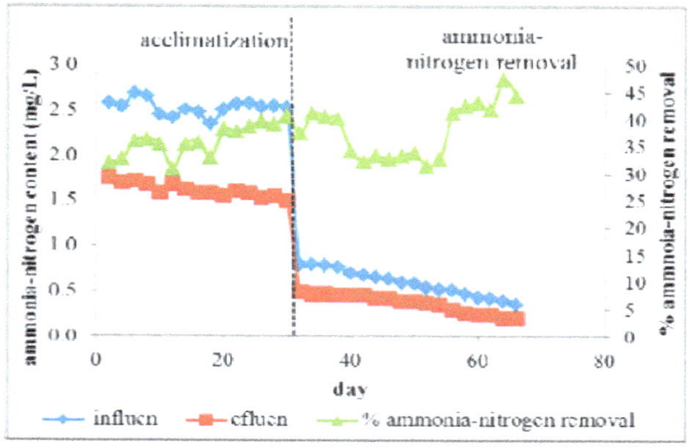

Figure 9. Ammonia-nitrogen removal for commercial media.

Through studies conducted by[5], the treatment of water using trickling filter yield high ammonia-nitrogen removal as high as 90%. The same removal efficiency gains also recorded through biological treatment using fluidized bed reactor using synthetic wastewater[7]. Meanwhile, the water treatment using rotating biological contactor, the percentage of ammonia-nitrogen removal efficiency was up to 88% in the treatment of industrial waste water[8].

Scanning Electron Microscopy Analysis (SEM)

Electron microscopy analysis was carried out on 3 samples of supporting medias. Figure 10 shows the surface of a supporting media on day 1 sample, Figure 11 shows the surface of a supporting media on day 15 samples were taken and Figure 12 shows the surface of a supporting media on the 27th day the sample was taken.

Through SEM, matrixes of biofilms germinate rapidly on both sides of media with the mechanism of the rapid growth of bacteria. Biofilm structure containing bacterial colonies that give biofilm surface looks rough and complex. At the beginning of the sampling, analysis showed the surface structure of coating media without biofilm who have not yet formed. This is because the media samples are still being tested in the early stages of wastewater treatment performance. However, along with increased performance during treatment, it was found that biofilm begin to form and stick rapidly on the surface structure of the supporting media until the take up of last supporting media.

Colony Formation Unit (CFU)

Figure 13 shows a graph of comparison CFU for hexafilter and commercial media. From the results, it was found that the amount of bacterial growth showed an increase over the period of performance of water treatment in both reactor containing hexafilter and commercial media. The number of bacteria in hexafilter increased from 2.3 x 107 to 4.7 x 107 CFU/ml while for commercial media, an increasing number of bacteria was from 1.9 x 107 to 4.2 x 107. On the 30th, the number increased further to 25.1 x 107 for the hexafilter and 23.5 x 107 for the commercial media.

Figure 10. Shows SEM sample. (a) Hexafilter. (b) Commercial media on day 1.

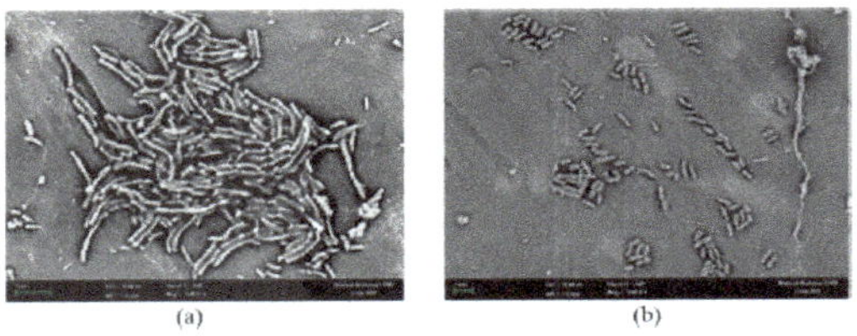

Figure 11. Shows SEM sample. (a) Hexafilter. (b) Commercial media on day 15.

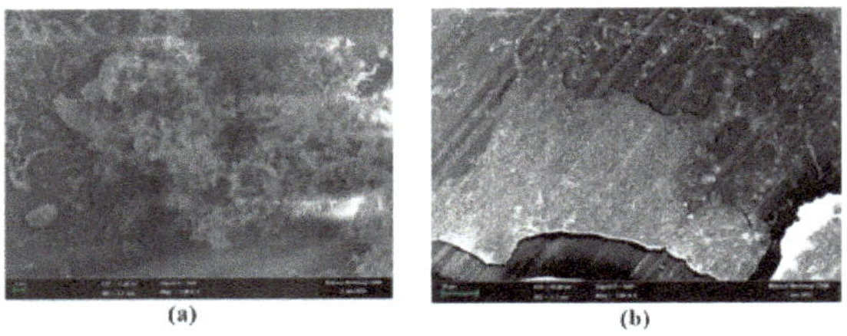

Figure 12. Shows SEM sample. (a) Hexafilter. (b) Commercial media on day 27.

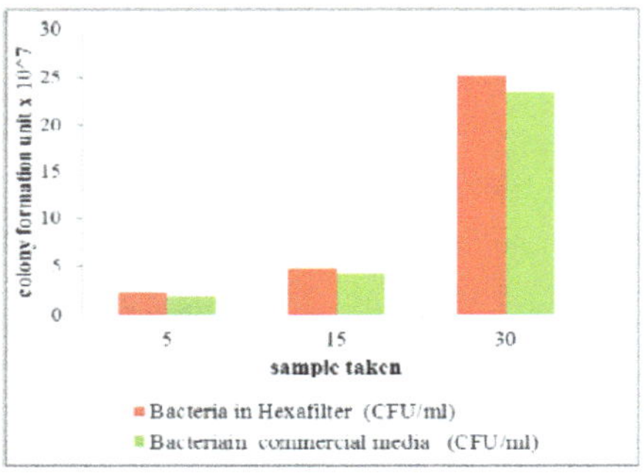

Figure 13. Colony formation unit.

CONCLUSIONS

Through the performance of wastewater treatment, both objectives for this scientific project to compare the performance of supporting media designed with commercial supporting media have been reached. During the performance of wastewater treatment, the average percentage of COD removal efficiency by hexafilter is 67.6% while the commercial media is 63.2%. This shows hexafilter media have higher removal efficiency. As for the percentage of ammonia-nitrogen removal, hexafilter recorded 40.6% which is slightly lower than the commercial media, which recorded 43.6%. Both supporting media for biofilm maintain optimum pH for growth of bacteria and higher DO concentrations.

From the results of this study, several conclusions can be highlighted. First, design of supporting media with a diameter of 25 mm and a height of 12 mm has been designed and tested its performance in the treatment of industrial wastewater. Comparison between supporting media designed are also made with commercial media, which shows through the experimental results of this scientific project, hexafilter is more effective in wastewater treatment and the experimental performance can be compared with existing commercial media.

ACKNOWLEDGEMENT

This research was financially supported by the Ministry of Education (MOE) through Grant No. FRGS/1/2014/ TK05/ UKM/02/1 and Universiti Kebangsaan Malaysia with Grant No. DLP-2015-01.

REFERENCES

1. Lo IMC. Characteristics and treatment of leachates from domestic landfills. Environment International. 1996; 22(4):433–42.
2. Diamadopoulos E, Samaras P, Dabou X, Sakellaropoulos GP. Combined treatment of landfill leachate and domestic sewage in a sequencing batch reactor. Water Science and Technology. 2007; 36 (2-3):61–8.
3. Orupold K, Tenno T, Henrysson T. Biological lagooning of phenols-containing oil shale ash heaps leachate. Water Research. 2000 Oct; 34:4389–96.
4. Flora EMCV, Suidan MT, Flora JRV, Kim BJ. Speciation and chemical interaction in nitrifying biofilms. Journal of Environmental Engineering. 1999 Sep; 125(9):871–7.
5. Matthews R, Winson M, Scullion J. Treating landfill leachate using passive aeration trickling filters; effects of leachate characteristics and temperature on rates and process dynamics. Science of Total Environment. 2009 Jan; 407(8):2557–64.
6. Selvakumar T, Ganesan M. Water quality improvements in Soil Aquifer Treatment (SAT) simulated soil columns. Indian Journal of Science and Technology. 2015 Oct; 8(28):1–6.
7. Vinod AV, Reddy GV. Simulation of biodegradation process of phenolic wastewater at higher concentration in a fluidized bed reactor. Biochemical Engineering Journal. 2005; 24:1–10.
8. Najafpour GD, Zinatizadeh AAL, Lee LK. Performance of a three stage aerobic RBC reactor in food canning wastewater treatment. Biochemical Engineering Journal. 2006 Jun; 30(3):297–302.

CHAPTER 11

BIOAUGMENTATION OF LATEX RUBBER SHEET WASTEWATER TREATMENT WITH STIMULATED INDIGENOUS PURPLE NONSULFUR BACTERIA BY FERMENTED PINEAPPLE EXTRACT

Nastee Kornochalert, Duangporn Kantachote, Sumate Chaipraрat, Somkiet Techkarnjanaruk

Keywords: Bioremediation Fermented, plant extract, Response surface, methodology Rotten-egg odor Rubber wastewater

INTRODUCTION

Since their founding in years 1994 and 1995 by the central government of Thailand, many of these small cooperative rubber sheet factories (CRSFs) are still functioning throughout Thailand [1]. One serious problem arising from the treatment of their wastewater is the associated noxious smell of rotten-egg gas (hydrogen sulfide, H2S). Open lagoons or natural oxidation ponds used for this treatment now are standard in these factories [1,2]. It is well recognized that H2S is a toxic gas which at high level is dangerous to human health and at low level causes nuisance odor [3, 4]. Devising a means to solve this problem by removing H2S requires the cooperative efforts of engineers and microbiologists. It is well recognized that phototrophic bacteria, especially purple nonsulfur bacteria (PNSB) are well capable of treating wastewaters due to their versatile metabolic pathways. They have a capability to grow as both photoautotrophs and photoheterotrophs under conditions of anaerobic or microaerobic-light, and chemoheterotrophs under conditions of anaerobic or aerobic-dark [5,6,7,8]. Some members of the PNSB such as Rhodopseudomonas, Rhodobacter and Rhodospirillum are known to reduce the H2S odor from facultative waste stabilization ponds [9,10].

In our previous work, a combination of fermented pineapple extract (FPE) as growth stimulator and a selected purple nonsulfur bacterium Rhodopseudomonas palustris P1 showed a high efficiency for treating latex rubber sheet wastewater to meet the Thai standard guidelines within only 3 d. It was also found that the mix ratio of bacterial culture as inoculum had a great influence during the reaction period [2]. It is, however, difficult for CRSFs to have an access to the functional PNSB culants to treat their wastewater. It was also observed that FPE, when added at an optimal amount, i.e. 0.13%, facilitated the treatment of non-sterile latex rubber sheet wastewater (RAW). This relatively small amount of FPE stimulated the growth of native/indigenous PNSB under conditions of anaerobic or microaerobic-light by providing simple substrates such as acetate and lactate that were easily consumed by PNSB [2,11,12]. This stressed the feasibility of preparing the PNSB inoculant culture with the help of a simple and easy to find FPE.

However, successful bioaugmentation depends on the use of the effective inoculants, either single culture or consortium, which are usually well adapted to each individual wastewater. This is mainly related to the survival and degrading ability of microorganisms introduced to a wastewater system as

Citation: Nastee Kornochalert, Duangporn Kantachote, Sumate Chaiprapat, Somkiet Techkarnjanaruk, Bioaugmentation of latex rubber sheet wastewater treatment with stimulated indigenous purple nonsulfur bacteria by fermented pineapple extract, http://dx.doi.org/10.1016/j.ejbt.2014.06.003.

dictated by various biotic and abiotic factors within the system. Therefore, autochthonous bioaugmentation is attractive to use for the enhancement of efficiency of wastewater system due to the fact that foreign microbes when used as inocula, are unable to avoid competition with normal flora.

The amount of added FPE into the wastewater medium decreases initial pH and increases chemical oxygen demand (COD) value, thereby optimum initial pH and COD values should be investigated as these are the key factors for stimulating growth of indigenous or native PNSB (PNSBsi) to make good inocula. The response surface methodology (RSM) which is a collection of mathematical and statistical techniques for the modeling and analysis of multivariate problems has been used to optimize the desired outputs [13]. This methodology is more practical as it arises from experimental methodology that includes interactive effects among the variables and eventually, it depicts the overall effects of the parameters in the process [14]. In the process of the wastewater treatments, RSM was reportedly effective in evaluating the interactive effects of operating parameters [15,16].

The aims of this study were firstly to stimulate indigenous PNSB (PNSBsi) in a RAW lagoon with FPE for re-inoculation in the treatment of rubber sheet wastewater, and secondly to determine the optimal values of the factors affecting the growth of indigenous PNSB and their efficiency to treat the rubber sheet wastewater.

MATERIALS AND METHODS

Rubber Sheet Wastewater Medium

Rubber sheet wastewater was collected from a lagoon of a CRSF at Pichit suburb, Songkhla Province, Thailand. The collected wastewater was filtered through cheesecloth into a 25 liter non-transparent plastic tank and stored in a cold room at 6 ± 2°C until use. The wastewater without autoclaving was supplemented with 0.05% NH4Cl as a supplementary nitrogen source to allow a reasonable growth of PNSB (based on our preliminary work). This medium was named RAW since the non-sterile rubber sheet wastewater was used.

Monitoring Parameters

The methods used in this study are described in the Standard Methods [17]. All effluent samples including RAW were placed in a cold room for 2 h to allow sedimentation. The supernatant (clear liquid near the water surface) was sampled for the measurement of settleable COD by the dichromate reflux method. Sulfide was measured in 3 forms, as total sulfide (TtS), dissolved sulfide (DsS) and un-ionized hydrogen sulfide (UHS: H_2S) using the iodometric method, while sulfate was measured using the turbidimetric method. A portable multi gas detector (MX 2100, Oldham, France) was used to measure H_2S in headspace of the treatment bottles. In this work, volatile fatty acids (VFAs) were analyzed by distillation method. A pH meter (Seven multi, Metller Toledo, USA) was used to measure pH and electrical conductivity (EC). The oxidation–reduction potential (ORP, Eh) probe (La Motte, USA) was used to measure the Eh values and the data were recorded after obtaining a constant value. The probe was checked frequently in a quinhydrone buffer solution following the method described by the manufacturer. Total acidity was determined by a titration method and presented as a 'lactic acid amount'. The actual amounts of lactic and acetic acids were determined using gas chromatography according to the method of Yang and Choong [18]. Viable cells count of PNSB was performed on GM (glutamate–malate agar) and incubated in anaerobic light conditions for 5 d [5] whereas yeasts, heterotrophic plate count (HPC) and lactic acid bacteria (LAB) were enumerated on potato dextrose agar (PDA), plate count agar (PCA) and de Man Rogosa and Sharp (MRS) agar for 3 d. Both LAB and yeasts were counted because these microbial groups play important roles in the production of the FPE.

Fermented Pineapple Extract

FPE was produced in our laboratory according to the method described by Kantachote et al. [19]. The fermentation process lasted for 2 months. At that time, it composed of 1.90% total acidity, 0.58% lactic acid and 0.15% acetic acid with a pH of 3.61 and an EC value of 3.51 mS cm-1. In addition to the nutrients, the population of HPC, yeasts and LAB were estimated in the region of 106 cfu mL-1 for each group. FPE was kept in a cold room until used.

Experimental design for studying the stimulation of indigenous PNSB from a lagoon with FPE to make an inoculum for treating RAW

RSM using the Box–Behnken Design (BBD) was chosen because relatively fewer experimental combinations of the variables were required to estimate potentially complex response functions with an acceptable reliability [20]. The BBD was used to optimize three input independent variables on the response of the amount of PNSBsi in RAW. The significant variables: initial COD, initial pH and amount of FPE were selected as the critical independent variables and designated as X_1, X_2 and X_3, respectively. The low, middle and high levels of each variable were coded as -1, 0, +1, respectively (Table 1) with the design matrix of a 17-trial experiment, established using a Design Expert 6 software (Stat Ease Inc. Minneapolis, USA). Variations on the values of each independent variable were designed according to the results of our preliminary work. The behavior of the system was explained by the following quadratic equation model.

$$Y = \beta_0 + \beta_1X_1 + \beta_2X_2 + \beta_3X_3 + \beta_{11}X_1^2 + \beta_{22}X_2^2 + \beta_{33}X_3^2 + \beta_{12}X_1X_2 + \beta_{13}X_1X_3 + \beta_{23}X_2X_3 \quad (1)$$

where Y is the predicted response; β_0 is the intercept; β_1, β_2 and β_3 are the linear coefficients, β_{11}, β_{22} and β_{33} are the square coefficients and β_{12}, β_{13} and β_{23} are the interaction coefficients. To predict the optimal point, a second order polynomial function was fitted to correlate the relationship between the independent variables and the response values (dependent variables; PNSBsi and ORP). The optimal conditions for stimulating PNSB (PNSBsi) were obtained by solving the regression equations and also by analyzing overlay interaction plots.

The experimental runs were performed in 120 mL serum glass bottles, containing 100 mL of RAW medium per bottle to achieve microaerobic conditions and all serum bottles were incubated under tungsten light conditions of 3500 lx for 2 d.

A Denki light meter (model DK-211) was used to measure light intensity. Three replicates were set for each experimental run. The following parameters: PNSB, COD, pH and ORP, were monitored at the start and at the end of the experiment (2 d incubation). The ORP was also monitored as a dependent variable as it might help explain the role of FPE on stimulating PNSB growth. To validate the optimum conditions of the 3 independent variables, confirmatory experiments were carried out under microaerobic light conditions. To determine the effect of FPE on PNSBsi growth in RAW, there were 2 sets of verified tests, one with sterile and one with non-sterile FPE, beside a control set without FPE addition. Parameters: pH, EC, ORP, VFAs, HPC, LAB and PNSBsi, were monitored at the beginning and end of the 2-d incubation. These experiments were conducted in triplicate and ANOVA (Tukey HSD post-hoc test) was used to analyze data in this verification test with the significance level of 0.05.

Table 1. The BBD matrix for the real variables along with the actual and predicted responses of the ORP and stimulated indigenous PNSBsi values in RAW under microaerobic light conditions for 2 d after adding FPE.

Run no.	Code of variables and levels			ORP (mV)		PNSBsi (log cfu mL^{-1})	
	X_1	X_2	X_3	Actual	Predicted	Actual	Predicted
1	1500	8	2	-188.5	-186.4	7.09	7.25
2	2000	8	1.5	-203.9	-205.12	7.53	7.52
3	2000	7	2	-230.7	-228.2	7.79	7.8
4	2000	7	2	-226.7	-228.2	7.79	7.8
5	2500	7	2.5	-205.3	-204.45	7.53	7.67
6	2500	7	1.5	-190.7	-190.13	7.2	7.2
7	2500	8	2	-188.9	-188.22	6.94	6.95
8	1500	7	1.5	-177.1	-177.95	7.19	7.05
9	2000	8	2.5	-206.9	-208.46	7.46	7.3
10	1500	7	2.5	-173	-173.5	7.1	7.1
11	2000	7	2	-227.3	-228.2	7.83	7.8
12	2000	7	2	-228.6	-228.2	7.8	7.8
13	2500	6	2	-198.7	-200.8	6.94	6.78
14	2000	7	2	-227.7	-228.2	7.81	7.8
15	1500	6	2	-158.8	-159.48	5.79	5.78
16	2000	6	1.5	-197.9	-196.34	6.06	6.21
17	2000	6	2.5	-204.1	-202.88	6.94	6.96

X_1: initial chemical oxygen demand, COD (mg L^{-1}), X_2: initial pH; X_3: FPE (%, V/V).

Experimental Design for studying the treatment of rubber sheet wastewater using stimulated PNSB inoculum and FPE

To determine the optimum conditions for the treatment of CRSF wastewater, the key factors most likely to affect the efficiency were investigated: PNSBsi numbers (X_1), FPE concentrations (X_2) and retention times (RT) (X_3). To fit a second order response surface, central composite design (CCD) is used. The levels of these 3 independent variables (X_1, X_2 and X_3) were studied at 5 coded levels; -1.682, -1, 0, +1 and +1.682 establishing the design matrix for a 20 trial experiment. Variable levels were selected based on the results obtained from preliminary work. Design Expert (version 6, Stat Ease Inc. Minneapolis, USA) was used for experimental design and statistical analysis. The quadratic model equation [Equation 1] was also used to explain the behavior of the system.

The optimum wastewater treatment conditions by PNSBsi were obtained by solving the regression equations and also by analyzing the response surface overlay contour plots. The quality of the fit of the model equations was expressed by the coefficient of determination, R_2, while regression coefficients were used to generate a contour map of the regression model. All experimental runs were conducted in 120 mL serum glass bottles and incubated in microaerobic light conditions previously described, and the following parameters were determined: COD, pH, PNSB, suspended solids (SS) and TtS. The effectiveness of the treatments was evaluated from the response values (COD, SS and TtS).

The results from the CCD were selected for obtaining the optimal percentage of FPE, inoculum size of PNSBsi and RT. The optimum conditions based on calculation from the CCD were confirmed. In order to explain the roles of PNSBsi, FPE, and a combination of starter cultures plus FPE, the experimental design was as follows: RAW (native control), RAW with optimal amount of FPE, RAW with optimal dose of PNSBsi, RAW with optimal amounts of both FPE and starter PNSBsi, and the other 4 sterile sets of each corresponding treatment were also conducted in the same way. After 4 d, the efficiency was assessed by the loss of COD, SS, TDS, sulfate ion (SO_4^{2-}), TtS, DsS, UHS (H_2S in wastewater) and H_2S in the head space. The amounts of HPC, LAB and PNSB were also counted. All data in this experiment were analyzed by ANOVA (Tukey HSD post-hoc test) with a significance level of 0.05.

RESULTS

Stimulation of Indigenous PNSB from a Lagoon with FPE for use as Inoculums

The experimental results of BBD were analyzed by regression analysis consisting of the linear, quadratic and interaction effects which generated the following regression equations with increasing Population of PNSBsi (Y_{PNSBsi}) and value of ORP (Y_{ORP}) at specific combination of three independent variables could be predicted by substituting the corresponding values of each variable in [Equation 2 or Equation 3], respectively. The predicted values from both equations for PNSBsi and ORP of each experimental run are presented in Table 1.

Table 2. ANOVA of the quadratic models for the values of stimulated indigenous PNSBsi and ORP in RAW incubated under microaerobic light conditions for 2 d after adding FPE

Source	Sum of squares	DF	Mean square	*F*-value	Prob > *F*	R^2
PNSBsi value						
Model	5.2744	9	0.5860	24.1706	0.0002	0.9688
Residual	0.1697	7	0.0242			
Lack of fit	0.1411	3	0.0470	6.5775	0.0502	Not significant
Pure error	0.0286	4	0.0072			
Corrected total	5.4441	16				
ORP value						
Model	7296.50	9	810.72	197.89	<0.0001	0.9961
Residual	28.68	7	4.10			
Lack of fit	19.05	3	6.35	2.64	0.1858	Not significant
Pure error	9.62	4	2.41			
Corrected total	7325.18	16				

PNSBsi and reducing ORP values as a function of COD (X_1), pH (X_2), and FPE (X_3).

$$Y_{PNSBsi} = -51.48 + 0.01X_1 + 12.02X_2 + 4.46X_3 - \left(1.64 \times 10^{-6}\right)X_1^2 - 0.67X_2^2 - \left(6.51 \times 10^{-4}\right)X_1X_2 - 0.48X_2X_3 \quad (2)$$

$$Y_{ORP} = 1421.52 - 0.61X_1 - 240.77X_2 - 156.34X_3 + \left(1.22 \times 10^{-4}\right)X_1^2 + 13.89X_2^2 + 44.43X_3^2 + 0.019X_1X_2 - 0.018X_1X_3 \quad (3)$$

ANOVA for the response surface model is summarized in Table 2. The statistical significance of model equation was evaluated by F-test. As the greater F-value indicates that the factors explain adequately the data variation about its mean and the estimated factors are real. In this study, F-values of 24.17 for PNSBsi model and 197.89 for ORP model, signify that the models were significant (Table 2). ANOVA at the low probability value (P N F) also demonstrated that the quadratic model of PNSBsi value (P N F = 0.0002) and ORP value (P N F b 0.0001) were significant. However, a lack of fit F-values of 6.5775 and 2.64 of these models suggested that it was not significantly relative to the pure error. Non-significant lack of fit would be appropriate for this experiment. These are 5% and 19% chances that a lack of fit F-value could occur due to the experimental errors in PNSBsi and ORP models, respectively. Based on these results, PNSBsi model was chosen to produce the response surface contour plots for the determination of optimal conditions. In addition, the coefficients of variation (R^2) were 0.9688 for PNSBsi population and 0.9961 for ORP value indicating a high correlation between the observed and predicted values from models [Equation 2 and Equation 3]. Thus, both equations could be used for predicting the amount of PNSBsi population and ORP value under conditions varied with only three independent variables in the experimental range. The regression coefficient in the response surface model for the linear, quadratic and interaction effects of the variables are presented along with the P-value in Table 3. The linear effect of COD (P b 0.05), pH (P b 0.01) and FPE (P b 0.05) was statistically significant for increasing PNSBsi population, whilst, the linear effect of COD (P b 0.01), pH (P b 0.01) and FPE (P b 0.05) was also statistically significant for reducing ORP value. Both regression equations show significant effects of interacting relationships between the COD vs. initial pH on the PNSBsi population (P b 0.01) and the ORP values (P b 0.01). However, a significant effect of interacting relationships between the initial pH versus the FPE concentration was observed only on the PNSB population (P b 0.05) while a significant effect of the interacting relationship between the COD versus the FPE concentrations was found only on the ORP (P b 0.01).

Table 3. Estimated regression coefficient and corresponding P-value for the values of stimulated indigenous PNSBsi and ORP in RAW incubated under microaerobic light conditions for 2 d after adding FPE

Term	PNSBsi value		ORP value	
	Coefficient	*P*-value	Coefficient	*P*-value
Constant	-51.47809	0.0002**	1421.52	<0.0001**
COD (X_1)	0.010655	0.0141*	-0.61	<0.0001**
pH (X_2)	12.02	0.0001**	-240.77	0.0015**
FPE (X_3)	4.46422	0.0487*	-156.34	0.0107*
COD × COD	-1.64E-06	0.0010**	0.00	<0.0001**
pH × pH	-0.66787	<0.0001**	13.89	<0.0001**
FPE × FPE	-0.42108	0.2079	44.43	<0.0001**
COD × pH	-6.51E-04	0.0041**	0.02	<0.0001**
COD × FPE	4.17E-04	0.2228	-0.02	0.0024**
pH × FPE	-0.47868	0.0180*	1.60	0.4552

* $P < 0.05$.
** $P < 0.01$.

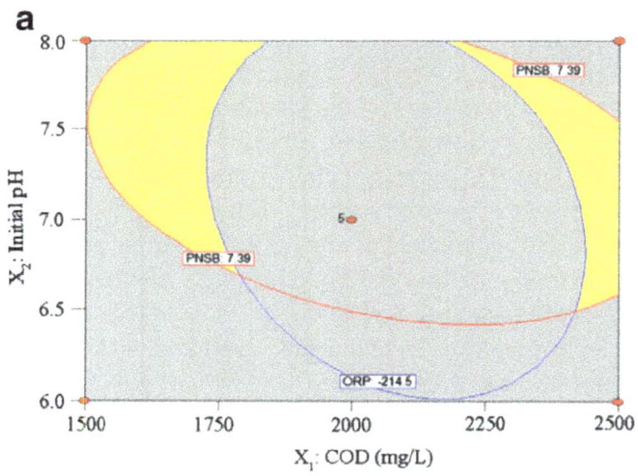

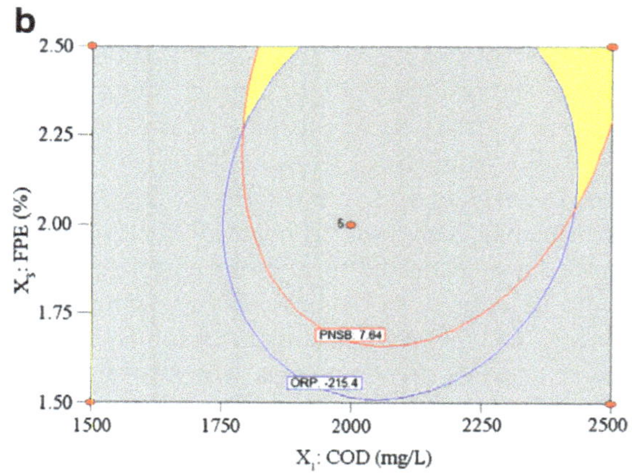

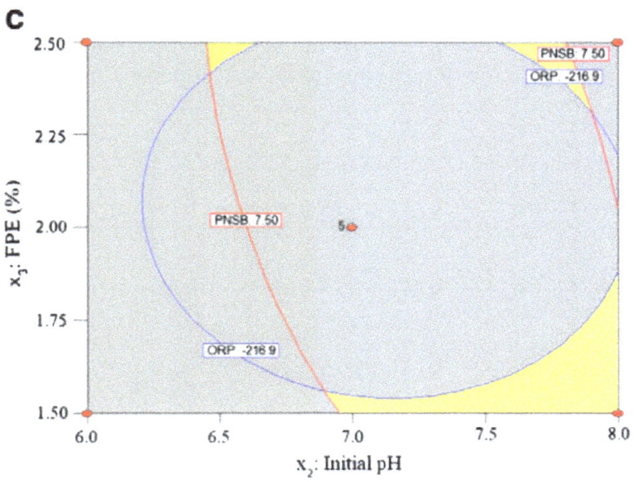

Figure 1. The overlay interaction plots of two variables: chemical oxygen demand (COD) and initial pH (a), COD and fermented pineapple extract (FPE) (b), initial pH and FPE (c) on oxidation–reduction potential (ORP) value and stimulating purple nonsulfur bacteria (PNSB) growth in non-sterile rubber sheet wastewater (RAW) under microaerobic light conditions over 2 d incubation.

The interacting relationship between 2 factors using the overlaid contour plot is presented by designation of another factor as zero level or mid-point. A 2% FPE was designed for the consideration of the interacting relationship between the COD and the initial pH on PNSBsi and ORP. It was found that the increase of the PNSBsi population with the decrease of ORP value relied on the following ranges of the examined variables: COD 1750–2400 mg L^{-1} and initial pH 6.5–8.0 (Fig. 1a). In a similar manner, the initial pH of 7 was used to consider the effect of COD and FPE showing that PNSBsi increased with the decrease of ORP when the COD was in the range of 1800–2400 mg L^{-1} and FPE was between 1.65 and 2.5% (Fig. 1b). When the COD was around 2000 mg L^{-1}, an increase of the PNSBsi with the decrease of ORP was observed at an initial pH and FPE in a range of 6.5–8.0 and 1.51–2.50% (Fig. 1c), respectively. The overlaid contour plot (Fig. 1) was also used to indicate the region where the optimization routine was searched for optimal points. The optimum conditions determined were COD concentration 2103 (mg L-1); initial pH 7.0 and FPE 2.0%. The predicted response of PNSBsi was 7.82 log cfu mL^{-1} while the actual value was between 7.79 and 7.83 log cfu mL-1 (run numbers 3, 4, 11, 12 and 14) (Table 1). These results were considered to be adequate and acceptable as the desirability obtained was 0.999. Hence, the optimal conditions for stimulating PNSB were 2000 mg L^{-1} COD, pH 7 and 2.0% FPE, and were therefore further verified, whilst the actual ORP values for those run numbers were in a range of -227 mV to -231 mV with the predicted value of -228 mV.

Table 4. Verification experiments of the stimulation of PNSB by adding sterile FPE and non-sterile FPE to RAW and incubating under microaerobic-light conditions for 2 d

Parameters	Treatments			
	RAW, 2000 mg L^{-1} COD, pH 7, t = 0	Control No addition of FPE	Verify test (2.0% FPE)	Verify test (Sterilized 2.0% FPE)
pH	7.0 ± 0.0^{a}	7.5 ± 0.0^{c}	7.2 ± 0.0^{b}	7.2 ± 0.0^{b}
EC (mS cm^{-1})	2.69 ± 0.05^{b}	2.47 ± 0.09^{a}	2.78 ± 0.05^{b}	2.97 ± 0.06^{c}
ORP (mV)	73 ± 1.1^{d}	-78 ± 1.7^{c}	-201 ± 0.4^{a}	-163 ± 0.1^{b}
VFAs (mg L^{-1})	53 ± 6^{d}	47 ± 6^{c}	23 ± 6^{a}	37 ± 6^{b}
HPC (log cfu mL^{-1})	5.42 ± 0.07^{a}	5.88 ± 0.08^{c}	5.27 ± 0.09^{a}	5.65 ± 0.08^{b}
LAB (log cfu mL^{-1})	0	0	0	0
PNSBsi (log cfu mL^{-1})	0^{a}	0^{a}	8.13 ± 0.06^{c}	7.69 ± 0.07^{b}

Different lowercase letters in each row indicate significant differences ($P < 0.05$).
EC: electrical conductivity; FPE: fermented pineapple extract; HPC: heterotrophic plate count; LAB: lactic acid bacteria; ORP: oxidation reduction potential; PNSB: purple nonsulfur bacteria; PNSBsi: stimulated indigenous purple nonsulfur bacteria; RAW: non sterile rubber sheet wastewater; VFA: volatile fatty acids.

Results of the verification experiments under microaerobic light conditions over 2 d test period with sterile and non-sterile FPE are shown in Table 4. Under the optimized conditions as calculation, significant differences were found for the amounts of HPC and PNSBsi, i.e. in a non-sterile FPE set, PNSBsi were 8.13 log cfu mL^{-1} and 7.69 log cfu mL^{-1} in a sterilized FPE set and no PNSB were detected in the control set. The lowest amount of the HPC was observed in a non-sterile FPE set which had the highest amount of PNSBsi. In contrast, there was no significant difference found in both verification sets for the following parameters: pH and LAB with the exception of ORP, VFAs and EC. The values of pH, VFAs and ORP including HPC in the control set were significantly higher than those found in both verification sets. However, no population of LAB was detected in any set.

Treatment of Rubber Sheet Wastewater using Stimulated PNSB and FPE

Results of CCD experiment are shown in Table 5 along with experimental and predicted values using RSM. COD, SS, and TtS were used as key responses to evaluate the efficiency of the treatment and the data obtained from all runs were analyzed by fitted regression models. The fitted regression models are as follows.

$$Y_{COD} = 4372.87 - 145.87X_1 - 957.70X_3 - 127.72X_1X_2 + 26.27X_1X_3 + 7.18X_1^2 + 2863.12X_2^2 + 84.17X_3^2 \quad (4)$$

$$Y_{SS} = 102.05 - 2.11X_1 - 21.61X_3 + 0.52X_1X_3 + 1.41X_3^2 \quad (5)$$

The models for the removals of COD, SS and TtS (Y_{COD}, Y_{SS} and Y_{TtS}) were significant by the F-test. The ANOVA for the quadratic models were highly significant as all had low p-value (P b 0.0001, 0.0008, b 0.0001 for COD, SS and TtS, respectively). The goodness of fit for COD, SS and TtS was confirmed by R^2 for COD, SS and TtS of 0.9369, 0.8947 and 0.9352, respectively, suggesting that only 6.31%, 10.53% and 6.48% of the total variation could not be explained by each respective model. In this study, the adj. R^2 values for COD, SS and TtS were 0.8801, 0.8000 and 0.8769, respectively, which are close to R^2 values. This is due to the fact that if there are many terms in the model and the sample is not very large, the adjusted R^2 might be noticeably smaller than the R^2. Based on the above results, the COD multiple regression model was selected to build the surface overlay contour plots and to derive the optimal condition for verification test.

Table 5. The central composite design for treating non sterile RAW with initial values in mg L^{-1} of 2024 COD, 57 SS and 13.56 TtS by PNSBsi inoculum under microaerobic light conditions.

Run no.	% PNSBsi	% FPE	RT (d)	COD (mg L^{-1})		SS (mg L^{-1})		TtS (mg L^{-1})	
	(X_1)	(X_2)	(X_3)	Actual	Predicted	Actual	Predicted	Actual	Predicted
1	2	0.75	4	770	953	32	38	8.67	8.98
2	4	0.625	5	673	596	30	25	8.67	8.36
3	8	0.625	3	724	825	32	33	8.89	8.88
4	4	0.875	3	1061	914	40	35	9.56	9.3
5	8	0.875	3	472	503	17	17	7.11	7.15
6	8	0.875	5	537	676	25	24	8	7.8
7[a]	6	0.75	4	288 (16)	289	24 (1.15)	24	6.89 (0.26)	6.77
8	6	0.5	4	660	797	27	30	8.44	8.63
9	6	0.1	4	867	794	23	25	7.78	7.97
10	10	0.75	4	394	275	15	14	7.33	7.39
11	4	0.625	3	1578	1393	52	49	9.33	9.27
12	8	0.625	5	976	868	28	26	8.44	8.3
13	6	0.75	2	1533	1504	48	48	8.89	8.85
14	4	0.875	5	1520	1582	47	45	9.11	8.99
15	6	0.75	6	886	979	27	33	8.22	8.63
R^2				0.9369		0.8947		0.9352	
Adj. R^2				0.8801		0.8000		0.8769	

[a] The experiment was repeated 6 times and the responses represented average values with their standard deviation in parenthesis.

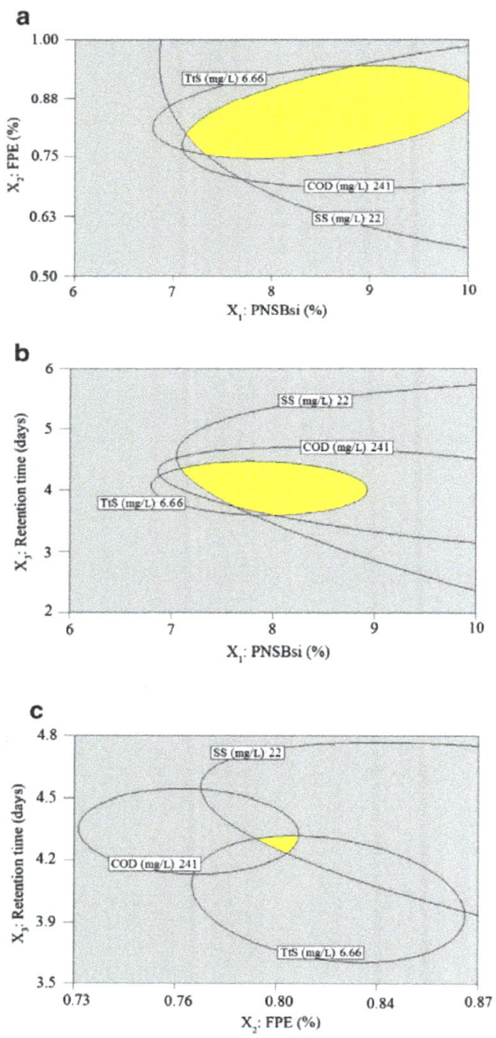

Figure 2. The overlay interaction plots of two variables: stimulated indigenous purple non sulfur bacteria (PNSBsi) and fermented pineapple extract (FPE) (a), PNSBsi and retention time (RT) (b), FPE and RT (c) on the efficiency to treat non sterile rubber sheet wastewater under microaerobic light conditions.

$$Y_{TtS} = 17.44 - 0.27X_1 - 0.36X_1X_2 + 0.61X_2X_3 + 0.03X_1^2 + 8.65X_2^2 + 0.17X_3^2 \quad (6)$$

Considering [Equation 4 and Equation 5], RT had a significant effect on reducing the COD and SS values (P b 0.05), and it also implies that the growth of the PNSBsi is dependent on the RT. Consequently, the PNSBsi had a significant influence on reducing the TtS [Equation 6]. The regression [Equation 4] shows the significant effects of interacting relationships between PNSBsi versus FPE and PNSBsi versus RT on the COD value. However, [Equation 5] shows that the interaction between PNSBsi vs. RT had a substantial impact on the SS value, whilst the TtS value was sensitive to the interaction between PNSBsi vs. FPE and FPE vs. RT [Equation 6]. Overlaid contour plots were used to present interacting relationships between 2 factors by providing another factor at a mid-point for the determination of optimum conditions. The interacting relationship between PNSBsi and FPE when the treatment was 4 d had decreasing values of COD, SS and TtS, when the levels of PNSBsi and FPE were in a range of 7.2–10% and 0.75–0.94%, respectively (Fig. 2a). When 0.75% FPE was the midpoint, the interacting relationship between PNSBsi and RT showed that when both factors were in a range of 7.2–8.9% and 3.7–4.3 d, there was a significant decrease of COD, SS and TtS values (Fig. 2b). The interacting relationship between FPE and RT when PNSBsi was 6% showed that values of COD, SS and TtS decreased when the FPE and RT levels were 0.79–0.81% and 4.2–4.3 d (Fig. 2c). The overlaid contour plot (Fig. 2) was also used to indicate the region where the optimization routine was found. The optimum conditions were determined as 7% PNSBsi population, 0.81% FPE and 4.22 d for the RT. The predicted responses of COD, SS and TtS were 241, 22 and 6.66 mg L-1 (corresponded to 88, 61 and 51%, respectively) and these were close to actual values of run number 7 as COD, SS and TtS were 288, 24 and 6.89 mg L-1 with the conditions of 6% PNSBsi, 0.75% FPE and 4 d (Table 5). The results were considered to be adequate and acceptable as the desirability was 0.910. However, for practical work, the optimal conditions were adjusted to 7% PNSBsi, 0.8% FPE and 4 d of RT with little changes of FPE and RT in the verification test.

3.3. The role of FPE and PNSBsi to treat rubber sheet wastewater The wastewater treatment efficiencies of FPE, PNSBsi and combination of FPE and PNSBsi (verification test) in a non-sterile set under microaerobic light conditions over 4 d are shown in Table 6. A native control set without the addition of FPE and PNSBsi produced the least efficient reduction of COD, SS and TtS levels and in this set PNSB were not detected. However, the addition of 0.8% FPE stimulated PNSB growth to 7.3 log cfu mL-1 and provided a higher efficiency for the reduction of COD than the native control. On the other hand, with the addition of only 7% PNSBsi, 8.11 log cfu mL-1 were achieved, with the reduction percentages of 85 for COD, 63 for SS and 52 for TtS. The best treatment efficiency was observed in a verification test with the addition of both 0.8% FPE and 7% PNSBsi as the reduction percentages were then 91 for COD, 75 for SS and 61 for TtS. However, H2S was not detected in any set except that RAW had an initial H2S level of 10 mg L-1. The highest amount of HPC was found in the native control set followed by a FPE set and the least in sets with PNSBsi. No detection of LAB was observed in any set. According to the results above, the verification test produced significantly higher efficiency than the predicted values derived from CCD experiment for COD (88%), SS (61%) and TtS (51%).

Results of a parallel sterile set to treat sterile RAW over 4 d are presented in Table 7. The abiotic control showed little degradation in RAW; however, with the addition of 0.8% FPE, the amount of HPC was 4.51 log cfu mL-1. Therefore, higher efficiencies for reducing COD, SS and TtS were found, but the most marked decrease of these levels were found in sets of verified and only addition of PNSBsi. A 7% PNSBsi inoculum provided PNSBsi (7.93 log cfu mL-1) and HPC (4.19 log cfu mL-1), and this effectively reduced COD 84%, SS 51% and TtS 49%. There were significant differences for the amounts of HPC and PNSBsi between the PNSBsi set and a verified set. Therefore, the verifying set produced the highest efficiency by reducing COD, SS, and TtS at 88%, 70% and 56%, respectively. The results of H2S in this study indicated that this gas was produced by HPC from 4 mg L-1 to 15 mg L-1 in a FPE set and it was completely removed by PNSBsi. However, a loss of H2S was also observed in an abiotic control.

Table 6. Characteristics of the effluents from the non-sterile sets after treatment by FPE, PNSBsi inoculants and results of the verification test.

Parameter	Non sterile	Control	Addition	Addition	Verification test,
mg L^{-1a}	RAW	Native control	0.8% FPE	7% PNSBsi	0.8% FPE + 7% PNSBsi
	T = 0	D 4	D 4	D 4	D 4
pH	7.03 ± 0.01^{a}	7.46 ± 0.03^{d}	7.31 ± 0.03^{b}	7.39 ± 0.01^{c}	7.35 ± 0.02^{bc}
COD	2722 ± 11^{e}	1196 ± 11^{d} (56)	1033 ± 11^{c} (62)	414 ± 11^{b} (85)	239 ± 11^{a} (91)
SS	43 ± 3^{d}	32 ± 3^{c} (26)	29 ± 5^{c} (33)	16 ± 6^{b} (63)	11 ± 3^{a} (75)
TDS	547 ± 12^{c}	464 ± 12^{b}	433 ± 42^{ab}	367 ± 31^{a}	380 ± 20^{a}
Sulfate	5.1 ± 0.2^{d}	3.9 ± 0.1^{c}	2.7 ± 0.1^{b}	2.1 ± 0.1^{a}	2.0 ± 0.2^{a}
TtS	13.78 ± 0.38^{d}	8.44 ± 0.38^{c} (39)	7.67 ± 0.67^{bc} (44)	6.65 ± 0.77^{ab} (52)	5.33 ± 0.67^{a} (61)
DsS	12.89 ± 0.38^{d}	10.22 ± 0.38^{c}	6.22 ± 0.38^{b}	5.78 ± 0.38^{ab}	5.11 ± 0.38^{a}
UHS	5.67 ± 0.17^{d}	2.45 ± 0.09^{c}	1.74 ± 0.11^{b}	1.39 ± 0.09^{a}	1.23 ± 0.09^{a}
H_2S	10 ± 2^{b}	0^{a}	0^{a}	0^{a}	0^{a}
HPC	8.3 ± 0^{d}	7.83 ± 0.1^{c}	7.63 ± 0.1^{b}	7.09 ± 0^{a}	7.13 ± 0^{a}
LAB	0	0	0	0	0
PNSB	0^{a}	0^{a}	7.3 ± 0^{b}	8.11 ± 0^{c}	8.39 ± 0^{d}

Numbers in parentheses are reduction percentages.
Different letters in each row indicate the significant differences ($P < 0.05$).
TDS: total dissolved solids; TtS: total sulfide; DsS: dissolved sulfide; UHS: unionized hydrogen sulfide (H_2S in wastewater); HPC: heterotrophic plate count; LAB: lactic acid bacteria; PNSB: purple nonsulfur bacteria.
[a] Unless otherwise stated and unit for microbial population is log cfu mL^{-1}.

DISCUSSION

The role of FPE for Stimulating PNSB Growth

It is well recognized that PNSB prefer to grow under a low oxygen tension with light conditions [21] and in this study it was confirmed that the lowest ORP (-228 mV) strongly promoted the growth of PNSB (Table 1). This result was supported by Izu et al. [21] who reported that maximum PNSB ratios of up to 80% of the total microbes were obtained using non-aeration conditions with ORP values of less than -200 mV. The ORP in the sediments of paddy fields is in the range of -200 to -250 mV was found as the most suitable condition for enhancing PNSB growth [22]. COD, initial pH and FPE are independent variables each, having significant influences on the ORP value (Table 3). This is because both COD and FPE factors are a source of nutrient and optimal pH that could stimulate the growth of normal flora and the consequent depletion of oxygen (ORP decreased). Hence, anaerobic conditions under light stimulated indigenous PNSB growth so they became the dominant organism. No PNSB were detected in a control set, without FPE, while under light conditions, 2.0% FPE had enough nutrients to stimulate PNSB growth (Table 4). This was because the FPE consisted of 1.90% organic acids and other nutrients with an EC value of 3.51 mS cm-1. In contrast, lower nutrients in the control set without added FPE in the light was enough to support some algal growth. However, why would FPE stimulate PNSB but have little effect on the HPC? It is possible that the VFAs in FPE in both verified sets were used preferentially as electron donors for the photosynthesis in the partially anaerobic light conditions as a lower amount of VFAs was found in both sets when compared with the control set (Table 4). It was found that LAB was not detected in any set (Table 4). This might be a result of very low number of LAB in the stored FPE since they could not survive the high competition for growth substrates from the heterotrophs and PNSB in the RAW under light conditions.

Table 7. Characteristics of effluents from sterile sets after treatment by FPE, PNSBsi inoculum and the verification test

Parameter	Sterile	Abiotic	Addition	Addition	Verification test,
mg L^{-1a}	RAW	Control	0.8% FPE	7% PNSBsi	0.8% FPE + 7% PNSBsi
	T = 0	D 4	D 4	D 4	D 4
pH	7.06 ± 0.01^{c}	7.33 ± 0.02^{d}	6.85 ± 0.02^{a}	6.96 ± 0.02^{b}	7.73 ± 0.02^{e}
COD	2684 ± 11^{e}	2632 ± 11^{d} (2)	1558 ± 11^{c} (42)	443 ± 11^{b} (84)	326 ± 11^{a} (88)
SS	37 ± 6^{b}	35 ± 5^{b} (3)	33 ± 3^{b} (11)	18 ± 3^{a} (51)	11 ± 5^{a} (70)
TDS	467 ± 12^{b}	459 ± 21^{b}	447 ± 12^{b}	440 ± 20^{b}	387 ± 12^{a}
Sulfate	4.8 ± 0.1^{d}	4.7 ± 0.1^{d}	2.5 ± 0.1^{c}	2.1 ± 0.2^{b}	1.9 ± 0.1^{a}
TtS	13.11 ± 0.38^{c}	12.65 ± 0.38^{c} (3)	10.22 ± 0.38^{b} (22)	6.67 ± 0.67^{a} (49)	5.78 ± 0.38^{a} (56)
DsS	12.44 ± 1.02^{b}	12.22 ± 0.38^{b}	7.78 ± 0.38^{a}	7.33 ± 0.67^{a}	6.44 ± 0.77^{a}
UHS	4.73 ± 0.39^{d}	3.42 ± 0.11^{c}	4.13 ± 0.47^{cd}	1.76 ± 0.16^{b}	0.84 ± 0.10^{a}
H_2S	4 ± 1^{b}	0^{a}	15 ± 3^{c}	0^{a}	0^{a}
HPC (log cfu mL^{-1})	0^{a}	0^{a}	4.51 ± 0.03^{d}	4.19 ± 0.03^{c}	4.10 ± 0.06^{b}
LAB (log cfu mL^{-1})	0	0	0	0	0
PNSB (log cfu mL^{-1})	0^{a}	0^{a}	0^{a}	7.93 ± 0.05^{b}	8.10 ± 0.02^{c}

Different letters in each row indicate the significant differences ($P < 0.05$).
For identification of abbreviations see Table 6.
[a] Unless otherwise stated and numbers in parentheses are reduction percentages.

According to BBD, the optimal conditions for promoting PNSB growth were a COD of 2000 mg L^{-1}, initial pH 7 and 2% FPE (Fig. 1 and Table 4). This can be explained by the fact that PNSB are photoorganotrophs therefore they prefer to grow in a condition that has a relatively high organic matter [2, 5, 10]. The 2.0% FPE addition in this study provided an optimal concentration of organic matter for proliferation of PNSB because a lower organic content would facilitate the growth of cyanobacteria as previously described in the control set. In contrast, too much organic matter may promote the growth of HPC and repress the growth of PNSB [2]. It is generally known that a neutral pH supports the growth of most microorganisms including PNSB [2,23]. Therefore, it was not surprising that PNSB became the dominant organism due to the provision of the most suitable conditions provided for the growth of photoheterotrophs under microaerobic light condition, which quickly switched to anaerobic conditions with the addition of FPE (ORP = -228 mV) (Table 1).

The role of PNSBsi for Treating Rubber Sheet Wastewater

Efficiency of the rubber sheet wastewater treatment is dependent on PNSBsi, FPE and RT. The results demonstrated that RT had a significant effect on the reduction of COD and SS values whereas the PNSBsi strongly reduced TtS [Equation 4, Equation 5, Equation 6]. RT is one of the major factors, which controls the degradation of carbonaceous wastes as it represents the length of time the microbial cells are in contact with the substrate, which directly dictates the efficiency of wastewater treatment [24]. In this case, it ensures that the PNSBsi had enough time to consume most of the substrate and nutrients over the RT of 4 d. There was a significant reduction of TtS caused by the PNSBsi because some members of the PNSB can use sulfide or other reduced forms of sulfur as an extra source of an electron donor to support photosynthesis [5, 10]. COD value was also dependent on the interactive relationship between PNSBsi vs. FPE and PNSBsi vs. RT [Equation 4], this again illustrated that FPE stimulated PNSB growth as previously described. The SS values were significantly governed by the interactive relationship between PNSBsi vs. RT as their biomass increased with time after consuming nutrients [Equation 5]. However, the PNSBsi gave a higher efficiency to reduce SS value than the HPC (Table 6 and Table 7). The reason is that PNSB have a growth yield ($Y_{biomass/substrate\ (COD)}$) in a range of 0.28–0.45 [25, 26], while, usually, HPC growth yields are around 0.40–0.60 [27]. The interactive relationships between PNSBsi vs.

FPE significantly decreased TtS while the FPE vs. RT interaction increased TtS [Equation 6]. This suggested that PNSBsi removed TtS while the FPE was involved with producing TtS as it supported the growth of the HPC. This corresponded with the results in Table 7. In contrast, H2S was not detected in the abiotic control after 4 d, although it was observed at time zero at 4 mg L^{-1} (Table 7). This indicated that the loss of rotten gas odor may be caused by the change of pH from 7.06 to 7.33 because a higher pH promotes the conversion of sulfide into its HSand also some may be lost by precipitation with small amount of metal ions existed in the medium. According to the overlaid contour plot, the adjusted optimal conditions for treating RAW were 7% PNSBsi, 0.8% FPE and 4 d treatment time, and this condition was verified with non-sterile and sterile FPE, including controls (Table 6 and Table 7). Results of the native control show that the addition of FPE stimulated PNSB growth which is consistent with the results shown in Table 4. As there was little change of most of the measured parameters in the abiotic control set (Table 7). The efficiency to treat RAW in the sets of non-sterile and sterile with no/addition of FPE/PNSBsi appeared to be governed by the microbes and the efficiencies were in the following order: the verified sets N PNSBsi N FPE N native control. As only 0.8% FPE was added into the sterile RAW, this allowed HPC growth to increase to 4.51 log cfu mL^{-1} (Table 7) and this could also treat RAW, although their activity was significantly lower than the treatments with the PNSBsi inoculum (Table 6 and Table 7). Addition of PNSBsi as inoculum also included some HPC population (Table 7). The removal efficiency to treat RAW was positively related to the additions of FPE and PNSBsi into RAW, although some normal flora present in the RAW may have involved (Table 6 and Table 7). The verified set in this study (Table 6) had higher efficiencies (91% COD, 75% SS and 61% TtS) to treat RAW that had an initial COD of 2722 mg L^{-1} than those predicted values (88% COD, 61% SS 51% TtS), with an initial COD of 2024 mg L^{-1} (Table 5). A possible reason for that is that the different batches of the wastewater used generally show high variations in both biotic and abiotic properties. The results in Table 7 showed that the sterile verified set had the removal efficiency of 88% for COD, 70% for SS and 56% for TtS, all of which were close to the predicted values. This means that biotic component in non-sterile verified set also supported RAW treatment.

The PNSBsi inocula not only easily adapted themselves to their original source (RAW) but they were also able to compete with the normal flora for treating RAW. This was evidenced by their higher efficiencies in the non-sterile condition than those in the sterile condition (Table 6 and Table 7). The results indicated that the use of autochthonous bioaugmentation (re-inoculation) for treating rubber sheet wastewater was a practical technique to be used by farmers who desire to use the treated RAW for the preparation of PNSB inoculum. The inoculum can be carried out onsite with a simple procedure of adding the appropriate amount of FPE into a portion of wastewater in a shallow pond nearby the plant under light condition. This application can sustain the need for inoculation and support a long term operation of this community level wastewater treatment ponds. However, the amount of UHS remained in the effluent must be solved and the use of selected PNSB strains might be possible to solve the problem, and this will be further studied.

CONCLUDING REMARKS

This work showed that the role of FPE in the stimulation of PNSB growth under light condition was to lower ORP leading to the reducing environment in the culture. A new approach for the preparation of indigenous PNSBsi inoculum from rubber sheet wastewater with the addition of optimal amount of FPE under microaerobic/anaerobic light conditions was developed. In addition, autochthonous bioaugmentation using PNSBsi proved to be effective for treating the wastewater.

ACKNOWLEDGMENTS

We thank the CRSF at Pijit suburb, Songkhla, Thailand, for providing the wastewater for this study. Thanks to Dr Brian Hodgson for the assistance with the English language.

REFERENCES

1. Chaiprapat S, Sdoodee S. Effects of wastewater recycling from natural rubber smoked sheet production on economic crops in southern Thailand. Resour Conserv Recycl 2007;51:577–90. http://dx.doi.org/10.1016/j.resconrec.2006.11.003.
2. Kantachote D, Kornochalert N, Chaiprapat S. The use of the purple non sulfur bacterium isolate P1 and fermented pineapple extract to treat latex rubber sheet wastewater for possible use as irrigation water. Afr J Microbiol Res 2010;4:2296–308.
3. Yalamanchili C, Smith MD. Acute hydrogen sulfide toxicity due to sewer gas exposure. Am J Emerg Med 2008;26:518.e5-e7. http://dx.doi.org/10.1016/j.ajem.2007.08.025.
4. Doujaiji B, Al-Tawfiq JA. Hydrogen sulfide exposure in an adult male. Ann Saudi Med 2010;30:76–80.
5. [5] Kantachote D, Torpee S, Umsakul K. The potential use of anoxygenic phototrophic bacteria for treating latex rubber sheet wastewater. Electron J Biotechnol 2005;8:314–23. http://dx.doi.org/10.2225/vol8-issue3-fulltext-8.
6. Okubo Y, Futamata H, Hiraishi A. Characterization of phototrophic purple nonsulfur bacteria forming colored microbial mats in a swine wastewater ditch. Appl Environ Microbiol 2006;72:6225–33. http://dx.doi.org/10.1128/AEM.00796-06.
7. Liang CM, Hung CH, Hsu SC, Yeh IC. Purple nonsulfur bacteria diversity in activated sludge and its potential phosphorus-accumulating ability under different cultivation conditions. Appl Microbiol Biotechnol 2010;86:709–19. http://dx.doi.org/10.1007/s00253-009-2348-2.
8. Madukasi EI, Dai X, He C, Zhou J. Potentials of phototrophic bacteria in treating pharmaceutical wastewater. Int J Environ Sci Technol 2010;7:165–74. http://dx.doi.org/10.1007/BF03326128.
9. Veenstra S, Al-Nozaily FA, Alatres GJ. Purple non-sulfur bacteria and their influence on waste stabilisation pond performance in the Yemen Republic. Water Sci Technol 1995;31:141–9. http://dx.doi.org/10.1016/0273-1223(95)00501-D.
10. Kim MK, Choi KM, Yin CR, Lee KY, Im WT, Lim JH, et al. Odorous swine wastewater treatment by purple non-sulfur bacteria, Rhodopseudomonas palustris, isolated from eutrophicated ponds. Biotechnol Lett 2004;26:819–22. http://dx.doi.org/10.1023/B:BILE.0000025884.50198.67.
11. Shi XY, Yu HQ. Conversion of individual and mixed volatile fatty acids to hydrogen by Rhodopseudomonas capsulata. Int Biodeterior Biodegrad 2006;58:82–8. http://dx.doi.org/10.1016/j.ibiod.2006.07.004.
12. Chen CY, Lu WB, Liu CH, Chang JS. Improved phototrophic H2 production with Rhodopseudomonas palustris WP3-5 using acetate and butyrate as dual carbon substrates. Bioresour Technol 2008;99:3609–16. http://dx.doi.org/10.1016/j.biortech.2007.07.037.
13. [13] Bezerra MA, Santelli RE, Oliveira EP, Villar LS, Escaleira LA. Response surface methodology (RSM) as a tool for optimization in analytical chemistry. Talanta 2008;76:965–77. http://dx.doi.org/10.1016/j.talanta.2008.05.019.
14. Bas D, Boyaci IH. Modeling and optimization I: Usability of response surface methodology. J Food Eng 2007;78:836–45. http://dx.doi.org/10.1016/j.jfoodeng.2005.11.024.
15. Ahmadi M, Vahabzadeh F, Bonakdarpour B, Mofarrah E, Mehranian M. Application of the central composite design and response surface methodology to the advanced treatment of olive oil processing wastewater using Fenton's peroxidation. J Hazard Mater 2005;123:187–95. http://dx.doi.org/10.1016/j.jhazmat.2005.03.042.
16. Wang G, Mu Y, Yu HQ. Response surface analysis to evaluate the influence of pH, temperature and substrate concentration on the acidogenesis of sucrose-rich wastewater. Biochem Eng J 2005;23:175–84. http://dx.doi.org/10.1016/j.bej.2005.01.002.
17. APHA, AWWA, WPCF. Standard method for examination of water and wastewater. 20th ed. Washington D.C.: American Public Health Association; 1998.

18. Yang MH, Choong YM. A rapid gas chromatographic method for direct determination of short-chain (C2–C12) volatile organic acids in foods. Food Chem 2001;75:101–8. http://dx.doi.org/10.1016/S0308-8146(01)00211-4.

19. Kantachote D, Kowpong K, Charernjiratrakul W, Pengnoo A. Microbial succession in a fermenting of wild forest noni (Morinda coreia Ham) fruit plus molasses and its role in producing a liquid fertilizer. Electron J Biotechnol 2009;12:1–11. http://dx.doi.org/10.2225/vol12-issue3-full12.

20. Annadurai G, Sheeja RY. Use of Box–Behnken design of experiments for the adsorption of experiments for the adsorption of verofix red using biopolymer. Bioprocess Eng 1998;18:463–6. http://dx.doi.org/10.1007/s004490050472.

21. Izu K, Nakajima F, Yamamoto K, Kurisu F. Aeration conditions affecting growth of purple nonsulfur bacteria in an organic wastewater treatment process. Syst Appl Microbiol 2001;24:294–302. http://dx.doi.org/10.1078/0723-2020-00027.

22. Overmann J, Garcia PF. The phototrophic way of life. The prokaryotes: Ecophysiology and biochemistry., Singapore: Springer; 2006. p. 32–85.

23. Panwichian S, Kantachote D, Wittayaweerasak B, Mallavarapu M. Isolation of purple nonsulfur bacteria for the removal of heavy metals and sodium from contaminated shrimp ponds. Electron J Biotechnol 2010;13:1–12

24. Bitton G. Wastewater microbiology. 3rd ed. Chichester, New York, USA: John Wiley and Sons; 2005.

25. He J, Zhang G, Lu H. Treatment of soybean wastewater by a wild strain Rhodobacter sphaeroides and to produce protein under natural conditions. Front Environ Sci Eng China 2010;4:334–9. http://dx.doi.org/10.1007/s11783-010-0239-5.

26. Lu H, Zhang G, Wan T, Lu Y. Influences of light and oxygen conditions on photosynthetic bacteria macromolecule degradation: Different metabolic pathways. Bioresour Technol 2011;102:9503–8. http://dx.doi.org/10.1016/j.biortech.2011.07.114. [

27. Heijnen JJ, Roels JA. A macroscopic model describing yield and maintenance relationships in aerobic fermentation. Biotechnol Bioeng 1981;23:739–63. http://dx.doi.org/10.1002/bit.260230407.

CHAPTER 12

FATE, BEHAVIOUR, AND IMPLICATIONS OF ZNO NANOPARTICLES IN A SIMULATED WASTEWATER TREATMENT PLANT

EFC Chaúque; JN Zvimba; JC Ngilal; N Musee

Keywords: Wastewater, activated sludge, nanoparticles, zinc oxide, OECD 303A

INTRODUCTION

Biological wastewater treatment processes employ a consortium of heterotrophic and autotrophic bacteria, essentially to degrade organic matter present in wastewater. Generally, the activated sludge process treats biodegradable organic material in domestic sewage as well as effluents from other sources such as pulp and paper mills, food industries, abattoirs, textile mills, edible oils, coal gasification wastes, petrochemical wastes, and oil refinery wastes (Henze et al., 2002; Metcalf and Eddy, 2002). The sorption of pollutants on activated sludge is among the fundamental processes for the removal of toxic substances including metals, synthetic organic chemicals, suspended solids, and pathogens in wastewater (Dobbs et al., 1989; Musee et al., 2007; Sheng et al., 2008). However, the sorption process may be ineffective as the bacteria used for wastewater treatment can be inhibited by toxic substances, e.g., heavy metals (Çeçen et al., 2010), thus adversely impacting the biologically-based treatment processes. The advent of nanotechnology has resulted in fast production and wide usage of engineered nanoparticles (ENPs) in consumer products and industrial applications. This has led to unintended release of ENPs into environmental systems at different stages of their product life cycles (e.g. manufacturing, use, etc.) (Koehler et al., 2008; Musee, 2011). For instance, zinc oxide (ZnO) ENPs are incorporated in numerous products including sunscreens, paints, cosmetics, dye-synthesized cells, plastic additives, catalysts and electronics (Woodrow Wilson, 2009; BCC Research, 2012; Piccinno et al., 2012); which in turn could lead to their release into natural and technical systems, e.g., wastewater treatment plants (WWTPs).

For example, ENPs released into wastewater from various sources eventually enter WWTPs (Kiser et al., 2009), and currently ENPs are considered among the rapidly emerging contaminants in municipal wastewater systems (Brar et al., 2010; Musee et al., 2011). Therefore, WWTPs are likely potential major point sources of ENP release into the environment, due to broad-scale and increasing usage of nanoproducts (Woodrow Wilson, 2015; BCC Research, 2012; Piccinno et al., 2012). Among the likely pathways of ENP release into the environment from WWTPs are water systems, soils, and air, through treated effluent, bio-solids, and plant-generated aerosols, respectively, (Limbach et al., 2008; Kiser et al., 2010; Musee, 2011; Westerhoff et al., 2011).

The effects of ZnO ENPs on aquatic organisms such as bacteria in different media have been investigated, and the

Citation: EFC Chaúque; JN Zvimba; JC NgilaI; N Musee, Fate, behaviour, and implications of ZnO nanoparticles in a simulated wastewater treatment plant, http://dx.doi.org/10.4314/wsa.v42i1.09.

observed toxicity was attributed to either soluble (dissolved) Zn^{2+} species, ZnO particulates, or both forms (Jiang et al., 2009; Wong et al., 2010; Thwala et al., 2013). Padmavathy and Vijayaraghavan (2008), using the disk diffusion method, and Premanathan et al. (2011), using the resazurin incorporation method, found the minimum inhibitory concentration (MIC) of nano ZnO against *Escherichia coli* to be 400 and 500 μg/mL, respectively. Moreover, Premanathan et al. (2011) indicated that the inhibitory activity was due to the generation of reactive oxygen species (ROS) which, in turn, induced apoptosis. However, a study by Liu et al. (2011) indicated that ZnO ENP toxicity to the biological populations in activated sludge (i.e. endogenous respiration, BOD biodegradation and nitrification) was solely due to soluble Zn^{2+} generated upon ZnO ENP dissolution. This finding points to the role of ENP dissolution for the generation of toxic effects on activated sludge bacterial communities. However, Musee and co-workers (2014) reported limited dissolution and impact of ZnO ENPs on bacterial viability under typical wastewater conditions.

In aqueous environments, natural organic matter (NOM) and other biomass can adsorb onto ENP surfaces (Zhang et al., 2009) and, in turn, enhance their stability, even at high ionic strength (Keller et al., 2010; Zhou and Keller, 2010). It has been reported that, for example, typical environmental concentrations of humic acids are sufficiently high to stabilize even high ZnO ENP concentrations, of up to 100 mg/L (Omar et al., 2014). However, the impacts of metal oxide ENPs on the treatment efficiency of WWTPs are largely unknown, and therefore, not adequately quantified (Brar et al., 2010). Moreover, there is evident discord in the published literature regarding the fate and behaviour of ZnO ENPs (Musee et al., 2014) as well as how this influences their toxicity (Liu et al., 2011).

The main goal of this study was to elucidate the fate and behaviour of ZnO ENPs in a simulated WWTP, developed as prescribed by the Organization for Economic Co-operation and Development (OECD) guideline 303A (OECD 303A, 2001). To achieve this goal, three specific objectives were set:

- To establish if transformation of ZnO ENPs occurred during the wastewater treatment process through characterization of ENP physicochemical parameters - e.g., morphology, formation of mineral phases, and association with organic matter - using XRD, transmission electron microscopy (TEM), and ultraviolet and visible spectroscopy (UV-Vis), before and after ZnO ENPs were dosed into influent wastewater
- Quantify the effect of ZnO ENPs on wastewater treatment processes by monitoring COD removal following the introduction of ENPs
- Through objectives (i) to (ii), establish the efficiency of the OECD WWTP model in the removal of COD and organic matter

MATERIALS AND METHODS

Collection and characterization of wastewater

Raw wastewater was withdrawn from a WWTP in Johannesburg (Johannesburg Water, Northern Wastewater Treatment Works; Gauteng, South Africa) that mainly collects and treats domestic sewage. The wastewater was stored at 4°C before use. Wastewater physical and chemical characteristics, namely, dissolved oxygen (DO) and chemical oxygen demand (COD) were measured using an oxygen meter (HANNA Instruments, Portugal), and a photometer (HACH, DR 3900, USA), respectively. The pH and conductivity were determined using pH meter and conductivity meters (METTLER TOLEDO Technologies, USA). All wastewater parameters were characterized as per the Standard Methods for the Examination of Water and Wastewater (Eaton et al., 2005). The wastewater physicochemical characteristics were continuously monitored (every 12 h for 77 d) in the influent as well as aeration chamber and effluent, before and after the ZnO ENPs were dosed into the simulated WWTP.

Zinc Oxide Nanoparticle Suspensions

Non-coated ZnO ENPs (#544906) were obtained from Sigma (Johannesburg, South Africa). TEM (JEOL Model JEM -2100F, Japan) and XRD (Philips, X'Pert PRO MPD, mineral powder diffraction) analyses of ZnO ENPs were carried out to establish their morphology and chemical state, respectively. XRD used to analyse ZnO ENP morphology and size in the sludge was equipped with monochromatic Cu Ka radiation (λ = 0.15406 nm). Diffraction patterns were recorded in the 2θ angular range from 10-80° with step sizes of 0.02°, at 40 kV and 40 A. Additional solid state analysis of the sludge was done using UV-Vis spectrophotometry (UV-2450 UV-Vis spectrophotometer, Shimadzu, Japan) coupled with UV-Probe and the patterns were collected for the 250 to 700 nm electromagnetic spectrum region using $BaSO_4$ as the reference material.

Three suspensions of ZnO ENPs with concentrations of 5, 10, and 20 mg/L were prepared by adding 0.1250, 0.2500, and 0.5000 g, respectively, of ZnO nanopowder to 1 L deionized water. Using a sonicator (Model 2000U, Ultrasonic Power Corp.) the suspensions were sonicated at 20 KHz for 30 min to break the aggregates before each suspension was added to the wastewater in the holding container. Each set concentration (5, 10, or 20 mg/L) was achieved by making the volume to 25 L using wastewater. Ice was added into the sonicator to minimize the possibility of ZnO ENP dissolution due to heat generated during the sonication process.

Simulated Activated Sludge Wastewater Treatment Plant Setup

The simulated activated sludge wastewater treatment plant (SAS WWTP) was constructed following the Organization for Economic Co-operation and Development (OECD 303A, 2001) specifications. The tests were conducted in aerobic digesters each designed to hold 3 L of activated sludge. Each model unit comprised of: influent holding container (25 L), a stirred and aerated tank reactor (aeration chamber), and a clarifier (settling vessel) simulating biological treatment using an activated sludge system (see supporting information in Appendix 1: Figs A1 and A2; schematic diagram and photograph of the simulated wastewater treatment plant), respectively.

Aeration Chambers

Each aeration chamber (3 L) was continuously stirred using IKA RW 16 basic stirrer to ensure thorough mixing of the substrate, essentially to mimic actual WWTP operational conditions. The aeration chamber was aerated at a flow rate of 0.29 L/min to maintain the dissolved oxygen above 2 mg/L using compressed air through a glass frit. In this study, 2 model units were used as test and control units. Both units (test and control) were fed with wastewater spiked with ZnO ENPs using Watson-Marlow (Falmouth, Cornwall, UK) 120S/DV pumps at 29 r/min. Deionized water was used as the control exposure media to offer a baseline comparison with the complex wastewater, and particularly to account for the ENP removal mechanism(s).

For the test unit, the influent wastewater was spiked with ZnO ENPs (at varied concentrations of: 5, 10, 20 mg/L) in 25 L containers and continuously stirred using IKA RW 20 digital stirrer at 1 800 r/min to keep the ENPs well dispersed in suspension. The use of ZnO ENPs in suspension was an attempt to mimic their actual pathway introduction into the WWTPs, and also to enhance reproducibility of the test conditions. Settling vessels of 1.5 L were used to separate treated effluent from the activated sludge. In accordance to standard practice, a portion of the sludge from the settling vessel was re-introduced as return activated sludge (RAS) into the aeration chamber to replenish the biomass, and to maintain the total suspended solids (TSS) in the range of 2 to 3 g/L of dry sludge. The aeration chamber with hydraulic retention time (HRT) of 6 h was designed to maintain nitrifying conditions, and the influent was introduced at feed flow rate of 0.50 L/h using peristaltic pumps. The study lasted for 77 d to establish the long-term effects of ZnO ENPs on the removal of organic matter from WWTPs, where COD removal was monitored continuously as a surrogate parameter, as previously reported (Musee et al., 2014).

Simulated Activated Sludge WWTP Operation

Using 425μm stainless steel mesh the raw wastewater was filtered before it was pumped into the operating units to remove big particulates, and to avoid clogging the tubing system. This operation mimicked the screening process of raw wastewater used in large-scale WWTPs. The simulated WWTP was acclimatized, stabilized, and optimized during the first 14 days followed by dosing with ZnO ENPs over 21 days for each concentration (5, 10, or 20 mg/L). Continuous sampling was carried out at the influent holding container, aeration chamber, settling vessel, and waste activated sludge chamber in order to monitor various wastewater physicochemical parameters. The effluent samples from the aeration chamber were filtered using 0.45 μm polyvinylidene difluoride (PVDF) syringe filter before performing COD and zinc analyses. On the other hand, the analysed influent samples were collected directly from the influent container. A similar procedure was followed in operating the control unit and analysing samples for the COD.

Determination of ZnO ENP Dissolution

To determine the dissolution of ZnO ENPs in wastewater, the influent and effluent wastewater were continuously collected after every 12 h for 77 d from the set up and acidified with 10 M HNO_3 to pH < 2, before being stored at 4°C until analysis. Additionally, the sludge was also continuously withdrawn, filtered (Whatman 41 filter paper), dried in the oven at 103-105°C, ground with mortar and pestle, and then stored in dry powder form before analysis. In this study, the digestion method used followed the procedure of Martin et al. (1994).

Zinc dissolution was evaluated using inductively coupled plasma optical emission spectrometry (ICP-OES) (SPECTRO ARCOS, Analytical Instruments GmbH, Germany) by measuring the concentration of zinc in the raw wastewater. First, the background zinc concentration in the blank solution was determined and subtracted from the concentration of Zn^{2+} in the wastewater dosed with ZnO ENP at 3 different concentrations. The ICP-OES was operated under forward power of 1 400 W, plasma argon flow rate of 13 L/ min, auxiliary argon flow rate of 2.00 L/min, and nebulizer argon flow rate of 0.95 L/min.

Aliquots from the test and control units were collected periodically to monitor the ZnO ENPs dissolution with time. The influence of ZnO ENP concentration on the wastewater treatment process was investigated as a function of COD removal efficiency.

The organic matter removal in the SAS WWTP was assessed through monitoring of COD removal after dosing ZnO ENP into the test unit containing influent wastewater and the control unit containing deionized water. Four ZnO ENP concentrations values were used in the study: 0, 5, 10, and 20 mg/L.

RESULTS AND DISCUSSION

Simulated Activated Sludge Wastewater Treatment Plant Performance

Physical Parameters of Commercial ZnO ENPs

The raw wastewater characteristics were monitored over time and yielded the following results: total dissolved solids (TDS) 344 ± 43 mg/L, pH 7 ± 0.4, chemical oxygen demand (COD) 543 ± 159 mg/L, conductivity 715 ± 52 μS/cm, salinity 0.3 ± 0.1 psu, and total suspended solids (TSS) 588 ± 39 mg/L. DO in the raw wastewater was 4 ± 0.4 mg/L, and was maintained above 2 mg/L in the aeration chamber throughout the experiment using compressed air to provide the bacteria with adequate oxygen supply.

Zinc oxide ENP Suspensions

XRD results for commercial ZnO ENPs (Fig. 1) confirmed that they were in pure phase, as evidenced by narrow spectral peaks with high intensity and hexagonal Wurtzite crystal structure. TEM images showed a size distribution for the ENPs in the range 10 to 130 nm (Fig. 2), consisting of a heterogeneous mixture of rods, cubes, regular, and irregular spheres.

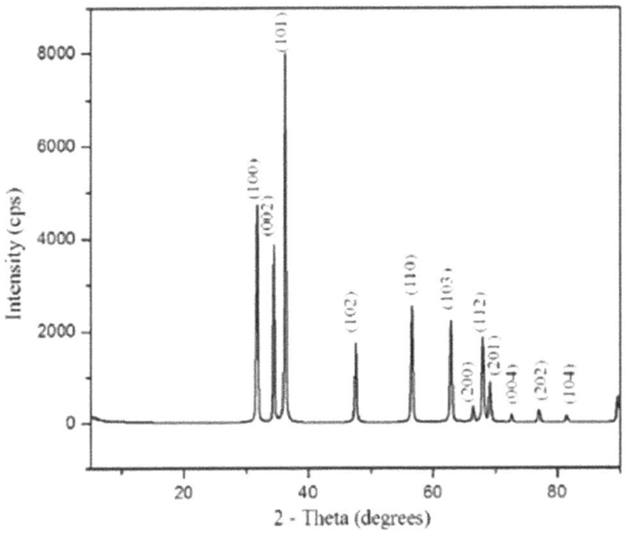

Figure 1. XRD pattern of ZnO ENPs (as purchased) prior to dosing into the effluent wastewater

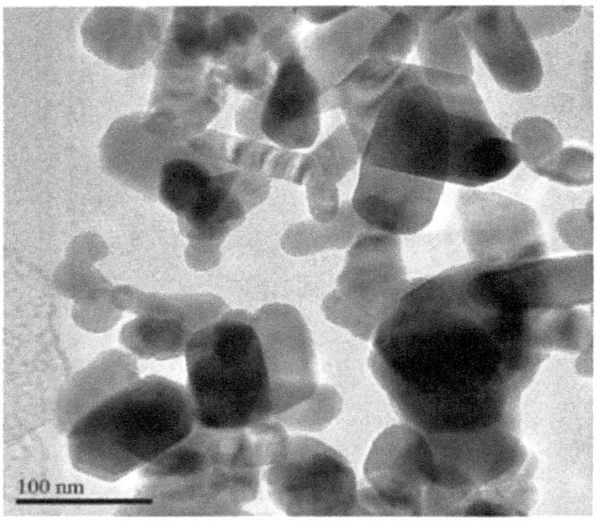

Figure 2. Transmission electron microscope (TEM) image of ZnO ENPs as purchased

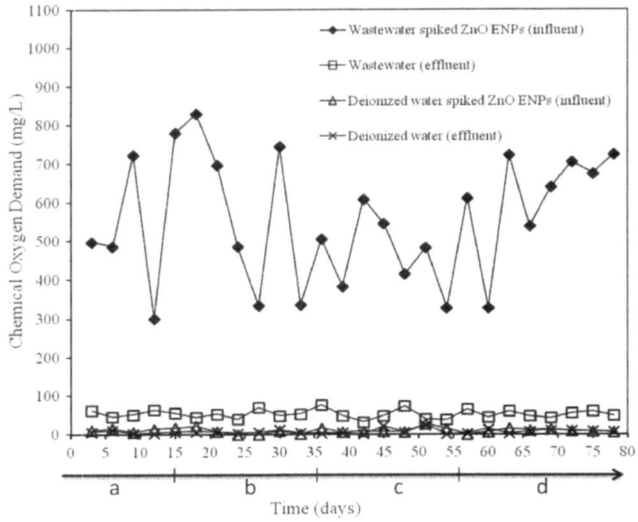

Figure 3. COD analysis for influent and effluent in the simulated wastewater treatment plant (test and control units) with continuous dosage of: a) 0 mg/L (bare sludge), b) 5 mg/L, c) 10 mg/L and d) 20 mg/L of ZnO ENPs

Effect of ZnO ENPs on COD Removal

As observed in Fig. 3, organic matter removal reached a steady state at about 50 mg/L COD in the effluent, irrespective of its initial concentration in the influent and ZnO ENP dosing concentration. The results in Fig. 3further indicate efficient organic matter removal with the average COD in the effluent ranging from 60-100 mg/L. Therefore, our results point to efficient removal of ZnO ENPs in the secondary treatment stage of WWTPs (cf. Sonune and Ghate, 2004). The average removal efficiency of COD observed was 91% in the test chamber; hence the simulated WWTP met the validity criteria as outlined in the OECD 303 A guidelines (OECD 303A, 2001). The COD removal results also suggest that dosing the influent with ZnO ENPs had no significant adverse impacts on the overall treatment efficiency of the stimulated WWTP, which is in agreement with earlier studies (Hou et al., 2013; Tan et al., 2015), especially at low dosing concentrations.

In light of our results and currently predicted relevant environmental concentrations of ZnO ENPs of around 24-300 µg/ L, where removal efficiency was assumed to be about 93% (Sun et al., 2014), it is unlikely that ZnO nanoparticles will impact on COD removal in WWTPs. However, as the production and use of ZnO ENPs increases their environmental concentrations will also increase with the possibility of causing potentially adverse effects. For example, Musee et al. (2014) reported negative impacts of ZnO ENPs at higher concentrations (100 mg/L): a 3-L bioreactor (simulated WWTP) with hydraulic residence time (HRT) of 6 h at a dose rate of 0.83 mg/min for 240 h yielded a mean COD removal efficiency of 71 ± 7%, which was much lower than that for the control (80 ± 5%). Similar inhibition of COD removal due to ZnO ENPs has also been reported for a batch membrane bioreactor, especially at high dose concentrations (Huang et al., 2013).

Notably, in the present study, though the simulated WWTP was exposed to high concentrations of ZnO ENPs, of up to 20 mg/L (levels unlikely to be found in actual treatment systems), the COD removal remained high. On the other hand, earlier toxicity studies showed that 10 mg/L of ZnO ENPs could induce significant growth inhibition of up to 90% and 20%, respectively, for *Bacillus subtilis* and *E. coli* (Adams et al., 2006). Thus, a number of reasons may account for high COD removal as the activated sludge in WWTPs entails large consortia of bacteria species; therefore, the potential impact of ZnO NPs on activated sludge based on the toxicity of ZnO NPs for pure bacteria is inadequate to account for these results.

Firstly, following ZnO ENP dissolution, cationic species could form complexes with natural organic matter, such as humic acid which is ubiquitous in wastewater and contains functional groups such as carboxylic (-COO^-) and phenolic (-ArO^-). These functional groups are known for their high complexation capacity with metal ions (Zhang et al., 2009). Thus, humic acid may cause the removal of zinc through processes such as: complexation, precipitation of zinc (e.g. through formation of insoluble $Zn(OH)^+$), and/or sorption of particulate ZnO ENPs (Applerot et al., 2009; Musee et al., 2014). In addition, neither ionic nor pure particulate forms of ZnO ENPs were present at high enough levels in the wastewater to cause adverse impacts by altering the microbial function in COD removal.

Secondly, ZnO ENPs may have agglomerated after being introduced into the influent as the dispersion of ENPs in wastewater is influenced by solution chemistry parameters, such as pH, ionic strength, and dissolved organic carbon. Kang et al. (2009) showed that higher conductivity and divalent cation concentrations (e.g., Mg^{2+} and Ca^{2+}) in wastewater effluents caused the ENPs be more aggregated than in freshwater systems, thereby influencing their plausible interactions with microbial populations. In this study, high conductivity of wastewater was observed. Agglomeration may have led to a drastic reduction in the bioavailable forms of ENPs (ionic and/ or particulate forms) essential to influence COD removal. For instance, it is likely that once the ENPs underwent agglomeration several processes could result, namely: (i) reduction in surface area essential for biointerfaces with bacteria (as evidenced by the increasing hydrodynamic size), (ii) reduction in the number of particles, and (iii) reduction in interfacial free energy (Oberdörster et al., 2005; Pettibone et al., 2008; Lowry et al., 2012). However, it should be noted that, to date, it is unclear as to the degree to which each of these potential mechanisms (i-iii) accounts for the diminished impact of ZnO ENPs on COD removal. Thus the mechanisms that may explain why the ZnO ENPs had minimal impact on the removal of COD in this study merit further investigation.

ZnO ENP removal and Plausible Mechanisms During SAS WWTP Processes

Residual zinc concentration in the effluent was 50-150 µg/L at 5 mg/L dosing concentration of ZnO ENPs (Fig. 4a). In the deionized water (control unit) residual Zn^{2+} was found to be approximately 6 times higher (~ 1 mg/L; Fig. 4b) than in the test unit. Dissolution of ENPs is dependent on the media chemistry (e.g. pH, organic matter, ionic strength, etc.) and inherent physicochemical properties (size, coating, surface chemistry, etc.) (Xia et al., 2008; Bian et al., 2011; Peralta-Videa et al., 2011).

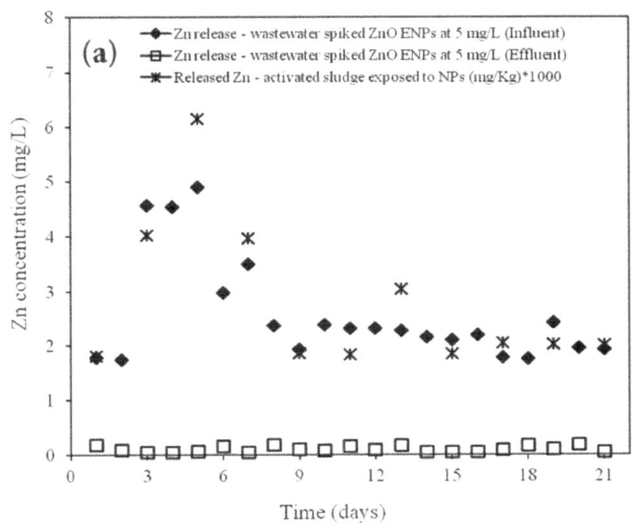

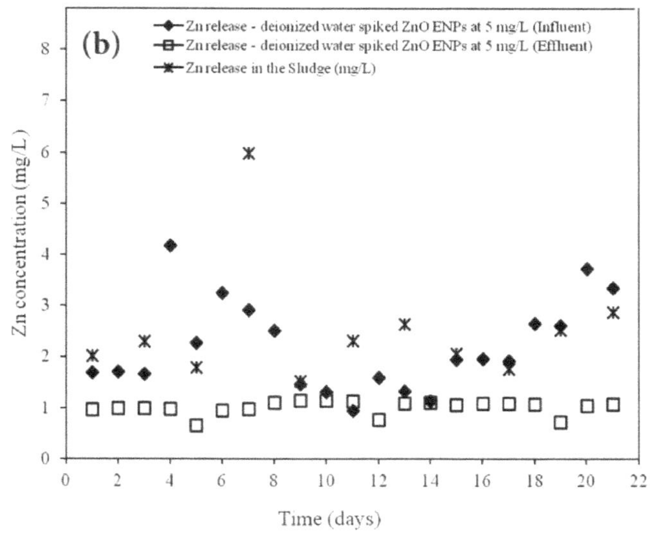

Figure 4. Effect of zinc release as a function of time after a dosage of 5 mg/L ZnO ENPs in: (a) wastewater and (b) deionized water

Results indicate that the organic matter played an important role in accounting for the significant differences (based on ANOVA statistical test at $p = 0.05$) in the amounts of zinc released by both the test and control media. As an example, the organic matter may have inhibited the chemical and physical 'speciation' of the ZnO ENPs through complexation processes and, in turn, led to a limited release of ionic species. In addition, the organic matter, through binding processes, may have adsorbed the solubilized metal species, causing a reduction in the quantities of metal ions observed in the effluent relative to the influent. Such scenarios point to different removal pathways of ZnO ENPs from the influent wastewater and the deionized water. For instance, as deionized water is devoid of organic matter, the ENP removal mechanism was possibly governed by other forms of abiotic factors, namely, pH and ionic strength, as adsorption was not feasible.

Effect of Concentration on ZnO ENP Removal

High levels of ENP removal from the wastewater could not be wholly linked to the influence of pH and ionic strength, although earlier findings have indicated that ZnO ENP stability is dependent on pH and ionic strength of wastewater (Chaúque et al., 2014). Additional influencing factors on ENP stability in aqueous media include natural organic matter (NOM). NOM is composed of humic substances (Becker et al., 2004), highly ubiquitous in wastewater, and offers a large surface area for contaminant adsorption (Metcalf and Eddy, 2002). Furthermore, NOM has been shown to enhance ENP stability in wastewater (Keller et al., 2010; Mu and Chen, 2011), and is therefore a significant factor influencing their removal in wastewater (Kiser et al., 2010).

For instance, NOM is known to stabilize ZnO ENPs (Bian et al., 2011) and, in turn, retard the dissolution rate. For ZnO ENPs in wastewater, it has been suggested that the retardation of dissolution was due to the hydrophobic nature of capric/ caprylic triglyceride dispersion formation (Lombi et al., 2012). Moreover, Liu et al. (2011) observed low zinc releases from wastewater due to the adsorption of Zn^{2+} ions onto the activated sludge immediately after their release from ZnO ENPs. To date, it is unclear which of the two mechanisms dominantly influences the fate and behaviour of ENPs as contaminants in wastewater, and/or whether they act concurrently in an antagonistic or synergistic fashion.

In this study, considering that the activated sludge pH was at 7 ± 0.4, the biomass was kept in the range of 2 000-3 000 mg/L of TSS - which is within the limits of typical WWTPs with biomass concentrations of 1 000-5 000 mg/L of TSS (Metcalf and Eddy, 2002) - and an average sludge retention time (SRT) of 6 d, provided favourable conditions for enhanced stabilization of ZnO ENPs by the NOM. The most plausible explanation is that the NOM played a dominant role in removing zinc ions from the influent wastewater. This is because humic acid is known to stabilize ZnO ENPs even at high concentrations (Omar et al., 2014), in addition to its ability to counteract the influence of high ionic strength, expressed as salinity (0.3 ± 0.1 psu) for the wastewater investigated in this study.

Low zinc concentrations (less than 3%) in the effluent, as shown in Fig. 4a, yielded similar observations to those of other researchers; e.g. Bolyard et al. (2013) and Mu and Chen (2011). These workers reported zinc release into the treated effluent as less than 2%, which accounted for the large percentages of zinc found in the activated sludge. Our findings suggest that the digestion process in the test chamber was responsible for the high percentage removal of ZnO ENPs, as confirmed by the observed low concentration of Zn^{2+} in the effluent. Furthermore, our results indicate that the release of zinc ions into the treated effluent exhibited a linear increase as the dosing ZnO ENPs concentration increased. Nevertheless, the Zn^{2+} released into the effluent was not high enough to adversely impact the organic matter removal, as ZnO ENP concentration was increased from 5 mg/L to 20 mg/L (Figs 5 and 6).

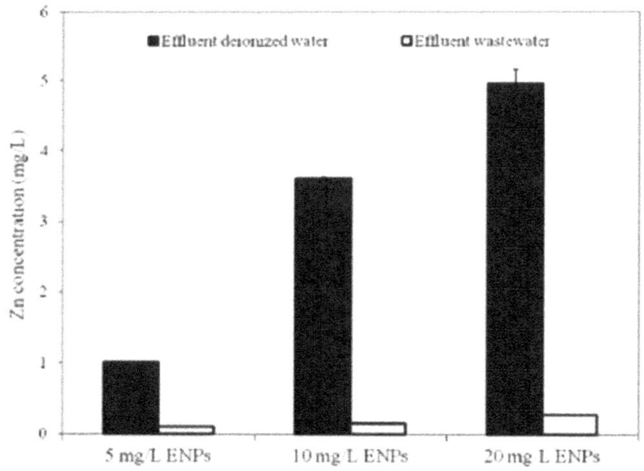

Figure 5. Comparitive zinc release in the effluent for the control (deionized water) and test (wastewater) after dosing with 5, 10, and 20 mg/L of ZnO ENPs

Spiking the influent with increasing concentrations of ZnO ENPs showed a substantial increase in the released Zn^{2+}. In this study, lower dosing concentrations exhibited higher ZnO NP dissolution compared to higher concentrations. The average concentrations of the released zinc ions were, respectively, 0.105 (2.1% of nominal ZnO), 0.150 (1.52%), and 0.270 (1.35%) mg/L at ZnO NPs concentrations of 5, 10, and 20 mg/L (Fig. 5). Our results were similar to those reported elsewhere for ZnO ENPs, e.g. Mu and Chen (2011), Xiong et al. (2011), and Thwala et al. (2013), whereas the dosing concentration increase led to enhanced aggregation (Maximova and Dahl, 2006), which is expected because of a higher probability of particle collisions - which ultimately leads to a decline in the dissolution rate (Xiong et al., 2011; Thwala et al., 2013).

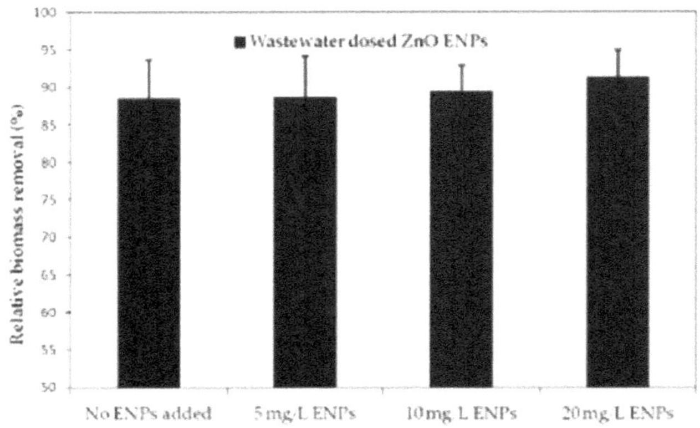

Figure 6. Biomass removal during the wastewater treatment exposed to different ZnO ENPs concentrations

Figure 6 shows biomass removal during wastewater treatment at different ZnO ENP dosing concentrations. The change in percentage biomass removal as the concentration of ZnO ENPs increased was found to be minimal. We had postulated that the ZnO ENP adsorption to the activated sludge would largely account for the removal mechanism. However, this does not appear to entirely account for these results. Our findings are, in some cases, in close agreement with that reported in the literature by various workers, while in other cases they differ. A few examples of studies related to biomass removal and release of Zn^{2+} ions into the treated effluent are discussed here to illustrate this observation:

Kiser et al. (2009) reported the sorption of titanium from TiO_2 ENPs by biomass during wastewater treatment; however, 10-100 μg/L Ti remained in the effluent. Limbach and colleagues (2008) observed the presence of 2-5 mg/L of cerium oxide in the effluent following exposure to 100 mg/L of CeO_2 ENPs in a model wastewater treatment process. Mu and Chen (2011) illustrated that a large amount of ZnO ENPs were adsorbed on the surface of activated sludge after a long-term exposure. Moreover, Liu and colleagues (2011) indicated that both the zinc ions and ZnO ENPs were adsorbed on activated sludge.

Studies by Klaine et al. (2008), Keller et al., (2010) and Bian et al. (2011) also showed that the removal of ENPs from aqueous media was dictated by chemical composition of the testing media chemistry (e.g. pH, organic matter, and ionic strength). Keller and co-workers (2010) reported enhanced stability of metal-oxides such as ZnO, TiO_2 and CeO_2 ENPs with organic matter adsorption to the particle surface providing a form of barrier to aggregation.

Analysis of ZnO ENPs in the sludge

In the previous section, results indicated that a large amount of the ZnO ENPs spiked into the influent were adsorbed onto the activated sludge, and removed from the effluent

(Figs 4a and 5). Therefore, only a small fraction of the ENPs was likely to be released into the treated effluent. However, the second route for the transport of the ZnO ENPs into the environment is via the sludge. The characteristics of ENPs in the sludge were therefore analysed using XRD (Fig. 7), and UV-Vis spectrophotometer coupled with UV-Probe (Fig. 8).

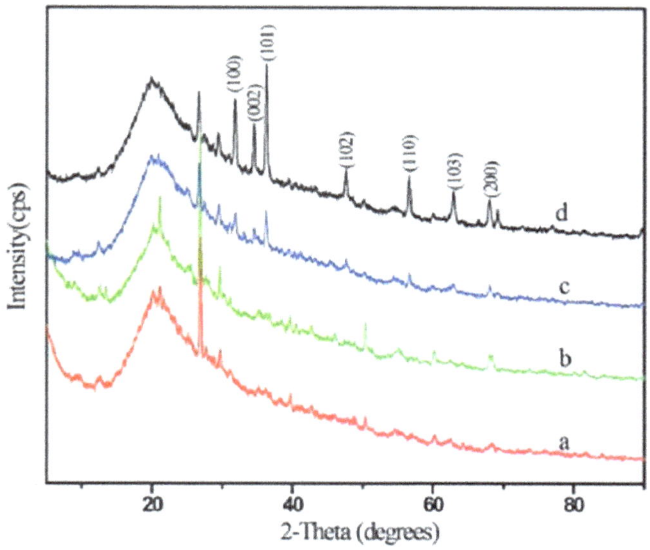

Figure 7. XRD for the ZnO ENPs in activated sludge after dosing influent at concentrations of (a) 0 mg/L (bare sludge), (b) 5 mg/L, (c) 10 mg/L, and (d) 20 mg/L

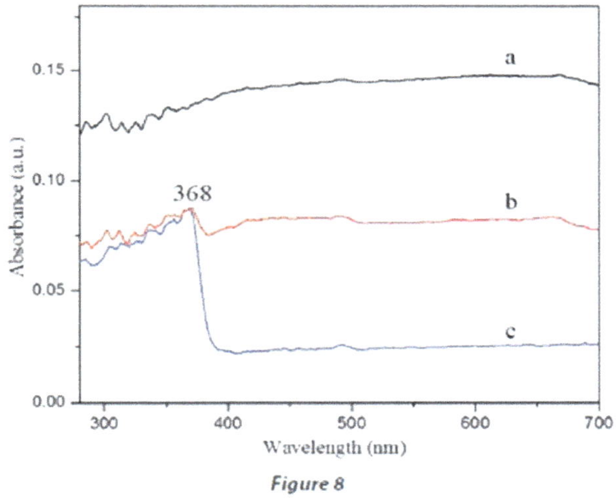

Figure 8. UV-Vis absorbance spectra of the sludge exposed to (a) 5 mg/L ZnO ENPs and (b) 10 mg/LZnO ENPs; compared to the spectrum of (c) ZnO ENPs before they were introduced into the wastewater

From the XRD spectra (Fig. 7, line b), no signal was detected in the sludge for the ZnO ENPs at low concentration of 5 mg/L, and this observation was confirmed by UV-Vis results (Fig. 8, line a). However, the XRD pattern showed marginal peaks after the dosage was increased to 10 mg/L and was characterized by low intensities (Fig. 7, line c), collaborated by UV-Vis results (Fig. 8, line b). For the UV-Vis, at 10 mg/L ZnO ENPs, an excitonic absorption peak at 368 nm was observed (Fig. 8, line b), though not as distinctive as in the case of ENPs before they were dosed into the influent (Fig. 1). Clear XRD patterns were observed at a dosage of 20 mg/L, and the peaks corresponded to characteristic polycrystalline hexagonal Wurtzite structure - signifying that the ZnO ENPs were in pure phase. The size distribution of ZnO ENPs in the sludge was computed based on XRD analysis and modified Scherrer's formula (Monshi et al., 2012):

$$L = \frac{k\lambda}{\beta \cos \theta} \tag{1}$$

where: L represents mean size of the particle distribution (nm), K is geometric factor (generally $K = 0.9$), λ is the wavelength of Cu *Ka* radiation (λ = 0.15406 nm), θ is the Bragg angle of diffraction, and ß is the full width at half maximum (FWHM) of the diffraction main peak at 26. The computed mean crystallite size distribution of sludge, determined for the influent dosed with 10 and 20 mg/L, was approximately 29.22 ± 2.47 nm.

In earlier work (Musee et al., 2014), where a high ZnO ENP concentration of 100 mg/L was used, sharper peaks were observed including the 002 diffraction peak - which signified one-dimensional nano-rod formation of ZnO ENPs. In the current study similar observation of sharp peaks was made at a relatively lower dosage of 20 mg/L ZnO ENPs.

Two key aspects merit highlighting. First, from our study, it was found to not be possible to detect very low concentrations of ZnO ENPs using the current analytical techniques, in this case ICP-OES, for determining zinc ions (Fig. 4) in complex matrixes such as wastewater, or XRD for determining zinc phases in the sludge (Fig. 7). Secondly, a comparison of XRD pattern results for ZnO ENPs (summarized in Figs 1 and 7, respectively), before being introduced into influent and after dosed in the influent, shows that their morphology did not change despite possible adsorption of the ENPs onto the sludge matrix.

Moreover, we observed a high efficiency of removal of ZnO ENPs (> 96%) from effluent wastewater. Therefore, the low concentrations of ZnO ENPs detected in the treated effluent indicated that it is unlikely that there is significant release and dispersion of ENPs into aquatic systems resulting from WWTPs as point sources. Furthermore, the low quantities of ENPs likely to be released through treated effluent into the environment can be further reduced by incorporating additional treatment steps such as the use of membranes.

Thirdly, it is important to take caution when generalizing findings about the fate and behaviour of ENPs, for example, based on studies on the same or other types of nanoparticles in in WWTPs. This is because increasing reports have shown that WWTPs exhibit non-uniformity of removal efficiencies for diverse types and forms of ENPs: e.g., 94% of CeO_2 (Limbach et al., 2008); 75-85% of TiO_2 (Kiser et al., 2009); 94% of CeO_2 (Gómez-Rivera et al., 2012); 97% of TiO_2; 95% of nC_{60}; and 88% of Ag (Wang et al., 2012); >95% of TiO_2 (Gartiser et al., 2013). These differences can be linked to the ENP type (e.g. surface properties and chemistry), and the nature of wastewater physical and chemical properties - where the latter varies widely due to the constituent components.

Due to elevated concentrations of ZnO ENPs in the sludge caution should be taken in utilizing such sludge for agricultural purposes or during disposal phase. As such, this calls for adoption of additional treatment steps to mitigate against possible dispersion of ENPs from various disposal mechanisms such as landfilling or incineration, as well as in agricultural applications. This has significant relevance in a country like South Africa, for instance, where over 80% to 97% of sludge from WWTPs (Musee, 2011) is used for agricultural purposes, and the rest disposed of through landfilling. Recent studies (Lombi et al., 2012; Wang et al., 2012) have also highlighted the likelihood of increased transfer of ENPs into the environment from WWTPs via different pathways, for instance, leaching from landfills - and therefore call for the adoption of mitigative approaches to guard against unknown long-term adverse effects of ENPs in aquatic systems.

Nitrification is an important process in numerous WWTPs and, therefore, likely to be sensitive to toxic effects of ENMs. Thus, to ensure the simulated wastewater treatment system results provide representative findings, both nitrification and denitrification processes will be considered in future work.

CONCLUSIONS

In this study, we carried out experiments to: (i) determine the fate and behaviour of ENPs in a simulated WWTP system following the OECD 303 A guideline, (ii) investigate plausible transformation of ZnO ENPs after entry into wastewater up to the point of discharge of the treated effluent, and (iii) examine the effect of increasing concentration of ZnO ENPs on the efficiency of organic matter removal from wastewater. The following observations were made and inferences have been drawn:

- ZnO ENPs were efficiently removed from wastewater with an insignificant portion being released into the environment through the treated effluent. A large percentage of the ENPs settled out into the sludge. However, the fraction of unreacted ENPs in the effluent may potentially be released into the environment. In order to mitigate any possible release of nanoparticles into the environment, additional removal methods may need to be put in place. Further investigations are necessary to collect more data on these processes.
- Under wastewater conditions most ENPs aggregated and were attached to the biomass and, therefore, were removed from influent through adsorption processes, for all of the concentration levels investigated.
- The currently available techniques for detection and quantification of metal-based ENPs, such as XRD and UV-Vis, were found to be inadequate since they could not detect ZnO ENPs in the sludge even after the influent was dosed with concentrations as high as 5 mg/L and above. Yet, a concentration of 5 mg/L ZnO ENPs is many times above the expected actual concentrations of metal-based nanoparticles in the environment (Musee, 2011; Gottschalk et al., 2013).

ACKNOWLEDGEMENTS

The study was supported by the Water Research Commission (WRC) (Project No. K5/2122) (EFCC, JNZ, JCN, NM) and Parliamentary Grant #EINI002 (NM, JNZ) from the CSIR. We would like to thank Siyasanga Mpelane (MINTEK) and Mathokone Jonas Shai (University of Johannesburg) for their assistance in providing the TEM and XRD services, respectively. We also acknowledge useful and insightful comments of anonymous reviewers of the manuscript.

REFERENCES

1. ADAMS LK, LYON DY and ALVAREZ PJJ (2006) Comparative ecotoxicity of nanoscale TiO_2, SiO_2, and ZnO water suspensions. *Water Res.* 40 3527-3532. http://dx.doi.org/10.1016/jwatres.2006.08.004
2. APPLEROT G, LIPOVSKY A, DROR R, PERKAS N, NITZAN Y, LUBART R and GEDANKEN A (2009) Enhanced antibacterial activity of nanocrystalline ZnO due to increased ROS-mediated cell injury. *Adv. Funct. Mater.* 19842-852. http://dx.doi.org/10.1002/adfm.200801081
3. BCC RESEARCH (2012) Nanotechnology: a realistic market research. NANO 31E, Wllesley, USA.
4. BECKER WC, O›MELIA CR, AU KK and YOUNG JS (eds) (2004) Using oxidants to enhance filter performance. American Water Works Association, Denver.
5. BIAN SW, MUDUNKOTUWA IM, RUPASINGHE T and GRASSIAN VHC (2011) Aggregation and dissolution of 4 nm ZnO nanoparticles in aqueous environments: Influence of pH, ionic strength, size, and adsorption of humic acid. *Langmuir* 27 6059-6068. http://dx.doi.org/10.1021/la200570n
6. BOLYARD SC, REINHART DR and SANTRA S (2013) Behaviour of engineered nanoparticles in landfill leachate.*Environ. Sci. Technol.* 47 8114-8122.
7. BRAR SK, VERMA M, TYAGI RD and SURAMPALLI RY (2010) Engineered nanoparticles in wastewater and wastewater sludge -Evidence and impacts. *Waste Manage.* 30 504-520.http://dx.doi.org/10.1016/j.wasman.2009.10.012
8. ÇEÇEN F, SEMERCI N and GEYIK AG (2010) Inhibition of respiration and distribution of Cd, Pb, Hg, Ag, and Cr species in a nitrifying sludge. *J. Hazardous Mater.* 178 619-627. http://dx.doi.org/10.1016/j.jhazmat.2010.01.130
9. CHAÚQUE EFC, ZVIMBA JN, NGILA JC and MUSEE N (2014) Stability studies of commercial ZnO engineered nanoparticles in domestic wastewater. *Phys. Chem. Earth* 67-69 140-144. http://dx.doi.org/10.1016/j.pce.2013.09.011
10. DOBBS RA, WANG L and GOVIND R (1989) Sorption of toxic organic compounds on wastewater solids: correlation with fundamental properties. *Environ. Sci. Technol.* 23 1092-1097. http://dx.doi.org/10.1021/es00067a004
11. EATON AD, CLESCERI LS, GREENBERG AE and FRANSON MAH (2005) *Standard Methods for the Examination of Water and Wastewater.* American Public Health Association, Washington, DC.
12. FRANKLIN N, ROGER NJ, APTE S, BATLEY G, GADD G and CASEY P (2007) Comparative toxicity of nanoparticulate ZnO, Bulk ZnO, and ZnCl2 to a freshwater microalga *(Pseudokirchneriella subcapitata):* The importance of particle solubility. *Environ. Sci. Technol.* 41 8484-8490. http://dx.doi.org/10.1021/es071445r
13. GARTISER S, FLACH F, NICKEL C, STINTZ M,

DAMME S, SCHAEFFER A, ERDINGER L and KUHLBUSCH TAJ (2014) Behaviour of nanoscale titanium dioxide in laboratory wastewater treatment plants according to OECD 303 A. *Chemosphere* 104 197-204.http://dx.doi.org/10.1016/jxhemosphere.2013.1L015

14. GÓMEZ-RIVERA F, FIELD JA, BROWN D and SIERRA-ALVAREZ R (2012) Fate of cerium dioxide (CeO_2) nanoparticles in municipal wastewater during activated sludge treatment. *Bioresour. Technol.* 108 300-304. http://dx.doi.org/10.1016/j.biortech.2011.12.113

15. GOTTSCHALK F, SUN TY and NOWACK B (2013) Environmental concentrations of engineered nanomaterials: Review of modeling and analytical studies. *Environ. Pollut.* 181 287-300.http://dx.doi.org/10.1016/j.envpol.2013.06.003

16. HENZE M, HARREMOES P, JANSEN J, LA C and ARVIN E (2002) *Wastewater Treatment: Biological and Chemical Processes* (3rd edn). Springer-Verlag, Berlin. ISBN: 3-540-42228-5. http://dx.doi.org/10.1007/978-3-662-04806-1

17. HOU L, XIA J, LI K, CHEN J, WU X and LI X (2013) Removal of ZnO nanoparticles in simulated wastewater treatment processes and its effects on COD and NH4-N reduction. *Water Sci. Technol.* 67 254-260. http://dx.doi.org/10.2166/wst.2012.530

18. HUANG F, WANG Z, MEI X and WU Z (2013) Effects of short-term presence of nanoparticles on MBR sludge properties and membrane fouling. Technol. *Water Treat.* 8 14-26.

19. JIANG W, MASHAYEKHI H and XING B (2009) Bacterial toxicity comparison between nano- and micro-scaled oxide particles. *Environ. Pollut.* 157 1619-1625. http://dx.doi.org/10.1016/j.envpol.2008.12.025

20. KANG M, MAUTER S and ELIMELECH M (2009) Microbial cyto-toxicity of carbon-based nanomaterials: implications for river water and wastewater effluent. *Environ. Sci. Technol.* 43 2648-2653. http://dx.doi.org/10.1021/es8031506

21. KELLER AA, WANG H, ZHOU D, LENIHAN HS, CHERR G, CARDINALE BJ, MILLER R and JI Z (2010) Stability and aggregation of metal oxide nanoparticles in natural aqueous matrices. *Environ. Sci. Technol.* 44 1962-1967. http://dx.doi.org/10.1021/es902987d

22. KISER MA, RYU H, JANG H, HRISTOVSKI K and WESTERHOFF P (2010) Biosorption of nanoparticles to heterotrophic wastewater biomass. *Water Res.* 44 4105-4114. http://dx.doi.org/10.1016/j.watres.2010.05.036

23. KISER MA, WESTERHOFF P, BENN T, WANG Y, PÉREZ-RIVERA J and HRISTOVSKI K (2009) Titanium nanomaterial removal and release from wastewater treatment plants. *Environ. Sci. Technol.* 43 6757-6763. http://dx.doi.org/10.1021/es901102n

24. KOEHLER A, SOM C, HELLAND A and GOTTSCHALK F (2008) Studying the potential release of carbon nanotubes throughout the application life cycle. *J. Clean Prod.* 16 927-937. http://dx.doi.org/10.1016/j.jclepro.2007.04.007

25. KLAINE JS, ALVEREZ PJJ, BATLEY GE, FERNANDES TF, HANDY RD, LYON DY, MAHENDRA S, MCLAUGHLIN MJ and LEAD JR (2008) Nanomaterials in the environment: Behavior, fate, bioavail-ability and effects. *Environ. Toxicol. Chem.* 27 (9) 1825-1851. http://dx.doi.org/10.1897/08-090.1

26. LAU BLT and HSU-KIM H (2008) Precipitation and growth of zinc sulfide nanoparticles in the presence of thiol-containing natural organic ligands. *Environ. Sci. Technol.* 42 (19) 7236-7241. http://dx.doi.org/10.1021/es801360b

27. LIMBACH LK, BEREITER R, MULLER E, KREBS R, GALLI R and STARK WJ (2008) Removal of oxide nanoparticles in a model wastewater treatment plant: Influence of agglomeration and surfactants on clearing efficiency. *Environ. Sci. Technol.* 42 (15) 5828-5833. http://dx.doi.org/10.1021/es800091f

28. LIU G, WANG D, WANG J and MENDOZA C (2011) Effect of ZnO particles on activated sludge: Role of particle dissolution. *Sci. Total Environ.* 409 2852-2857. http://dx.doi.org/10.1016/j.scitotenv.2011.03.022

29. LOMBI E, DONNER E, TAVAKKOLI E, TURNEY TW, NAIDU R, MILLER BW and SCHECKEL KG (2012) Fate of zinc oxide nanoparticles during anaerobic digestion of wastewater and post-treatment processing of sewage sludge. *Environ. Sci. Technol.* 46 (16) 9089-9096. http://dx.doi.org/10.1021/es300839e

30. LOWRY GV, GREGORY KB, APTE SC and LEAD JR (2012) Transformations of nanomaterials in the environment.*Environ. Sci. Technol.* 46 (13) 6893-6899. http://dx.doi.org/10.1021/es300839e

31. MARTIN TD, CREED JT and BROCKHOFF CA (1994) Sample preparation procedure for spectrochemical determination of total recoverable elements. Revision 2.8, Method 200.2. United States Environmental Protection Agency Environmental Monitoring Systems Laboratory -Office of Research and Development, Cincinnati.

32. MAXIMOVA N and DAHL O (2006) Environmental implications of aggregation phenomena: current understanding.*Curr. Opin. Colloid. Interface Sci.* 11 246-266. http://dx.doi.org/10.1016/j.cocis.2006.06.001

33. METCALF and EDDY (2002) *Wastewater Engineering: Treatment and Reuse* (4th edn). McGraw Hill, New York.

34. MONSHI A, FOROUGHI MR and MONSHI MR (2012) Modified Scherrer equation to estimate more accurately nanocrystallite size using XRD. *World J. Nano Sci. Eng.* 2 154-160.http://dx.doi.org/10.4236/wjnse.2012.23020

35. MU H and CHEN Y (2011) Long-term effect of ZnO nanoparticles on waste activated sludge anaerobic digestion. *Water Res.* 45 5612-5620. http://dx.doi.org/10.1016/j.watres.2011.08.022

36. MU H, ZHENG X, CHEN Y, CHEN H and LIU K (2012) Response of anaerobic granular sludge to a shock load of zinc oxide nanoparticles during biological wastewater treatment. *Environ. Sci. Technol.* 46 (11) 5997-6003. http://dx.doi.org/10.1021/es300616a

37. MUSEE N, TRERISE MA and LORENZEN L (2007)

Post-treatment of distillery wastewater after UASB using aerobic techniques. *S. Afr. J. Enol. Vitic.* 28 (1) 50-55.

38. MUSEE N (2011) Simulated environmental risk estimation of engineered nanomaterials: a case of cosmetics in Johannesburg City. *Hum. Exp. Toxicol.* 30 (9) 1181-1195.http://dx.doi.org/10.1177/0960327110391387

39. MUSEE N, THWALA M and NOTA N (2011) The antibacterial effects of engineered nanomaterials: implications for wastewater treatment plants. *Environ. Monit.* 13 (5) 1164-1183.http://dx.doi.org/10.1039/c1em10023h

40. MUSEE N, ZVIMBA JN, SCHAEFER LM, NOTA N, SIKHWIVHILU LM and THWALA M (2014) Fate and behaviour of ZnO- and Ag-engineered nanoparticles and a bacterial viability assessment in a simulated wastewater treatment plant. *J. Environ. Sci. Health A 49* (1) 59-66. http://dx.doi.org/10.1080/10934529.2013.824302

41. OBERDÖRSTER G, OBERDÖRSTER E and OBERDÖRSTER J (2005) Nanotoxicology: an emerging discipline evolving from studies of ultrafine particles. *Environ. Health Perspect. 113* (7) 823-839. http://dx.doi.org/10.1289/ehp.7339

42. OECD 303A (2001) Aerobic sewage treatment - activated sludge units. Guidelines for the testing of chemicals. OECD, Paris.

43. OMAR FM, AZIZ HA and STOLL S (2014) Aggregation and disaggregation of ZnO nanoparticles: Influence of pH and adsorption of Suwannee River humic acid. *Sci. Total Environ.* 468-469 195-201. http://dx.doi.org/10.1016/j.scitotenv.2013.08.044

44. PADMAVATHY N and VIJAYARAGHAVAN R (2008) Enhanced bioactivity of ZnO nanoparticles-an antimicrobial study. *Sci. Technol. Adv. Mater.* 9 035004. http://dx.doi.org/10.1088/1468-6996/9/3/035004

45. PERALTA-VIDEA JR, ZHAO L, LOPEZ-MORENO ML, DE LA ROSA G, HONG J and GARDEA-TORRESDEY JL (2011) Nanomaterials and the environment: A review for the biennium 2008-2010. *J. Hazardous Mater. 186* 1-15. http://dx.doi.org/10.1016/).jhazmat.2010.11.020

46. PETTIBONE JM, CWIERTNY DM, SCHERER M and GRASSIAN VH (2008) Adsorption of organic acids on TiO_2nanoparticles: Effects of pH, nanoparticle size, and nanoparticle aggregation. *Langmuir* 24 6659-6667. http://dx.doi.org/10.1021/la7039916

47. PICCINNO F, GOTTSCHALK F, SEEGER S and NOWACK B (2012) Industrial production quantities and uses of ten engineered nanomaterials in Europe and the world. *J. Nanopart. Res. 14* 1109-1119. http://dx.doi.org/10.1007/s11051-012-1109-9

48. PREMANATHAN M, KARTHIKENYAN K, JEYASUBRAMANIAN K and MANIVANNAN G (2011) Selective toxicity of ZnO nanoparticles toward Gram-positive bacteria and cancer cells by apoptosis through lipid peroxidation. *Nanomed. Nanotechnol. Biol. Med.* 7184-192. http://dx.doi.org/10.1016Ai.nano.2010.10.001

49. SHENG GP, ZHANG ML and YU HQ (2008) Characterization of adsorption properties of extracellular polymeric substances (EPS) extracted from sludge. *Colloid Surf. B Biointerfaces* 62 83-90. http://dx.doi.org/10.1016/jxolsurfb.2007.09.024

50. SONUNE A and GHATE R (2004) Developments in wastewater treatment methods. *Desalination 167* 55-63. http://dx.doi.org/10.1016/).desal.2004.06.113

51. SUN T, GOTTSCHALK F, HUNGERBUHLER K and NOWACK B (2014) Comprehensive probabilistic modelling of environmental emissions of engineered nanomaterials. *Environ. Pollut. 185* 69-76. http://dx.doi.org/10.1016/j.envpol.2013.10.004

52. TAN M, QIU G and TING Y-P (2015) Effects of ZnO nanoparticles on wastewater treatment and their removal behavior in a membrane bioreactor. *Bioresour. Technol.* 185 125-133. http://dx.doi.org/10.1016/j.biortech.2015.02.094

53. THWALA M, MUSEE N, SIKHWIVHILU L and WEPENER V (2013) The oxidative toxicity of Ag and ZnO nanoparticles towards the aquatic plant *Spirodela punctuta* and the role of testing media parameters. *Environ. Sci.: Process. Impacts* 151830-1843. http://dx.doi.org/10.1039/c3em00235g

54. WANG Y, WESTERHOFF P and HRISTOVSKI KD (2012) Fate and biological effects of silver, titanium oxide, and C60 (fullerene) nanomaterials during simulated wastewater treatment processes. *J. Hazardous Mater.* 201-20216-22. http://dx.doi.org/10.1016/).jhazmat.2011.10.086

55. WESTERHOFF P, SONG G, HRISTOVSKI K and MEHLIKA AK (2011) Occurrence and removal of titanium at full scale wastewater treatment plants: implications for TiO_2 nanomaterials. *J. Environ. Monit.* 131195-1203. http://dx.doi.org/10.1039/c1em10017c

56. WONG SWY, LEUNG PTY, DJURIŠIČ AB and LEUNG KMY (2010) Toxicities of nano zinc oxide to five marine organisms: influences of aggregate size and ion solubility. *Anal. Bioanal. Chem.* 396 609-618. http://dx.doi.org/10.1007/s00216-009-3249-z

57. WOODROW WILSON INTERNATIONAL CENTRE FOR SCHOLARS (2015) A nanotechnology consumer products inventory. Project on Emerging Nanotechnologies. URL: www.nanotech-project.org (Accessed 31 March 2015).

58. XIA T, KOVOCHICH M, LIONG M, MADLER L, GILBERT B, SHI H, YEH JI, ZINK JI and NEL AE (2008) Comparison of the mechanism of toxicity of zinc oxide and cerium oxide nanoparticles based on dissolution and oxidative stress properties. *Am. Chem. Soc.* 2 (10) 2121-2134. http://dx.doi.org/10.1021/nn800511k

59. XIONG D, FANG T, YU T, SIMA X and ZHU W (2011) Effects of nano-scale TiO_2, ZnO and their bulk counterparts on zebrafish: Acute toxicity, oxidative stress and oxidative damage. *Sci. Total Environ.* 409 1444-1452. http://dx.doi.org/10.1016/j.scitotenv.2011.01.015

60. ZHANG Y, CHEN Y, WESTERHOFF P and CRITTENDEN J (2009) Impact of natural organic matter and divalent cations on the stability of aqueous nanoparticles. *Water Res.* 43 4249-4257.http://dx.doi.org/10.1016/j.watres.2009.06.005

61. ZHOU D and KELLER AA (2010) Role of morphology in the aggregation kinetics of ZnO nanoparticles. *Water Res.*44 2948-2956. http://dx.doi.org/10.1016/j.watres.2010.02.025

CHAPTER 13

MICROBIAL ASSESSMENT AND EFFECT OF SEASONAL CHANGE ON THE REMOVAL EFFICIENCY OF FAB BASED SEWAGE TREATMENT PLANT

Dilafroza Wani, Ashok K. Pandit and Azra N Kamili

Keywords: Microbial evaluation; faecal coliform; faecal streptococci; total coliforms

INTRODUCTION

Wastewater disposal is done both on land and in water body by most urban centres without pretreatment. Wastewater collection in most urban centres is done by means of sewerage system which is carried to the treatment plants and finally disposed off into the environment. Wastewater treatment must be made mandatory for all sizes of urban centres [1]. Wastewater treatment is aimed at removing impurities present in wastewater in the form of floating material, suspended solids, biodegradable organics and elimination of pathogens. The two fundamental reasons for treating wastewater is the prevention of pollution of water sources and protection of public health by safeguarding water supplies against the spread of diseases [2].

Microbial pollutants can also serve as indicators of water quality. For instance total and faecal coliforms, and the entercocci - faecal streptococci are the indicator organisms currently used in the public health arena as they indicate fecal contamination [3] and presence of enteric pathogens in surrounding water [4]. Indicator organisms are always used to determine the relative risk of the possible presence of a particular pathogen in wastewater [6]. Coliform bacteria include all aerobic and facultative anaerobic, gram-negative, nonsporeforming, rod-shaped bacteria that ferment lactose with gas formation, and comprises mainly species of the genera Citrobacter, Enterobacter and Klebsiella and also includes Faecal coliforms, of which Escherichia coli is the predominant species [6] and Faecal streptococci. These are found naturally in the soil, as well as in faeces. Usually between 60% and 90% of total coliforms are faecal coliforms. *E. coli* are a particular species of bacteria that may or may not be pathogenic but are ubiquitous in the human intestinal tract. Faecal streptococci is an indicator in tropical conditions and especially used to compare the survival with that of Salmonellae.

The increased population and urbanization in Srinagar city has resulted in an increase in the discharge of domestic sewage into the water bodies especially Dal lake, due to which the lake has undergone tremendous changes in its chemical, biological and trophic status. So there is an immediate need for trapping these pollutants before releasing into the lake. It was in this background that wastewater treatment plant was installed at the periphree of the Dal lake at Hazratbal to treat the wastewater coming from Hazratbal area and University campus. In past, no extensive study has been carried out to assess the efficiency and quality of these STPs. Because of the associated dangers of sewage, present study was carried out to investigate the impact of the wastewater effluent discharged and to estimate the removal efficiency of STPs around the Dal lake. In addition, seasonal change in the removal efficiency rate was evaluated.

Citation: Wani D, Pandit AK and Kamili AN: Microbial assessment and effect of seasonal change on the removal efficiency of FAB based sewage treatment plant. journal of Environmental Engineering and Ecological Science 2013, 2:1. http://dx.doi.org/10.7243/2050-1323-2-1.

MATERIAL AND METHODS

Study Area

The study was conducted on Hazratbal Sewage Treatment Plant (34° 08' 06"N, 74° 50'29"E) located in the vicinity of Dal lake with a design capacity of 7.5 MLD. The wastewater treatment plant receives domestic sewage from the Hazratbal area of Srinagar city and treat it using the FAB based biological treatment systems. The final treated effluent of Hazratbal sewage treatment plant is discharged into the Dal lake.

The water samples were collected at the inlet and outlet of sewage treatment plant on seasonal basis for a period of 12 months, in white plastic containers, which were previously sterilized with 70% alcohol and rinsed with distilled water. At the sites, the containers were rinsed thrice with the wastewater before being used to collect the samples.

Microbiological examination of samples was conducted as promptly as possible (within 24 hours) after collection or were stored at 4°C in a refrigerator until use. A multiple tube fermentation technique or Most Probable Number (MPN) technique [7] was used to determine bacterial indicators as Total Coliforms (TC), Faecal Coliforms (FC) and Faecal Streptococci (FS) according to standard methods as described in APHA [7]. Multiple tube fermentation method used in present work included measurement of total plate count and MPN of coliform. After incubation for 24 hours at 35°C results were recorded when acid and gas liberated in durham tubes with change of colour to yellow. Lactose broth was used as a medium for the growth of coliform. Number of positive test tubes with acid (yellow colouration) and gas production were matched with the McCardy's Statistical Table and the most probable number (MPN) of coliforms present in 100ml of each sample was thus determined. Different pathogenic bacteria were enumerated by serial dilution method. 0.1 ml aliquots of each dilution (10^{-1} to 10^{-7}) directly from tubes was inoculated in triplicate on FC agar, Mac Conkey agar and FS agar by spread-plate method for enumeration of Faecal coliform, *E. coli* and Faecal streptococci respectively. The culture plates were incubated at 37°C for 24h. The removal efficiency of bacterial indicators was calculated using the following formula [8,9].

$$\text{Removal efficiency} = \frac{\text{MPN/100ml in influent- MPN/100ml in effluent}}{\text{MPN/100ml in influent}} \times 100$$

Results

The results showed significant differences ($p<0.05$) in mean values of microbial load (Faecal coliform, *E. coli* and Faecal streptococci) between wastewater (raw) and treated wastewater (final) from the from sewage treatment plant. The total coliform was significantly lower (*t-test*=13.6, $p<0.05$) in effluent (7.62±0.10 MPN/100mL) in comparison to raw (8.78±0.13 MPN/100mL). In case of other microbial indicators like faecal coliform, faecal streptococci and *E.coli* the data showed significantly ($p<0.05$) lower values for effluent in comparison to the raw sewage ($p<0.05$). Average values of Faecal coliform, *E. coli* and Faecal streptococci concentrations in the inlet samples were found to be 7.83±0.07 MPN/100mL, 7.63±0.14 MPN/100mL and 6.85±0.11MPN/100mL respectively. Similarly, FC, *E. coli* and FS concentrations were found to vary in the effluent samples (Table 1).

Table 1: Comparison of average raw and effluent microbial load (±SD) in hazratbal sewage treatment plant with statistical analysis

Parameter	Raw	Effluent	*t-test*	df	*p- value*
Total coliform	8.78±0.13	7.62±0.10	13.6	6	p<0.05
Faecal coliform	7.83±0.07	6.78±0.06	14.5	6	p<0.05
E. *coli*	7.63±0.14	6.82±0.03	15.2	6	p<0.05
Faecal streptococci	6.85±0.11	6.08±0.05	12.8	6	p<0.05

The comparison of Percentage efficiency rate showed significant differences among different microbial parameters. Removal efficiency rate was greater for Faecal coliform (72.33%) and least in Faecal streptococci (38.7%; Table 2).

Table 2: Log concentration of influent and effluent log removal (%) efficiency of microbial load on seasonal basis

Table 2. Log concentration of influent and effluent log removal (%) efficiency of microbial load on seasonal basis.

Season	Total coliform			Faecal coliform			E. *coli*			Faecal streptococci		
	Raw	Effluent	log removal (%) efficiency	Raw	Effluent	log removal (%) efficiency	Raw	Effluent	log removal (%) efficiency	Raw	Effluent	log removal (%) efficiency
Summer	8.72	7.49	1.25 (54.85)	7.81	6.72	1.11 (72.33)	7.58	6.79	0.80 (61.44)	6.86	6.06	0.82 (43.93)
Autumn	8.89	7.62	1.28 (51.10)	7.85	6.74	1.15 (68.09)	7.65	6.86	0.81 (62.12)	6.94	6.08	0.86 (44.35)
Winter	8.62	7.73	0.92 (48.15)	7.74	6.86	0.90 (56.94)	7.48	6.85	0.63 (51.44)	6.70	6.04	0.66 (38.70)
Spring	8.90	7.66	1.30 (50.39)	7.93	6.80	1.15 (68.36)	7.82	6.81	1.04 (59.67)	6.93	6.15	0.78 (40.597)

The results show significant variation in different parameters of microbial load among different seasons ($p<0.05$). The efficiency rate of total coliform is significantly higher in summer (54.85, *Chi-square*=11.2, $p<0.05$) and least in winter (48.15). Other microbial parameters also show significant variation among different seasons (Table 2 and Table 3). When mean values of microbial load (Faecal coliform, *E. coli* and Faecal streptococci) were compared by using box plot showed least mean values in winter in comparison to other seasons (Figure 1).

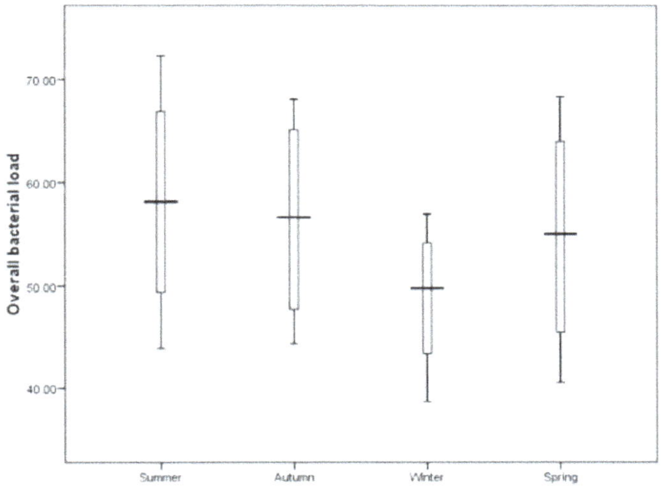

Figure 1: Seasonal variation of bacterial load in Hazratbal STP.

Table 3: Seasonal variation in efficiency rate of microbial load in Hazratbal sewage treatment plant with Chi-Square test

Season	Chi square	P- value
Summer	11.2	P<0.05
Autumn	7.5	P<0.05
Winter	10.1	P<0.05
Spring	11.3	P<0.05

DISCUSSION

It is evident from the data that microbial load showed significant differences between the inlet and outlet, water samples nevertheless, these variations do not match as per Indian national standards. As per Indian standards, Total coliform, Faecal coliform and Faecal streptococci should be less than 10000MPN/100ml, 2500MPN/100ml, 500MPN/100ml respectively for discharge into some water body. The high coliform count in raw sewage obtained in our results may be an indication that the sewage is comprised of faecal matter coming from household latrines (APHA, 1998). The presence of pathogenic bacteria in treated wastewater effluent is a potential public health hazard, as this water source is directly discharged in receiving water bodies and may be used by communities for multiple purposes. At the inlet physico-chemical parameter values are relatively high which causes microbial biomass development in particular increase faecal coliform, faecal streptococci [10]. So, it is essential to include tertiary treatment step in STPs so that complete disinfection of effluent is done [11].

Efficiency rate showed significant differences between raw and effluent in all the microbial parameters. Hazratbal sewage treatment plant uses FAB based technology which is a recent technique for the removal of pollutants from municipal sewage. The comparatively low efficiency rate in this sewage treatment plant may be because of the fact that primary sedimentation is not carried out in this sewage treatment plant and as such efficiency rate is affected [12]. The reduction of microbes also depends on protozoan predation, settlement of suspended solids, inactivation due to sunlight, activity of bacteria and environmental conditions [13]. It is clear from our results that large amount of microbial load is retained even after the purification process [14,15 and 16]. However, it is to be mentioned that these investigated STPs were using different types of technologies including Activated Sludge Process (ASP), Extended Aeration (EA), BIOFORE, Trickling filter and Oxidation Pond for the treatment of sewage [16].

Seasons seem to play an important role in predicting the removal efficiency of this STP. The greater efficiency rate in summer and autumn season could be due to the efficient working capability of mechanical parts during these months. Moreover feasible temperature in these months brings about the inactivation of microbes. The reverse is the case in winter months when harsh climatic conditions may reduce the efficiency of machine parts and thus will reduce efficiency rate. So it can be concluded that temperature plays an important role in working capability of Hazratbal wastewater treatment plant. According to Central Pollution Control Board (2008 report), the effluent concentration of Total coliforms was observed lower in winters whereas the highest effluent TC concentration was noted during monsoon period [17]. In case faecal coliforms, higher effluent concentration was during summer and monsoon period and lowest in winters whileas almost similar FC removal was reported during all the seasons. These findings coincide with our results.

CONCLUSION

From the study it was concluded that efficiency rate of Hazratbal sewage treatment plant did not reach to the level of standards and international standards and as such discharge of effluent from this sewage treatment plant into the lake is not suitable prior to complete disinfection. It is further recommended that authorities should include tertiary treatment processes such as rapid sand filtration, UV disinfection, artificial lagoons, wetlands, adequate contact time etc. to increase the removal efficiency of this STP. It is also recommended that wastewater effluents should be routinely monitored to ensure that strict adherence to effluent discharge standards is met.

ACKNOWLEDGEMENT

We are greatly indebted to LAWDA (Lakes and Waterbody Development Authority) for their cordial help during my period of work.

REFERENCES

1. Central Public Health and Environmental Engineering Organisation (CPHEEO) Ministry of Urban Development Government of India, 2005. Status of Water Supply, Sanitation and Solid Waste Management in Urban Areas. 2000
2. Akpor OB and Muchie M : Environmental and public health implications of wastewater quality. *African Journal of Biotechnology.* 2011, 10:2379-2387.
3. Geldreich, EE: Water and Sewage Works Ref. No. R-98.1967
4. Momba, MNB, Mfenyana, C: Inadequate treatment of wastewater: A source of coliform bacteria in receiving surface water bodies in developing countries- Case Study: Eastern Cape Province of South Africa. In: Water

Encyclopedia- Domestic, Municipal and Industrial Water Supply and Waste Disposal. Lehr JH, Keeley J (eds). John Wiley & Sons Inc. 2005, pp. 661-667.

5. Paillard D, Dubois V, Thiebaut R, Nathier F, Hoogland E, Caumette P and Quentin C: Occurrence of Listeria spp. in effluents of French urban wastewater treatment plants. *Appl Environ Microbiol* 2005, 71:7562-6.

6. Hurst CJ, Knudsen GR, Mclnerney MJ, Stetzenbach LD and Walter MV (eds): *Manual of Environmental Microbiology*. ASM Press, Washington DC. 1997

7. Americal Public Health Association (APHA): Standard Methods for the Examination of Water and Waste water, 19th ed. American Water works Association and Water Pollution Control Federation Publication, Washington, D.C. 1998.

8. Kantachote D, Kantasorn, D and Cherdchan S: Treatment fficiency in wastewater treatment plant of Hat Yai Municipality by quantitative removal of microbial indicators. *Songklanakarin J. Sci. Technol.* 2009, 31:567-576.

9. Shalaby IMI, Altalhy AD and Mosallam HA: Preliminary field study of a model plant for sewage water treatment using Gravel Bed Hydroponics Method. *World Applied Sci. J.* 2008, 4: 238-243.

10. Rajeb AB, Kallali H, Saidi N, Abidi S, Jedidi N and Hassen A: Physicochemical and Microbial Caracteristics Performency in Wastewater Treated Under Aerobic Reactor. *American Journal of Environmental Sciences.* 2011,7: 254-262.

11. Koivunen J, Siitonen A and Heinonen-Tanski H: Elimination of enteric bacteria in biological-chemical wastewater treatment and tertiary filtration units. *Water Res* 2003, 37:690-8.

12. EPA. Washington, DC: «Primer for Municipal Waste water Treatment Systems.»Document no. EPA 832-R-04-001. 2004

13. Katsoyiannis A and Samara C: Persistent organic pollutants (POPs) in the sewage treatment plant of Thessaloniki, northern Greece: occurrence and removal. *Water Res* 2004, 38:2685-98.

14. Cohen J and Shuval HI: Coliforms, faecal coliforms and faecal streptococci as indicators of water pollution. *Water, Air and Soil Pollution,* 1972, 2:85-95.

15. Payment P, Plante R and Cejka P: Removal of indicator bacteria, human enteric viruses, Giardia cysts, and Cryptosporidium oocysts at a large wastewater primary treatment facility. *Can J Microbiol* 2001, 47:188-93.

16. Jamwal P, Jamwal P, Mittal A K and Mouchel J M: Efficiency evaluation of sewage treatment plants with different technologies in Delhi (India). *Environ Monit Assess* 2009, 153:293-305.

17. Central Pollution Control Board: Status of water quality in india- 2007. Ministry of Environment and Forests. 2008

CHAPTER 14

CLEANING PROCESS RESEARCH OF MVR HIGH CONCENTRATION SALTY WASTEWATER TREATMENT SYSTEM

Na Wang, Wenqiang Jiang and Libin Fang

Keywords: High-concentrated salty wastewater; MVR system; cleaning process

INTRODUCTION

High-concentrated salty (inorganic salt) wastewater is a form of pollution which is difficult to treat. It is generated in numerous industrial sectors, and the discharge of such wastewater without prior treatment will pollute the environment [1-3]. In particular, wastewater that contains dissolved salts (namely contains Cl^-, SO_4^{2-}, Na^+, Ca^{2+}, etc.) could be recycled if these salts are directly discharged. Hence, the treatment and recovery of such wastewater are necessary. At present, these types of wastewater are usually treated using evaporation techniques, such as multi-stage flash (MSF), multi-effect evaporator (MEE) and reverse osmosis (RO) methods. Among them, MEE is considered to be more effective than other processes [4, 5]. However, the MEE method requires a large amount of external heating vapor, the system and its operation are complex [6]. So the mechanical vapor re-compression (MVR) is an alternative method for the treatment of high-concentrated salty wastewater in numerous chemical enterprises.

Although the MVR system has been widely studied, it has not been widely applied in wastewater process. The MVR system used in the wastewater distillation with reliable and long-term operation is limited in China [7]. One of the most pressing problems is the absence of design optimization process. Therefore, the purposes of this study were to develop a comprehensive design cleaning process of the effective mechanical vapor compression process, and to analyze the performances focusing on the salt content, discharge conditions of mother liquor and the solvent recovery by flashing technology.

The design processes of the MVR high concentrated salty wastewater treatment system are important parts in actual production or research. In this study, the factors that influence the mechanical vapor compression system are discussed more comprehensively. The influencing factors include the accumulation of organic matter in the heater, and the organic solvent waste water which cannot be recycled and so on.

Moreover, the performance of the system was investigated through using sodium chloride salt wastewater. The following sections include a description of the optimized process, the determination of experimental conditions of the system and an analysis of the system.

PROCESS DESCRIPTION AND PROCESS IMPROVEMENT

Present Situation of the Enterprise Production Wastewater

According to the actual survey, when the pesticide chemical plant is in the process of production preparation, the amount

Citation: Na Wang, Wenqiang Jiang and Libin Fang, Cleaning Process Research of MVR High Concentration Salty Wastewater Treatment System, DOI: 10.1051/matecconf/20152504008.

of its wastewater emission is about 423.6 tons a day in 2014. And the high-concentrated salty wastewater is about a third of the total discharge quantity. This part of the wastewater mainly contains methanol, toluene, ethanol, acetonitrile, sodium chloride, macromolecular heterocyclic reaction by-products and so on. The wastewater appears to be red, containing pharmaceutical intermediates. It is a kind of typical refractory wastewater. The main pollutants include residual polyethylene monomer by polymerization and all kinds of polymerization additives such as dispersant, initiator, termination agent and reaction and decay products and so on. The water quality is as follows: COD is as high as 20000–35000 mg/L, among which the NH_3-N is in the range of 626–1260 mg/L, the pH value is in the range of 5.0– 6.5 and the NaCl salt content is more than 18%.

High-concentrated salty wastewater refers to the total waste water containing inorganic salt mass fraction which is more than 3.5% [8, 9]. The distillation-based water treatment is a conventional technology in the field of wastewater treatment, especially in the process of high salinity wastewater treatment, which contains salts with Cl^-, $SO4^{2-}$, Na^+, Ca^{2+} and the mass concentration which is more than 1% [10, 11]. The existence of a large amount of salt can lead to the bacteria cell wall dehydration. Practice has proved that when the content of Cl^- in water is more than 15g/L, the biochemical process becomes very difficult, and the concentration of Cl^- is 14–16g/L in enterprises' wastewater production. If we continue to increase the content of Cl^-, the biochemical processes is very bad, so we must remove the salts in the wastewater in advance in order to improve the effect of biochemical means. These types of wastewaters are usually treated using evaporation techniques. The characteristic of MVR is the reuse of the energy of vapor produced in the evaporator. And the efficiency of MVR is higher than that of MSF and MEE [12].

Our government attaches great importance to the development of MVR wastewater treatment technology, and has put the technology into the national strategic emerging industries. "The complete recycling for zero discharge water vapour recompression crystallization" project has been listed as the "National Strategic Technology Plan for Newly-emerged Industry by the Support of Chinese Academy of Sciences" in one of the themes of energy conservation and environmental protection industries [13]. Our government has provided policy and project investment guarantee to establish the domestic independent research and development of MVR crystallization recovery system engineering research platform, and the development of industrial wastewater MVR zero emissions crystal full recovery technology.

Recently, the MVR system is mainly used in the process of high salinity wastewater treatment. Matsuda et al. [14] used the actual operation data of the treatment process in a heavy chemical industrial field as an industrial application, and the initial feasibility study was undertaken by applying energy conservation and environmental protection from an industrial point of view.

MVR before Optimization Processing System

The sewage in the workshop storage tanks is staged 1– 2 days later by the pump into the sewage transfer tank. Two reconcile the tanks can be alternately used every other day in order to make the sewage mix even. Reconciling the tank is mainly to add liquid alkali pH adjustment. The liquid is filtrated and then into the MVR system. Then after becoming concentrated crystallization in MVR system, it will be concentrated into the centrifuge and directly recycled into reconciled tanks away from the mother liquor. Finally, the concentration of salt crystals will be regarded as a product for sale. We steam out of the condensate to send to the sewage treatment station after Fenton oxidation into biochemical treatment. The specific process is expressed in Figure 1.

As the previous description, the MVR system can precipitate NaCl crystals from the wastewater and quantitative water into the Fenton oxidation and biochemical treatment. Therefore, this system is environmentally friendly. Recently, this kind of MVR system has been mainly used in the process of salt crystals. The major part of literature study of the MVR system is the crystallization of ammonium sulfate [15], sodium chloride [16] and other salts. However, this MVR is not an energy-saving choice for the treatment of high-concentrated salty wastewater if the previously described single-effect MVR system is used. Under certain conditions the mother liquor recycled, the chromaticity of the centrifuged salt increases as the organics of mother liquor increases. At the same time, the higher liquor organics leads to higher power consumption to maintain the heat-transfer temperature difference. Therefore, a certain number of liquor circulation needs to be quantitatively discharged in order to maintain the quality of salt. The main problems are divided into four aspects about the MVR system in high-concentrated salty wastewater treatment process, and the main problems are presented as follows:

(1) The solvent content in the wastewater becomes more and more, and the solvent content of being more than 10% will subsequently lead to the multi-effect evaporation steamed effluent COD which becomes high, thus affecting water quality of subsequent pretreatment effluent.

(2) Salt content in the waste water was larger. And the mixed wastewater including sodium chloride, potassium chloride and organic halogen contains salt after evaporation, causing multi-effect evaporation at the scene of the salt. Small amounts of workshop intermediates, whose appearance is grey and red-brown organic impurities, are unable to deal with. They can only become waste.

(3) After the mother liquor of high-concentrated wastewater which is separately discharged through using four-effect evaporator concentrate is processed, the treated wastewater is incorporated by low concentration waste water treatment processes. However, in the process of high-strength wastewater, the main problem is that the high solvent content causes the multi-effect evaporator scaling phenomenon to be more serious, affecting the amount of change heat treatment. What's worse, more waste oil and tar-like substances can cause blockage of multi-effect evaporator heat exchanger tube.

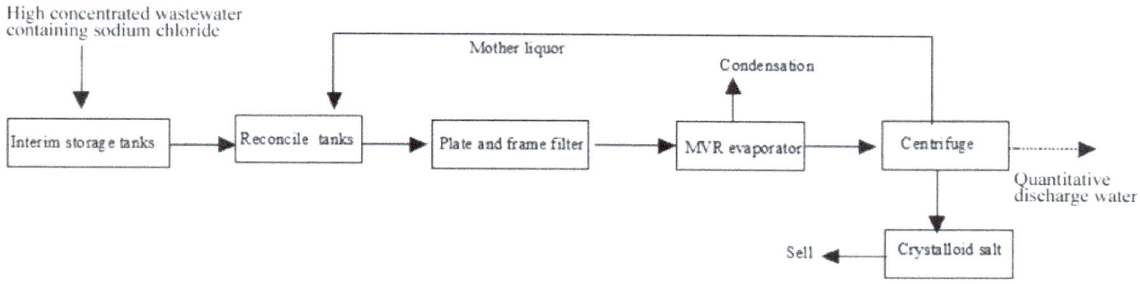

Figure 1. MVR before optimization processing system

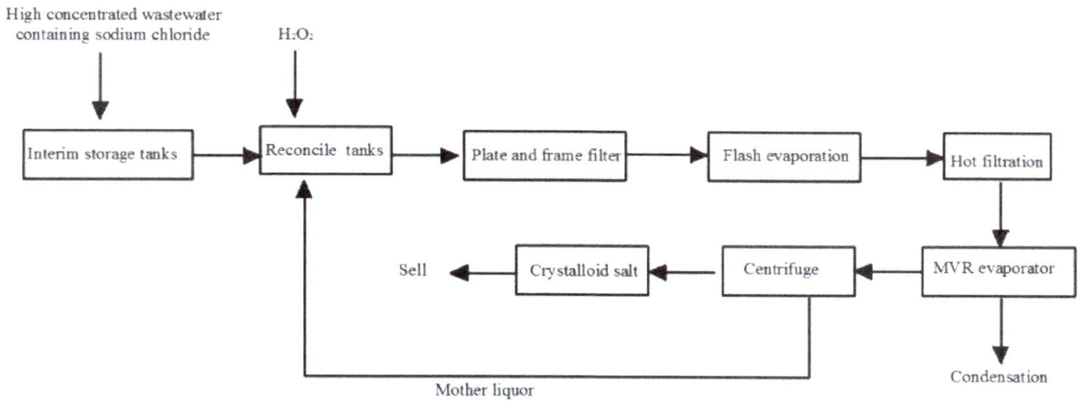

Figure 2. MVR optimized processing system

So it is necessary to optimize the traditional process from the aspect of process technology. Liang, et al. [17] reported on the MVR evaporation method of saline wastewater treatment process. Yu Haichen, et al. [18] thoroughly investigated the improved MVR process to improve the efficiency of wastewater treatment. In addition, they found that MVR can be combined with other systems such as MSF [19-22] and MEE [21-23] systems.

MVR Optimized Processing System

According to the above defects and considering the salt chromaticity which is the accumulation of organic matter in the heater, the final decision is that adding "H_2O_2 oxidatio + filter + flash evaporation +hot filter" technology. The operation is simple, and it can greatly reduce the energy consumption. After being optimized, the high salt wastewater treatment process is shown in Figure 2.

The building of the MVR optimized processing system is based on the relation of energy saving, consumption and reduction of pollution and efficiency in each part of the system. Compared with the actual working conditions, the processing system reaches a good desired effect: (1) zero discharge of mother liquid of high-concentrated organic impurities; (2) recycling salt chromaticity drop, and the salt content which is over 95% has reached the delivery requirements; (3) a large part of the solvents are recycled after fully utilization, and it simultaneously reduces the evaporator MVR system operation load, hence greatly reducing the company operation cost. Therefore, this system is environmentally friendly.

THE CHOICE OF OPTIMUM CONDITION

Optimum H_2O_2 dosage and Reaction Time

In the workshop, we take 4 groups of 500 mL high-concentrated salty wastewater samples, and the measured water quality is as follows: 22027.04 mg/L COD, 671.6 mg/L NH3-N, pH 8.5, including our 22% of sodium chloride salt, and the methanol accounting for 23.7% in total. Then according to the proportion of 5L, 10L, 15L and 20L water which are added to H_2O_2, those reactions which are respectively for 2.0h, 4.0h, 6.0h, 8.0h and 24 hours under the condition of pH 8.5 are simultaneously evaporated out of salt. The experimental results are shown in Figures 3 and 4:

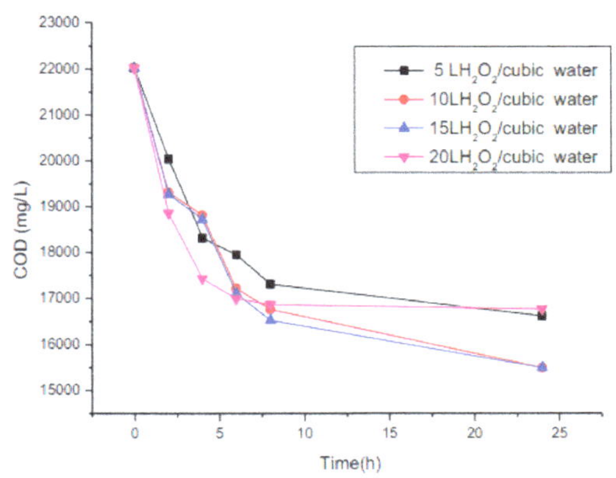

Figure 3. H_2O_2 dosage and reaction time on the influence of the waste water COD

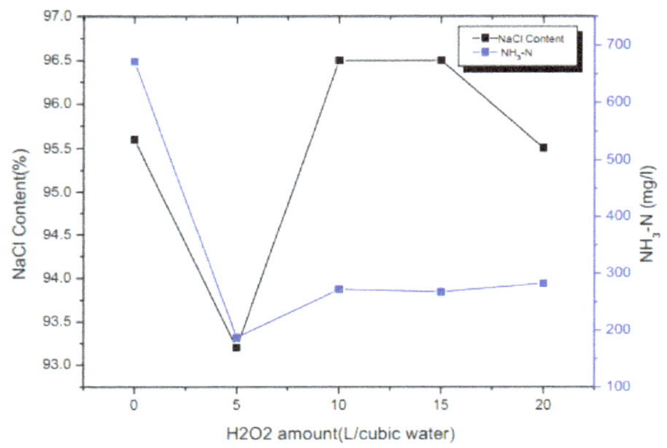

Figure 4. H_2O_2 amount of influence on the NH_3-N and salt content

Figure 3 shows the results for the influence of the waste water COD on account of H_2O_2 dosage and reaction time. As being also evident in Figure 3, the waste water COD has a certain degree of being reduced with the increase of the reaction time. The degradation of waste water COD and NH_3-N rate is increased as the response time increases, and the amount of H_2O_2 is 10L; 15L/water COD degradation rate is highest and the qualified rate of salt is the best. As shown in Figure 5, the influence of H_2O_2 amount on the NH_3-N and salt content is evident when the amount of hydrogen peroxide under different conditions is added. The figure can be easily found that the sodium chloride salt content is peaked when the amount of hydrogen peroxide is 10L/cubic water. So it is optimum to choose the H_2O_2 amount which is 10L H_2O_2/cubic water. And the longer the reaction time is, the better the effect can be. It can be judged according to the color and turbidity of water, and it needs to react for at least 8 hours. However, considering the two which reconcile the tanks can be alternately used together, so the reaction time we choose is 24 hours.

Optimum pH Value Conditions

For the same sample, according to the proportion, 10L H_2O_2/ cubic water under the condition of pH 5.5, 6.5, 7.5, 8.5, 9.5 and 10.0 respectively responses in 24 hours. After the flash process and MVR treatment, we compare the reaction condition, and the experimental results are shown in Figure 5:

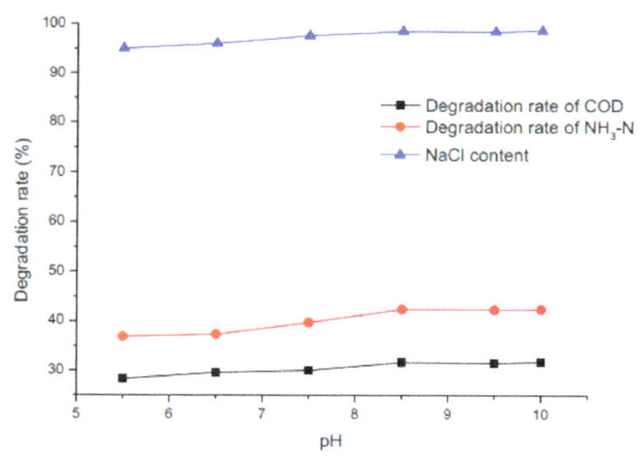

Figure 5. Degradation rate versus the oxidation reaction at different pH value condition

Figure 5 shows the degradation rate versus the oxidation reaction at different pH value condition. The pH value condition is respectively chosen from 5.5 to 10. It can be seen from Figure 3 that the degradation rate is slightly increased with the increase of pH value. The results also show that it is similar to the hydrogen peroxide treatment effect under the condition of acid and alkaline conditions, but the effect of reaction under condition of alkaline reaction is slightly better than the effect of acid condition. Through the comparison of the color salt and COD reduction rate and the comprehensive consideration of color in water sample, we choose to directly add the hydrogen peroxide to do oxidation processing under alkaline conditions. Then we can find that there are a lot of flocs when mother liquor reacts with hydrogen peroxide under the alkaline condition. And it can be used to neutralize acidic waste water and won't make cost rise. Therefore, reaction pH value we choose is 8.5.

Influence of Adding Flash Device for Waste Water Treatment

The raw water is filtered into the flash system after oxidation, and then we control the flash system according to the different solvents with different control temperature by 1–3% of the distillate water. The flash water contains light component (this topic is to steam out of methanol), and the pH value of adjusted water is neutral. A large number of free ammonia exhalations are steamed out with light component, which enter the MVR system's content is reduced. Therefore, even the centrifugal mother liquor which was repeatedly reused will also not affect the quality of salt; what's more, it can greatly reduce the high concentrated emission of mother liquor at the same time. The specific test results are shown in Table 1.

Table 1. Adding flash system for wastewater treatment effect

Monitoring point	State	COD(mg/L)	NH_3-N(mg/L)	NaCl content of the first out (%)	NaCl content after 35 times reuse of mother liquor (%)
Raw water	Raw water tanks	15506.40	327.2	--	--
MVR inlet	Without flash before	15506.40	327.2	--	--
	After add the flash	9020.42	96.74	--	--
MVR outlet	Without flash before	12783.62	236.22	83.2	67.2
	After add the flash	1472.22	23.46	98.2	96.2

According to the results of calculation and analysis above, after adding flash system, it can not only recover solvents but also make the MVR import water COD and NH_3-N be respectively reduced by 41.83% and 70.4%. Meanwhile, the effluent COD and NH_3-N MVR system were respectively reduced by 88.5% and 90.1% than before without flash. It can also greatly reduce the operation load of MVR evaporation system at the same time. Also, it can be seen that the centrifugal NaCl salt content is above from 80% to 98% at the first time, but salt quality of mother liquor stay above 95% for several times. So it can avoid the mother liquor to be discharged in order to maintain the quality of the salt. The literature shows that the flash system can achieve higher energy efficiency than that of the conventional processes.

CONCLUSION

A comprehensive optimization process is developed to overcome the problems in traditional MVR system in practical production. The overall pH value condition, response time and oxidant dosage were given in the optimization process.

The performance of the system was investigated using high salinity wastewater containing NaCl. This process by adding "H_2O_2 oxidation + filter + flash evaporation + hot filter" technology is on the basis of originally traditional MVR treatment system, and the optimum technological conditions were determined through the experiment: The pH value is 8.5; the oxidation time is 24.0h; H_2O_2 amount accounting for 1% of the total wastewater under the condition of outlet test MVR system respectively fell 88.5% and 90.1% than the traditional process of effluent COD and NH_3-N removal rate. The MVR system is used for the sodium chloride salt qualified rate which is increased from 83.2% to 98.2%. On the other hand, this process can avoid the high concentrated outside of mother liquor by flashing evaporation recycling volatile solvent. The optimization process results were found to be consistent with practical industrial data which was published. Compared with the new process MVR system, the operation load of MVR system was lower.

The experimental and practical industrial data indicated that the optimized MVR system is more energy-conserving and environmentally-friendly than traditional system when they are used to treat sodium chloride wastewater. It can not only effectively improve the clean production level of this company, but also have a very important practical guiding significance for the high-concentrated salty wastewater treatment of chemical enterprise.

REFERENCES

1. G. Parthasarathy, & R.F. Dunn, 2004. Graphical strategies for design of evaporation crystallization networks for environmental wastewater applications, Adv. Environ. Res. 8: 247-265.
2. A. Fakhru'l-Razi, A. Pendashteh, L.C. Abdullah, D.R.A. Biak, S.S. Madaeni, & Z.Z. Abidin, 2009. Review of technologies for oil and gas produced water treatment. Hazard. Mater.170: 530-551.
3. L. Di Palma, N. Verdone, A. Chianese, M. Di Felice, C. Merli, E. Petrucci, & G. Veriani, 2002. Treatment of wastewater with high inorganic salts content, Environ. Eng. Sci. 19: 329-339.
4. O. Lefebvre, & R. Moletta, 2006. Treatment of organic pollution in industrial saline wastewater: a literature review, Water Res. 40 3671-3682.
5. D.F. Zhao, J.L. Xue, S. Li, H. Sun, & Q.D. Zhang, 2011. Theoretical analyses of thermal and economical aspects of multi-effect distillation desalination dealing with high salinity wastewater, Desalination 273:292-298.
6. Lin Liang, Dong Han, Ran Ma, & Tao Peng, 2013. Treatment of high-concentration wastewater using double-effect mechanical vapor recompression, Desalination 314: 139–146.
7. Yasu Zhou, Chengjun Shi, & Guoqiang Dong, 2014. Analysis of a mechanical vapor recompression wastewater distillation system, Desalination 353: 91-97.
8. Woolard C R, & Irvine R L. 1995.Treatment of hypersaline wastewater in the sequencing batch reactor. Water Research, 29: 1159-1168.
9. Shaomei Nong, Handong Li, & Shuzeng Zhang, et al. 2007. Treatment technology research progress of hyper saline wastewater. Jiangsu Environment. 21(2008) 72-76 (in Chinese).A.A. Mabrouk, A.S. Nafey, H.E.S. Fath, Thermoeconomic analysis of some existing desalination processes, Desalination 205: 354-373.
10. X. Zhao, & W. Li, et al., 2011.Suppression of liquid film rupture during falling film evaporation for high salinity wastewater, J. Environ. Eng. 5: 726-730 (in Chinese).
11. Q. Xie, Y.H. Li, & Y.N. Zhu, 2004. Study on biological treatment process of high salinity wastewater, Environ. Eng. 22: 15-17 (in Chinese).
12. F.N. Alasfour, & H.K. Abdulrahim, 2011.The effects of stage temperature drop on MVC thermal performance, Desalination 265: 213-221.
13. Guoxiao Li, & Liping Lu, 2013. MVR marinating all alkali run summary, China Chlor-Alkali (4)10-12.
14. K. Matsuda, K. Kawazuishi, & Y. Hirochi, 2010. Advanced energy saving in the reaction section of the hydro-desulfurization process with self-heat recuperation technology, Appl. Therm. Eng. 30: 2300-2306.
15. Jian Wang. 1999. The impact of WTO on China's organic silicon industry. Silicone materials and applications, 1-3(in Chinese).
16. [16]D. G. Nicola, 2001.Sol-Gel-Derived Silicon-Boron Oxycarbide Glasses Containing Mixed Silicon Oxycarbid (SiCxO4-x) and Boron Oxycarbide (BCyO3-y) Units. J. Am. Ceram. Soc, 2160-2164.
17. Liag L, Han D, & Ma R, et al. 2013.Treatment of hign concentration wasterwater using double - effect mechanical vapor recompression. Desalination, 314: 139-146.
18. Haichen Yu, Peiwen Zhong, & Wei Gao, et al. 2013. Zero Discharge Process Design and Synthesis of Chemical high salt wastewater. Urban Environment & Urban Ecology. 26: 44-46(in Chinese).
19. F.N. Alasfour, & H.K. Abdulrahim, 2011. The effects of stage temperature drop on MVC thermal performance, Desalination 265: 213-221.
20. A.A. Mabrouk, A.S. Nafey, & H.E.S. Fath, 2007. Analysis of a new design of a multi-stage flash–mechanical vapor compression desalination process, Desalination 204:482-500.
21. A.A. Mabrouk, A.S. Nafey, & H.E.S. Fath, 2007. Thermoeconomic analysis of some existing desalination processes, Desalination 205:354-373.
22. A.S. Nafey, H.E.S. Fath, & A.A. Mabrouk, 2006. A new visual package for design and simulation of desalination processes, Desalination 194: 281-296.
23. A.S. Nafey, H.E.S. Fath, & A.A. Mabrouk, 2008. Thermoeconomic design of a multi-effect evaporation mechanical vapor compression (MEE–MVC) desalination process, Desalination 230: 1-15.

CHAPTER 15 PHYSICO-CHEMICAL TECHNOLOGIES FOR NITROGEN REMOVAL FROM WASTEWATERS

Andrea G. Capodaglio, Petr Hlavínek, Massimo Raboni

Keywords: Nitrogen removal; physico-chemical technologies; wastewater treatment

INTRODUCTION

Human activities (e.g. communities, industrial plants, intensive livestock and farming, uncontrolled waste disposal) discharge many pollutants in water, including nitrogen [Urbini et al., 2015]. Nitrogen is one of the causes of eutrophication, while some N-compounds (e.g., ammonia, ammonium, nitrite, nitrate, chloramines) can be harmful for human health. Nitrogen removal from wastewater can be achieved by biological or physico-chemical processes.

Biological processes are widely applied to the treatment of both municipal and industrial wastewaters. They are based on the action of heterotrophic bacteria that, under anoxic conditions, carry out the biochemical reduction of nitrate to nitrogen gas. At present, the dominant technology is pre-denitrification in activated sludge systems. Research advances in this field are mainly oriented to (i) evaluate the inhibiting effects of dissolved oxygen on denitrification rates [Raboni et al., 2014b; Torretta et al., 2014]; (ii) limit, as much as possible, the presence of dissolved oxygen in the denitrification reactor [Urbini et al., 2015]; (iii) highlight denitrification efficiency limitations in plants serving small communities due to strong fluctuations of the fed loads [Raboni et al., 2013a; 2014a; c]; (iv) evaluate the effects of the hydrodynamic behavior of the denitrification reactor (complete-mixing or plug-flow) on denitrification performance [Eramo et al., 1994; Farabegoli et al., 2003]. With reference to industrial wastewaters, studies which highlight the inhibition effect of high ammonia concentrations on microbial activity [Li and Zhao, 1999] suggest the use of alternative solutions based on physico-chemical processes.

Currently, a broad spectrum of physico-chemical technologies for the removal of nitrogen from wastewater is available: ammonia air and steam stripping; ammonia vacuum distillation; ammonia precipitation as struvite; ammonia and nitrate removal by selected ion exchange; breakpoint chlorination; chloramine removal by selected activated carbon; ammonia adsorption on charcoal; chemical reduction of nitrate; advanced oxidation processes to convert ammonia and organic-N into nitrogen gas or nitrate. The above-mentioned technologies are individually examined in this study to show both their applicability and actual viability in full-scale plants.

TECHNOLOGIES REVIEW

Ammonia Stripping and Distillation

Ammonia removal from highly concentrated wastewaters can be accomplished by air or steam stripping. Ammonia stripping is the best known with the most extensive applications in concentrated wastewater treatment (e.g. landfill leachate, supernatants of anaerobic digestion processes, specific flows of the petrochemical industry), while very few applications in sewage treatment are known. In ammonia stripping, lime or some other caustic substance is generally added to the wastewater until pH reaches 10.8-11.5 standard units, converting ammonium hydroxide ions to ammonia gas according to equilibrium reaction:

Citation: Andrea G. Capodaglio, Petr Hlavínek, Massimo Raboni, Physico-chemical technologies for nitrogen removal from wastewaters, http://dx.doi.org/10.4136/ambi-agua.1618.

$$NH_4^+ + OH^- \leftrightarrows H_2O + NH_3\uparrow \qquad (1)$$

The equilibrium is governed by pH and temperature, according to the equation describing the reaction's ionization constant (Keq)

$$K_{eq} = \frac{[NH_4^+][OH^-]}{[NH_3]} = 1.8 \cdot 10^{-5} \quad \text{at } 25°C \qquad (2)$$

The equilibrium in water of the two nitrogenous forms (ammonia and ammonium ion) is well represented in Figure 1.

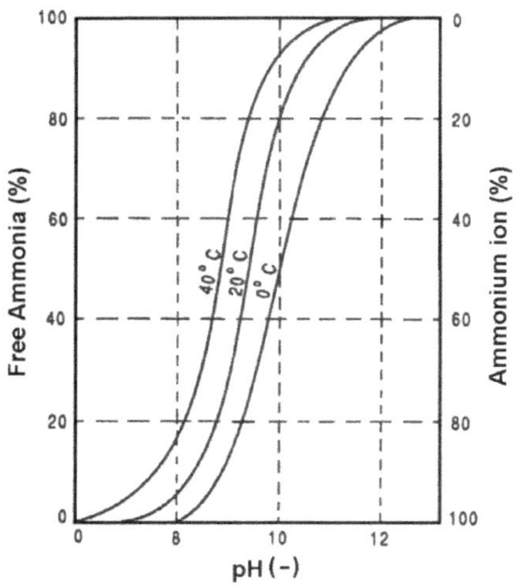

Figure 1. Free ammonia fraction as a function of temperature and pH

Efficiency of ammonia stripping depends mainly on pH and temperature, in addition to the dimensioning criteria of the process (e.g. liquid rate, air/liquid ratio, packing height, packing characteristics).

Air stripping is more common and more often applied than steam stripping, which is a quite similar process, with the difference that it requires operating temperatures higher than 95°C (Tchobanoglous et al., 2003). Steam stripping is usually adopted when ammonia recovery is economically attractive.

It should be considered that the resulting free dispersion of ammonia into the atmosphere may be unacceptable in many areas due to air quality concerns or regulations; in such cases, air stripping units are coupled with absorption towers which capture the released ammonia. In modern facilities, absorption onto sulfuric acid solutions allows the production of ammonium sulfate, which could constitute a marketable product (base fertilizer).

Figure 2 shows typical flow diagrams of the stripping process with air and steam.

Stripping is a relatively simple operation, as it only requires that pH and temperature remain stable enough. Its disadvantages are mainly related to the formation of $CaCO_3$ scale on the tower packing, resulting in a progressive loss of stripping performance, with the need of frequent chemical cleaning operations (Viotti and Gavasci, 2015). Furthermore, air stripping cannot be performed in freezing conditions when icing of the lower packing strata may occur, resulting in a dramatic reduction of process performance. Closed loop systems are designed to avoid this risk. Air stripping is often used in the treatment of municipal solid waste (MSW) landfill leachate (or groundwater polluted by the same leachate). An example of the process is represented in Figure 3.

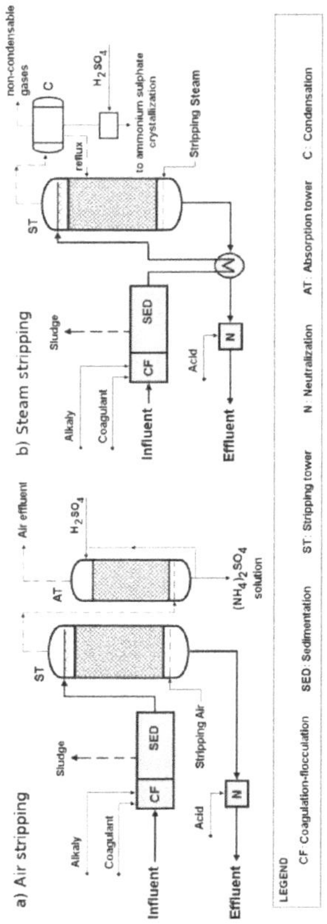

Figure 2. Typical flow diagrams of the stripping processes with air (a) and steam (b)

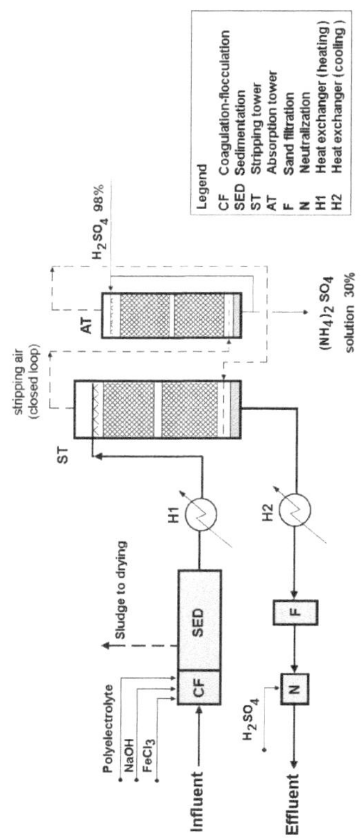

Figure 3. Diagram of a physico-chemical process for leachate treatment, with wastewater pre-heating and a closed-loop air stripping process

The system consists of: coagulation-flocculation at pH > 11; water heating at 38°C; ammonia air stripping; ammonia recovery by absorption with a sulfuric acid solution. Raboni et al. (2013b) obtained a reduction of ammonia from 199 mg L-1 (as N) down to 22.5 mg L-1 (as N). The plant showed severe scaling issues, with an increase in weight of the single filling components (Pall rings) up to seven times their original weight. The progressive scaling produced an 18% loss of ammonia removal efficiency, caused by the reduction of both the packing interfacial surface and the air/water ratio due to head loss increase. Periodical (every six months) chemical washing of the stripping towers with a 30% HCl solution was carried out to avoid scaling build-up. Closed-loop air recirculation allowed the water temperature inside the column to be maintained in the range of 25-38°C, even during winter, but caused a gradual decrease of the solution's pH within the tower. This was due to the acidic characteristics of the stripping air coming from the ammonia absorption tower. Subsequent experiences proved the actual possibility of achieving high ammonia removal efficiencies (greater than 95%) by operating the stripping towers at 38°C with liquid rates below 4.5 m3 m-2 h-1 and air/water ratios above 1,500 Nm3 m-3 (Viotti and Gavasci, 2015).

Vacuum distillation is another viable solution for ammonia recovery (US-EPA, 2013). This process can be justified in case of highly concentrated streams and when waste heat is available. Vacuum flash distillation processes (Figure 4) have been applied to ammonia recovery from centrate of manure digested sludge, using as heat source the biogas generated by the anaerobic digestion (Orentlichter et al., 2009).

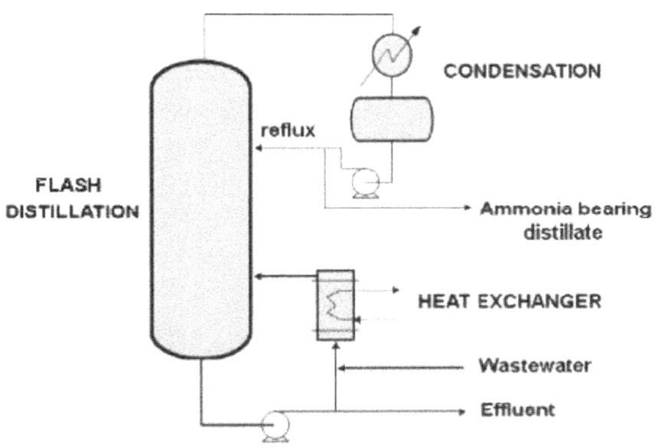

Figure 4. Diagram of a typical flash distillation process for ammonia recovery.

By operating the process at pH = 9.5 and 60°C under vacuum, an ammonia removal of approximately 1,000 mgNH3-N L-1 to less than 100 mgNH3-N L-1 can be achieved (Orentlichter et al., 2009). Other combinations of temperature and pH achieve similar high free ammonia fractions, allowing an effective stripping process. The resulting ammonia-rich distillate is a marketable product or it can be converted into an ammonium sulfate solution. If desired, such a solution can be further concentrated, using multiple vacuum distillation stages. According to pilot-scale results obtained with digested sludge liquor and landfill leachate at various concentrations, the process behaves according to a first order kinetics with respect to ammonia concentration, so it can be expected to provide 90% removal over a wide range of influent ammonia concentrations (Orentlicher, 2012).

Ammonia Precipitation as Struvite

Precipitation of NH4+-N as Magnesium Ammonium Phosphate (MAP), also named struvite (Figure 5), has been studied and implemented for different types of wastewaters, such as tannery effluents of leather industries (Tünay et al., 1997), digester supernatant (Siegrist et al., 1994; 1996; Celen and Turker, 2001) and in industrial wastewater treatment plants, from coke plants and nitrogen works to pig farms (Zdybiewska and Kula, 1991; Nelson et al., 2003), sludge liquor (Schulze-Rettmer, 1999) and landfill leachate (Prater, 2014). The reaction is as follows:

$$Mg^{2+} + PO_4^{3-} + NH_4^+ + 6H_2O \leftrightarrows MgNH_4PO_4 \cdot 6H_2O\downarrow \quad (3)$$

MAP has low water solubility (0.169 g L-1 at 25°C) and can be easily separated from the water phase. Since the molecular weight of MAP is 245 g mol-1, 17.5 g MgNH4PO4•6H2O are theoretically formed as precipitate when 1 g of NH4+-N is removed. Because MAP has a similar composition of Mg, P and N found in guano, it can be used in lieu of commercial fertilizers.

The basic chemical reactions for MAP formation are reported as follows (Li et al., 1999):

$$MgCl_2 \cdot 6\,H_2O + Na_2HPO_4 \cdot 12\,H_2O + NH_4^+ \leftrightarrows MgNH_4PO_4 \cdot 6H_2O\downarrow + 2NaCl \quad (4)$$

$$MgO + H_3PO_4 + NH_4^+ \rightarrow MgNH_4PO_4 \cdot 6H_2O\downarrow + H_2O \quad (5)$$

$$Ca(H_2PO_4)_2 \cdot H_2O + MgSO_4 \cdot 7H_2O + NH_4^+ \rightarrow MgNH_4PO_4 \cdot 6H_2O\downarrow + CaSO_4\downarrow \quad (6)$$

Struvite precipitation is controlled by pH and temperature, and can be influenced by impurities (e.g. calcium). Crystal formation proceeds by nucleation from crystal embryos followed by their growth, and their precipitation will not occur without nucleation (Kim et al., 2007).

Figure 5. Electron microscopy imaging of MAP precipitates.

Zhang et al. (2009) showed that ammonium in landfill leachate could be removed at the optimum pH of 9.5. The molar ratio $Mg^{2+}:NH_4^+:PO_4^{3-}$ was controlled at 1.15:1:1 to remove ammonium effectively and avoid higher concentrations of PO43− in the effluent.

Magnesium chloride ($MgCl_2$) is the most commonly used reagent in these cases. However, Huang et al. (2010) used magnesite as a low-cost magnesium source for the treatment of rare-earth elements separation process wastewaters. Since magnesite solubility is low, prior to use it was decomposed to magnesia, which has a higher reaction rate. The cost of adding magnesium salts is the major economic constraint to the application of struvite crystallization for nutrient recovery. Lee et al. (2003) report the removal of nitrogen and phosphate from wastewater by addition of bittern, the salt produced by evaporation of seawater which is a byproduct of salt manufacture. Bittern is a low-cost source of magnesium ions because contains mostly magnesium chloride (approximately 32 g L-1, or 27 times that of seawater) with smaller amounts of other inorganic compounds. The effectiveness of bittern as a precipitant, in relation to magnesium chloride and seawater, was evaluated at different pH's, with results slightly lower than those of alternative Mg^{2+} sources.

The most-used industrial reactor technologies for struvite precipitation are fluidized bulk reactors and stirred tanks, the latter being the easiest to set-up and operate. Fluidized bulk reactors are harder to control since they need steady flow to keep the bulk fluidized. Some proprietary industrial processes are currently being commercialized. Figure 6 shows the diagram of a process for struvite precipitation with a fluidized reactor applied in a full scale plant in USA (Garcia-Belinchón et al., 2013; Hazen and Sawyer, 2010).

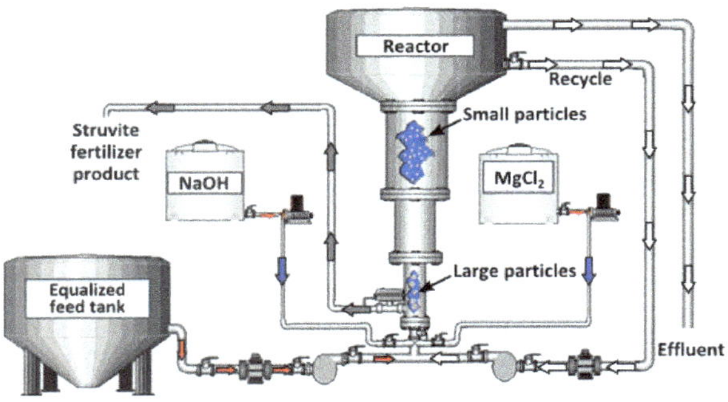

Figure 6. Diagram of the process for struvite precipitation and recovery.

NaOH and MgCl2 are added to the concentrate liquor (supernatant of a sludge digester with high N and P content). The resulting solution is fed to the fluidized reactor. The upper part consists of a clarification stage while the lower cylindrical parts are the stages in which struvite precipitates. Small MAP particles aggregate progressively up to form large particles which drop on the bottom of the reactor, where they are collected and reused as fertilizer.

Ion Exchange for Ammonia and Nitrate Removal

Ion exchange using zeolites is a technologically simple process for ammonia removal. Zeolites have a porous structure that can accommodate a wide variety of cations (e.g. Na^+, K^+, Ca^{2+}, Mg^{2+} and others). Zeolite minerals differ in Si and Al content. Natural zeolites (aluminosilicates) particularly used are clinoptilolite and modernite. Clinoptilolite, the most common and abundant high-siliceous zeolite (Si:Al ratio equal to 5.7) is usually employed as ion exchange medium. It has a two-dimensional channel system that allows the mineral to act as a molecular sieve. For this reason it has high sorption and ion-exchange capacity, ion exchange selectivity, catalytic activity and structural temperature stability up to 700-750°C (Zabochnicka-Świątek and Malinska, 2010). Synthetic zeolites are also produced and they have some key advantages over their natural analogs, as they are manufactured in a uniform, phase-pure state. The main raw materials used to manufacture zeolites are silica and alumina, which are among the most abundant mineral components on earth; therefore the potential to supply synthetic zeolites is virtually unlimited.

The mechanism of ammonia-nitrogen removal by clinoptilolite is based on cationic ion exchange. But ammonia can be also removed through adsorption in the structural pores of the zeolite. The efficiency of ammonia removal is strongly affected by temperature: the higher the temperature, the higher the removal efficiency achieved. Regarding pH, the removal capacity of clinoptilolite is approximately constant between 4 and 8, diminishing rapidly outside this range (US-EPA, 1971a; b). Clinoptilolite and other zeolite minerals easily adsorb potassium, ammonia, calcium, sodium, magnesium and other ions. Therefore, high cationic concentration of the solution will reduce ammonia removal capability.

In demonstration studies on municipal wastewaters containing about 20 mgNH_3-N L^{-1}, an average ammonia removal of 95.7% was reached (US-EPA, 1971a; b). Another important advantage of zeolite applied to water treatment is its high porosity when compared to other minerals: this results in good hydrodynamic properties (head loss through zeolite filters is 1.5 2.0 times smaller than that in sand filters) and adsorption properties; furthermore its high capacity allows for adsorption of other contaminants (e.g. phytoplankton and bacteria).

Both natural and synthetic zeolites can be regenerated with solutions containing high concentrations of calcium ions. The regenerant is a relatively small volume of liquid with high ammonia concentrations. Ammonia-rich eluate can be stripped, recovered by chemical absorption or destroyed by electrolysis producing chlorine which reacts with the ammonia to produce nitrogen gas (Figure 7).

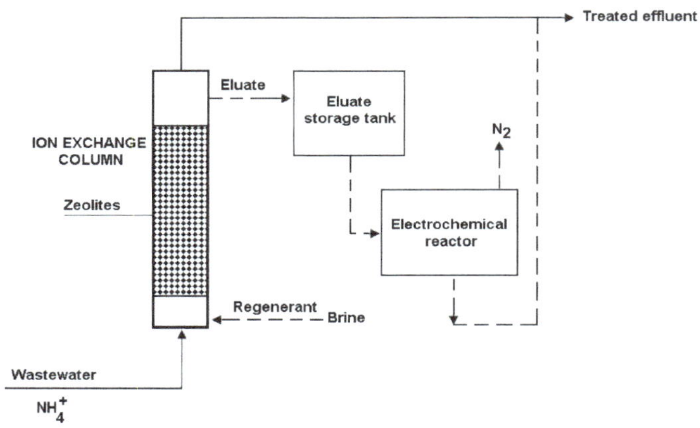

Figure 7. Simplified diagram of a ion exchange process on a zeolite bed for ammonia removal. Ammonia in the eluate is decomposed to nitrogen gas by an electrochemical reaction

Rozic et al. (2000) investigated the removal of ammonium ions from aqueous solutions using a combination of natural clay and zeolite. Clays and zeolites are both hydrated aluminosilicates, but they have different crystalline structure. Zeolites consist of symmetrically stacked alumina and silica tetrahedra forming an open and stable three dimensional honeycomb structure with negative charge, while clays are extremely fine allumino-silicates crystals or particles. Because of their fineness, clays exhibit the chemical properties of colloids and structurally appear as "sandwiches" of tetrahedral and octahedral sheet structures. Rozic et al. (2000)found that zeolite removes more than 50% (in weight) of ammonia nitrogen, while clay must be added in a liquid colloidal state in order to increase the specific surface area and to obtain maximum efficiency. A further process is required to be performed on the effluent as a consequence of the adding of colloidal clay.

Jorgensen and Weatherley (2003) evaluated the effects of organic contaminants in synthetic wastewater, showing that in most cases their presence enhances the uptake of ammonium onto the ion exchangers. Rahmani et al. (2004) demonstrated successfully removal of ammonia from synthetic solutions using Semnan clinoptilolite samples from northern Iran. They showed that the total exchange capacity of such zeolite for three different mesh sizes was between 17.31 to 18.98 $mgNH^{4+}$ g^{-1} of dry weight. Zeolite particles with mesh 40 gave an ammonia adsorption capacity higher than other, coarser ones, indicating that smaller particle sizes have higher ion-exchange capacity thanks to the greater available surface area. They also demonstrated that the ammonia adsorption capacity of the Semnan clinoptilolite remained constant after multiple regenerations with NaCl.

The ion exchange process is also effective at removing nitrates. Applications of this type are mainly in the field of water supply using synthetic resin beads. Since conventional anion exchange resins have a higher affinity for sulfate and other anions, nitrate-selective strong base anion resins with triethylamine or tributylamine functional groups were developed.

Ethyl groups force nitrate selectivity over sulfate demonstrating to remove over 90% of nitrates present in water. Such resins are regenerated with NaCl or KCl. PuroliteTM type resins have also shown the ability to remove arsenic and nitrate (Wang et al., 2011). A model of the process is reported in the literature (Kim and Benjamin, 2004).

Chemical Oxidation of Ammonia

Ammonia and Chloramines Removal by Activated Carbon

Activated carbon (AC) is commonly used for the adsorption of ammonia gas, while the aqueous phase adsorption is generally considered ineffective due to ammonia's strong affinity with water. Adsorption of aqueous ammonia is obtained with bamboo charcoal carbonized at 400°C, subsequently treated with diluted sulfuric acid to enhance its adsorption properties (Asada et al., 2006). Its adsorption capacity is evaluated through an adsorption isotherm, in about 8.5 mg ammonia g-1 charcoal in correspondence of an ammonia equilibrium concentration of 5 mg L^{-1}.

Formation of chloramines, and their subsequent removal by activated carbon, is well known. Chloramines are formed by adding chlorine or its by-products to a water with some ammonia content. The reactions are:

$$HOCl + NH_3 \rightarrow NH_2Cl + H_2O \text{ (monochloramine)} \quad (7)$$

$$HOCl + NH_2Cl \rightarrow NHCl_2 + H_2O \text{ (dichloramine)} \quad (8)$$

$$HOCl + NHCl_2 \rightarrow NCl_3 + H_2O \text{ (nitrogen trichloride)} \quad (9)$$

The formation of the specific chloramines depends on both the pH and the $Cl^2:NH^{4+}$ molar ratio (Tchobanogolous et al., 2003). At pH 7-8 and $_{Cl2}:NH^{4+} > 1.8$ dichloramine is dominant; viceversa, with $Cl_2:NH^{4+} < 0.7$ the monochloramine is largely prevalent.

Chloramines possess disinfectant power, like chlorine, but they have a negative impact on the taste and odor of water, and some toxicological impact on human health. Therefore, their removal by AC is commonly practiced in drinking water treatment plants and, in a few cases, also in sewage treatment plants for the protection of receiving waters' ecosystems.

The reactions of AC with residual chlorine and dichloramine can be summarized as follows:

$$AC + 2\,Cl_2 + 2\,H_2O \rightarrow 4\,HCl + CO_2 \quad (10)$$

$$AC + 4\,NHCl_2 \rightarrow 2N_2 + CO_2 + 8H^+ + 8Cl^- \quad (11)$$

Hence, AC does not really adsorb chloramines but acts as a catalyst for their chemical breakdown. This catalytic reaction involves formation of an oxide of carbon intermediate (CO*). Fine mesh sizes of AC remove chloramines more efficiently than coarse ones, since they have greater surface area, and allow faster access to the catalytic sites. Also, selected AC with increased sites for CO* formation improves chloramine removal. When using new AC, an initial dosing with chlorine to pre-oxidize the carbon may result in a more effective chloramine removal (WQA, 2013). In order to improve AC's adsorption capacity of chloramines and chlorine, new types of surface-enhanced, catalytic AC were manufactured. These catalytic carbons are marketed with a peroxide number (rate of hydrogen peroxide decomposition), in addition to the traditional iodine number. With these new products the minimum empty bed contact time (EBCT) has been reduced to 3 min, while for classical ACs the minimum EBCT's need to be greater than 10 min.

Breakpoint Chlorination for Ammonia Removal

Breakpoint chlorination is a widely used process that oxidizes ammonia to nitrogen gas, thus removing it from wastewaters. This process is practical only as an effluent polishing technique, not for the removal of high levels of nitrogen. When chlorine is added to wastewater containing ammonia-nitrogen, it initially reacts with hypochlorous acid to form chloramines. Continued addition of chlorine after "breakpoint" (occurring when free chlorine residual is formed) converts chloramines to nitrogen gas. The overall breakpoint reaction is given as:

$$NH_4^+ + 1.5\,HOCl \rightarrow 0.5\,N_2 \uparrow + 1.5\,H_2O + 2.5\,H^+ + 1.5\,Cl^- \quad (12)$$

Stoichiometrically, a Cl2:NH3-N weight ratio of 7.6:1 is required to oxidize ammonia to nitrogen gas. Figure 8shows a theoretical breakpoint curve.

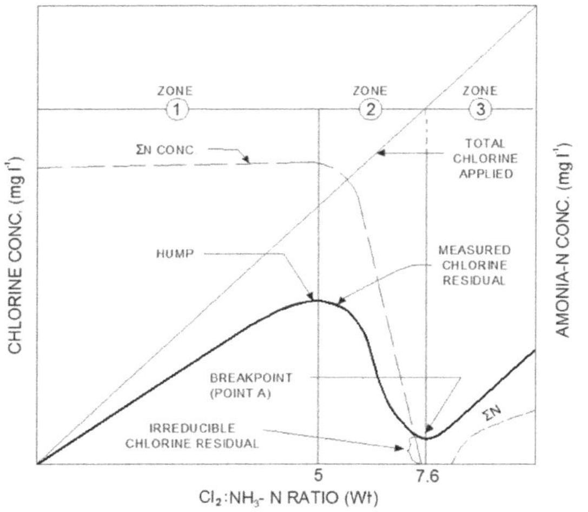

Figure 8. Theoretical breakpoint curve

Zone 1 is associated with the reactions of chlorine and ammonia to form monochloramine; Zone 2 is associated with an increase in dichloramine and the disappearance of NH_3; Zone 3 is associated with the appearance of free chlorine after the breakpoint. Ammonia concentration begins to decline at about Cl_2:NH_3 ratio equal to 5, until it becomes nearly completely absent at the breakpoint. It is generally accepted that, as applied chlorine increases from 0 up to the "hump" in the curve (Zone 1), monochloramine is being formed. The declining portion of the breakpoint curve is where dichloramine is formed, and then decomposes to yield nitrogen gas, resulting in a decline of the ammonia nitrogen concentration.

Design of the mechanical components of a breakpoint chlorination system is quite simple. The first reaction between ammonia and chlorine occurs rapidly, and no special design features are necessary, except providing for complete, uniform mixing of the chlorine with the wastewater. A minimum contact time of 10 minutes is a good practice. Sizing the chlorine-producing and feed device depends on the influent ammonia concentration, as well as on the degree of treatment that the wastewater has received. As the quality of influent to the breakpoint process improves, the required amount of chlorine decreases, and approaches the theoretical amount required to oxidize ammonia to nitrogen. If insufficient chlorine to reach the breakpoint is added, there is no nitrogen gas formation, and formed chloramines have to be destroyed prior to discharge.

Design should include provisions to continuously monitor the effluent for free chlorine residual following its addition, and to pace the rate of chlorine feed to maintain a set-point free chlorine residual. When using breakpoint chlorination, attention should be paid to the possible formation of excess acid (that some wastewaters do not neutralize), and to the elimination from the effluent of the active chlorine residual by dechlorination. Theoretically, 14.3 mg L^{-1} of alkalinity as $CaCO_3$ is required to neutralize the acid (HCl) produced by the oxidation of 1 mg L^{-1} of ammonia-nitrogen gas. To maintain pH above 6.3, at least twice this theoretical alkalinity is needed. Except for wastewaters already possessing high alkalinity, and treatment systems using lime coagulation prior to chlorination, design normally includes the means to feeding an alkaline chemical to the wastewater stream, to maintain pH in the proper range. The pH should be monitored carefully and continuously. In fact, a simple pH increase will not correct the problem: if pH rises above 8, breakpoint will be reached only very slowly. If NaOCl is used instead of Cl_2, the alkalinity requirement is reduced by 75%. Dechlorination may be required as a companion process to breakpoint chlorination (WEF, 2009).

Advanced Oxidation Processes

Many advanced chemical oxidation processes for the conversion of ammonia and N-organic into nitrogen gas or nitrate have been proposed and applied: ozonation, peroxone oxidation, catalytic wet air oxidation, photocatalityc oxidation, and electrochemical oxidation.

Ozonation is one of the methods for the oxidation of low concentration of ammonia into nitrate. The reaction can be expressed as:

$$NH_3 + 4\,O_3 \rightarrow H^+ + NO_3^- + H_2O + 4\,O_2 \qquad (13)$$

However, since the persistence of ozone in water is quite short (its half-life is about 15 min at 298 K, and pH = 7), its rate of reaction with ammonia is slow.

A more advanced oxidation using ozone and hydrogen peroxide mixtures as combined oxidants (known as peroxone oxidation) may also be effective for ammonia removal from wastewaters.

The combined kinetics of traditional ozonation and peroxone (hydrogen peroxide mixtures) oxidation of ammonia in alkaline solutions was also studied (Chiang-Hai et al., 1997). The ozonation reaction is governed by the direct oxidation of ammonia with ozone molecules. The overall kinetics is second order, with first order each in ozone and ammonia.

Khuntia et al. (2012) found that the ozonation using a pilot plant with micro-bubbles instead of a traditional system increased the removal efficiency of ammonia thanks to (i) a larger gas–liquid interfacial area; (ii) slower bubble rising velocity; (iii) longer persistence; (iv) smaller amount of ozone required due to the generation of hydroxyl radicals responsible for pollutants oxidation. In fact, the solubility of ozone in water is several times higher than that of oxygen. Therefore, when a mixture of ozone and oxygen is passed into water in the form of micro-bubbles, ozone preferentially dissolves in water. Li et al. (2009) reported that free hydroxyl radicals are generated from collapsing oxygen micro-bubbles under strongly acidic conditions. As a consequence, the possibility of ammonia oxidation by these radicals increases. The authors also found that aqueous ammonia was effectively oxidized by ozone micro-bubbles under alkaline conditions (i.e., pH > 7), even at low concentrations (e.g., 1 mgNH_3 L^{-1}). Micro-bubble-aided ozonation is a fast process and can be used for water treatment at large scales, and at a wide pH range. After ozonation is complete, ozone self decomposes to oxygen without introducing any new pollutant into the water, and leaving only the products of oxidation.

Catalytic wet air oxidation (CWAO) consists of oxidizing pollutants with oxygen at high pressure (5-200 bar) and temperature (125-320°C) in the presence of a catalyst (e.g. oxides, noble metals). This process is applied to the treatment of liquid waste particularly rich in complex organic compounds. Several applications to chemical and petrochemical waste liquors are known. Among the nitrogenous organic compounds that can be treated by CWAO are aniline and its azoic derivatives: nitrotoluenesulfonic acid, nitrobenzene, dinitrotoluene, quinoline, nitrophenol, fenuron, acenaphtene, acetonitrile, carbamide, nitromethane, etc. CWAO is able to achieve the complete mineralization of organic pollutants (into CO_2, N_2 and H_2O), but in some cases the efficiency of the process is controlled at a lower level, in order to simply increase effluent's biodegradability. In general, also most of the non-catalytic techniques (WAO, O_3/H_2O_2, O_3/UV) can be applied, but to obtain significant conversions of nitrogenous organic products, they need to operate with a large excess of oxidants. CWAO processes have been applied to the elimination of ammonia with a good efficiency, particularly when associated with the use of cobalt (II) oxide (CoO) as a catalyst at temperatures up to 260°C. (Aguilar et al. (2003)) studied the oxidation of aqueous ammonia to nitrogen by CWAO process with AC as catalyst, confirming that ammonia conversion to nitrogen is favored by hydrogenated ACs in which the most acidic groups had been eliminated.

Photocatalytic oxidation of aqueous ammonia (into NO_2^- and NO_3^-) over microwave induced Titanate ($Na_xH_{2-x}Ti_3O_7$) NanoTubes (TNTs) was investigated (Ou et al., 2008). Photocatalytic oxidation of NH_3/NH^{4+} over TiO_2 is a promising process due to its inherent stability and high photocatalytic potential. TNTs have received much attention due to their one-dimensional nanostructure and versatile applications, including solar cell photocatalysis, and electroluminescent hybrid devices. Some researchers consider TNTs effective photocatalysts for pollutant' degradation (Tsai and Teng, 2004). A selective photocatalytic oxidation of NH_3 to N_2 on platinized TiO_2 in water was also investigated, achieving 80% ammonia removal efficiency (Lee et al., 2002).

Electrochemical oxidation of ammonia has attracted attention in recent years since it addresses both clean energy supply (CO_2 free) and environmental protection. In general, ammonia can be transformed to other nitrogen forms by direct anodic oxidation, or indirect oxidation in an electrochemical process. Direct oxidation at a reasonable reaction rate according to the chemical reactions 14, 15, and 16 implies many advantages; however, large scale applications have been strongly hindered by insufficient performance, and by the high cost of electrodes. In an indirect oxidation process, strong oxidants (HClO and •OH) are first produced in the bulk of solution via electrochemical reactions. Subsequently, these oxidants destroy the ammonia by oxidation reactions.

(Anode reaction) $$2\ NH_{3(aq)} + 6OH^- \rightarrow N_2 + 6H_2O + 6e^- \quad (14)$$

(Cathode reaction) $$6H_2O + 6e^- \rightarrow 3H_2 + 6OH^- \quad (15)$$

(Overall reaction) $$2NH_{3(aq)} \rightarrow N_2 + 3H_2 \quad (16)$$

Active components of electrodes include pure Pt, Pt-based alloys, and Pt-free metal oxide, such as $Ni/Ni(OH)_2$, IrO_2, etc. Despite the high catalytic activities of Pt and Pt-alloys for electrochemical oxidation of ammonia, the high cost of these electrodes limits their application at an industrial scale. Hence, Pt-free electrodes provide an alternative for accelerating its application. Electrochemical oxidation of ammonia over a Pt-free, Ru–Ir/TiO2 anode under continuous flow conditions was evaluated (He et al., 2015). The commercial anode was effective for ammonia removal by electro-oxidation at neutral pH in a continuous electrochemical reactor, and provided 80% removal of ammonia in a packed bed reactor, under optimum conditions, proving to be a suitable, economic alternative for the treatment of ammonia wastewater with high chloride concentration.

Chemical Reduction of Nitrate

Nitrate is a weak oxidizing ion; therefore, in order to react with reducing agents and produce nitrogen gas or ammonia, it requires the presence of a catalyst or improved thermodynamic conditions (pressure and temperature). For this purpose, reducing agents that have been experimentally applied under (i) acidic condition are: formic acid, iron metal, methanol and ammonium ion; (ii) alkaline conditions are: aluminum, zinc, iron, iron (II), ammonia, hydrazine, glucose, and hydrogen (Fanning, 2000). More recently, a chemical process capable of converting nitrate ions into environmentally neutral nitrogen gas was proposed. Nitrate-contaminated water is put in contact with sulfamic acid and zinc at pH 2.0 3.5. Sulfamic acid acts as reducing agent of nitrous ion, while Zn promotes nitrate reduction. The advantages of this process are its low cost, the absence of limitations by nitrate concentration levels and its lack of sensitivity to other contaminants (Seungmoon et al., 2006). Thus it may be applied to high strength wastewater from optoelectronic, fertilizer, pharmaceutical, food and dairy industries. Chemical denitrification depends on the pH as well as the Zn, sulfamic acid and nitrate concentrations (i.e. high Zn concentration increases the nitrate removal rate).

CONCLUSIONS

Physico-chemical processes have long been widely applied for the removal of nitrogen from wastewater. Among them, ammonia air stripping is perhaps the most widespread. It is typically applied to the treatment of wastewaters with high ammonia concentration (i.e. digested sludge liquor; landfill leachate; specific wastewater streams from chemical and petrochemical industries). In leachate treatment, ammonia removal efficiencies higher than 95% can be achieved. Recently, the process is often coupled to ammonia absorption onto sulfuric acid, yielding a solution that could constitute a marketable product (base fertilizer). Stripping is a relatively simple operation. Its disadvantages are mainly related to the formation of calcium carbonate scale in the packing of stripping towers and the icing of the lower packing strata in freezing conditions, resulting in the progressive loss of performance. Solutions can be the packing chemical cleaning with strong acidic solutions and closed loop air recirculation, respectively. In a few cases, ammonia removal and recovery is achieved by steam stripping or vacuum distillation processes. Breakpoint chlorination (ammonia converted into nitrogen gas) and activated carbon adsorption of chloramines are long known for their applications, mainly in the field of surface water potabilization. Currently, great interest is attached to ammonia precipitation as struvite (magnesium ammonium phosphate) from concentrated wastewater (e.g. digested sludge liquor; landfill leachate). Magnesium chloride is the most commonly used reagent, but other low cost magnesium salts can be used with some efficiency loss. A few proprietary industrial processes are currently being commercialized. Ion exchange processes for ammonia and nitrate removal have certain diffusion, but they are mainly confined to (clean) water treatment. Clinoptilolite is often applied for ammonia removal, while specifically manufactured strong anion-based resins with triethylamine or tributylamine functional groups have proven to be effective in nitrate removal, achieving over 90% efficiency. Other resins have demonstrated the ability to simultaneously remove arsenic and nitrate. Great attention is currently given to advanced chemical oxidation processes for the conversion of ammonia and N-organic into nitrogen gas or nitrate. Some of these processes (e.g. ozonation; peroxone oxidation; catalityc wet air oxidation) have already been applied in real scale applications, while others (e.g. photocatalytic oxidation, electrochemical oxidation; nitrate chemical reduction), although very promising, are still confined to the level of applied research.

REFERENCES

1. AGUILAR, C.; GARCIA, R.; SOTO-GARRIDO, G.; ARRIAGADA, R. Catalytic wet air oxidation of aqueous ammonia with activated carbon. Applied Catalysis B: Environmental, v. 46, n. 2, p. 229-237, 2003. http://dx.doi.org/10.1016/S0926-3373(03)00229-7
2. ASADA, T.; OHKUBO, T.; KAWATA, K.; OIKAWA, K. Ammonia adsorption on bamboo charcoal with acid treatment. Journal of Health Science, v. 52, n. 5, p. 585-589, 2006. http://dx.doi.org/10.1248/jhs.52.585
3. CELEN, I.; TÜRKER, M. Recovery of ammonia as struvite from anaerobic digester effluents. Environmental Technology, v. 22, n. 11, p. 1263-1272, 2001. http://dx.doi.org/10.1080/09593332208618192
4. CHIANG-HAI K.; YUAN, F.; HILL D. O. Kinetics of oxidation of ammonia in solutions containing ozone with or without hydrogen peroxide. Industrial & Engineering Chemistry Research, v. 36, p. 4108-4113, 1997. http://dx.doi.org/10.1021/ie9702082
5. CULP, R. L.; CULP, G. L. Advanced wastewater treatment. New York: Van Nostrand Reinhold, 1971. 310 p.
6. ERAMO, B.; GAVASCI, R.; MISITI, A.; VIOTTI, P. Validation of a multisubstrate mathematical model for the simulation of the denitrification process in fluidized bed biofilm reactors. Water Science and Technology, v. 29, n. 10-11, p. 401-408, 1994.
7. FANNING, J. C. The chemical reduction of nitrate in aqueous solution. Coordination Chemistry Reviews, v. 199, n. 1, p. 159-179, 2000. http://dx.doi.org/10.1016/S0010-8545(99)00143-5
8. FARABEGOLI, G.; GAVASCI, R.; LOMBARDI, F.; ROMANI, F. Denitrification in tertiary filtration: application of an up-flow filter. Journal of Environmental Science and Health - Part A Toxic/Hazardous Substances and Environmental Engineering, v. 38, n. 10, p. 2169-2177, 2003. http://dx.doi.org/10.1081/ESE-120023349
9. GARCIA-BELINCHÓN, C.; RIECK, T.; BOUCHY, L.; GALÍ, A.; ROUGÉ, P.; FÀBREGAS, C. Struvite recovery: pilot-scale results and economic assessment of different scenarios. Water Practice and Technology, v. 8, n. 1, p. 119-130, 2013. http://dx.doi.org/10.2166/wpt.2013.013
10. HAZEN AND SAWYER ENIRONMENTAL ENGINEERS & SCIENTISTS. New process makes asset of phosphorus and nitrogen from wastewater. 2010. Available in: <http://www.hazenandsawyer.com/news/new-process-makes-asset-of-phosphorus-and-nitrogen-from-wastewater>. Access in: May 2015.
11. HE, S.; HUANG, Q.; ZHANG, Y.; WANG, L.; NIE, Y. Investigation on Direct and Indirect Electrochemical Oxidation of Ammonia over Ru–Ir/TiO2 Anode. Industrial & Engineering Chemical Research, v. 54, n. 5, p. 1447–1451, 2015. http://dx.doi.org/10.1021/ie503832t
12. HUANG, H. M.; XIAO, X. M.; YANG, L. P.; YAN, B. Removal of ammonium as struvite using magnesite as a source of magnesium ions. Water Practice & Technology, v. 5, n. 1, 2010. http://dx.doi.org/10.2166/wpt.2010.007
13. JORGENSEN, T. C.; WEATHERLEY, L. R. Ammonia removal from wastewater by ion exchange in the presence of organic contaminants. Water Research, v. 37, n. 8, p. 1723-1728, 2003. http://dx.doi.org/10.1016/S0043-1354(02)00571-7
14. KIM, D.; RYU, H. D.; KIM, M. S.; KIM, J.; LEE, S. I. Enhancing struvite precipitation potential for ammonia nitrogen removal in municipal landfill leachate. Journal of Hazardous Materials, v. 146, n. 1-2, p. 81-85, 2007. http://dx.doi.org/10.1016/j.jhazmat.2006.11.054
15. KIM, J.; BENJAMIN, M. M. Modeling a novel ion exchange process for arsenic and nitrate removal. Water Research,v. 38, n. 8, p. 2053-2062, 2004. http://dx.doi.org/10.1016/j.watres.2004.01.012
16. KHUNTIA, S.; MAJUMDER, S. K.; GHOSH, P. Removal of ammonia from water by ozone microbubbles. Industrial & Engineering Chemical Research,v. 52, n. 1, p. 318-326, 2012. http://dx.doi.org/10.1021/ie302212p
17. LEE, J.; PARK, H.; CHOI, W. Selective photocatalytic oxidation of NH3 to N2 on platinized TiO2 in water. Environmental Science & Technology, v. 36, n. 24, p. 5462-5468, 2002. http://dx.doi.org/10.1021/es025930s
18. LEE, S. I.; WEON, S. Y.; LEE, C. W.; KOOPMAN, B. Removal of nitrogen and phosphate from wastewater by addition of bittern. Chemosphere, v. 51, n. 4, p. 265-271, 2003. http://dx.doi.org/10.1016/S0045-6535(02)00807-X
19. LI, X. Z.; ZHAO, Q. L. Inhibition of microbial activity of activated sludge by ammonia leachate. Environmental International, v. 25, n. 8, p. 961-968, 1999. http://dx.doi.org/10.1016/S0160-4120(99)00068-9
20. LI, X. Z.; ZHAO, Q. L.; HAO, X. D. Ammonium removal from landfill leachate by chemical precipitation. Waste Management, v. 19, n. 6, p. 409-415, 1999. http://dx.doi.org/10.1016/S0956-053X(99)00148-8
21. LI, P.; TAKAHASHI, M.; CHIBA, K. Enhanced free-radical generation by shrinking microbubbles using a copper catalyst. Chemosphere, v. 77, n. 8, p. 1157-1160, 2009. http://dx.doi.org/10.1016/j.chemosphere.2009.07.062
22. NELSON, N.O.; MIKKELSEN, R. L.; HESTERBERG D. L. Struvite precipitation in anaerobic swine lagoon liquid: effect of pH and Mg:P ratio and determination

of rate constant. Bioresource Technology, v. 89, n. 3, p. 229-236, 2003. http://dx.doi.org/10.1016/S0960-8524(03)00076-2

23. ORENTLICHER, M.; BROWN, S.; HALL, C. Centrate ammonia reduction with ARP(r): pilot data with synthetic and actual wastewaters. In: WEF/WERF NUTRIENT REMOVAL SPECIALTY CONFERENCE, 2009, New York. Proceedings... New York: WEF; WERF, 2009. p. 546-555.

24. ORENTLICHER, M. Overview of nitrogen removal technologies and application/use of associated end products. In: GOT MANURE TRADESHOW AND CONFERENCE, March 2012, New York. Proceedings... New York: Cornell University, 2012. p. 27-29.

25. OU, H. H.; LIAO, C. H.; LIOU, Y. H.; HONG, J. H.; LO, S. L. Photocatalytic oxidation of aqueous ammonia over microwave-induced titanate nanotubes. Environmental Science & Technology,v. 42, n. 12, p. 4507-4512, 2008. http://dx.doi.org/10.1021/es703211u

26. PRATER. J. Improved production of magnesium ammonium phosphate (struvite) from landfill leachate - Final report. Stevens Point: University of Wisconsin, 2014. 98 p.

27. RABONI, M.; TORRETTA, V.; URBINI, G. Influence of strong diurnal variations in sewage quality on the performance of biological denitrification in small community wastewater treatment plants (WWTPs). Sustainability, v. 5, n. 9, p. 3679-3689, 2013a. http://dx.doi.org/10.3390/su5093679

28. RABONI, M.; TORRETTA, V.; VIOTTI, P.; URBINI, G. Experimental plant for the chemico-physical treatment of groundwater polluted by MSW leachate, with ammonia recovery. Revista Ambiente & Agua, v. 8, n. 3, p. 22-32, 2013b. http://dx.doi.org/10.4136/ambi-agua.1250

29. RABONI, M.; TORRETTA, V.; VIOTTI, P.; URBINI, G. Pilot experimentation with complete mixing anoxic reactors to improve sewage denitrification in treatment plants in small communities. Sustainability,v. 6, n. 1, p. 112-122, 2014a. http://dx.doi.org/10.3390/su6010112

30. RABONI, M.; TORRETTA, V.; VIOTTI, P.; URBINI, G. Calculating specific denitrification rates in pre-denitrification by assessing the influence of dissolved oxygen, sludge loading and the mixed-liquor recycle. Environmental Technology,v. 35, n. 20, p. 2582-2588, 2014b. http://dx.doi.org/10.1080/09593330.2014.913690

31. RABONI, M.; GAVASCI, R.; URBINI, G. UASB followed by sub-surface horizontal flow phytodepuration for the sustainable treatment of the sewage generated by a small rural community. Sustainability,v. 6, n. 10, p. 6998-7012, 2014c. http://dx.doi.org/10.3390/su6106998

32. RAHMANI, A. R.; MAHVI, A. H.; MESDAGHINIA, A. R.; NASSERI, S. Investigation of ammonia removal from polluted waters by Clinoptilolite zeolite. International Journal of Environmental Science & Technology,v. 1, n. 2, p. 125-133, 2004. http://dx.doi.org/10.1007/BF03325825

33. ROZIC, M.; CERJAN-STEFANOVIC, S.; KURAJICA, S.; VANCINA, V.; HODZIC, E. Ammoniacal Nitrogen Removal from Water by Treatment with Clays and Zeolites. Water Research,v. 34, n. 14, p. 3675-3681, 2000. http://dx.doi.org/10.1016/S0043-1354(00)00113-5

34. SCHULZE-RETTMER, R. The simultaneous chemical precipitation of ammonium and phosphate in the form of magnesium-ammonium-phosphate. Water Science & Technology, v. 23, n. 4-6, p. 659-667, 1999.

35. SEUNGMOON, L.; SANJEEV, M.; JUNG-HWA, J.; KWINAM, P.; JIN-WON, P. Development of physicochemical nitrogen removal process for high strength industrial wastewater. Water Research,v. 40, n. 5, p. 975-980, 2006. http://dx.doi.org/10.1016/j.watres.2006.01.018

36. SIEGRIST, H.; GAJCY, D.; SULZER, S.; ROELEVED, P.; OSCHWALD, R.; FRISCHKNECHT, H. et al. Nitrogen elimination from digester supernatant with magnesium-ammonium-phosphate precipitation. In: Chemical water and wastewater treatment II. Berlin Heidelberg: Springer, 1994. p. 457-465.

37. SIEGRIST, H. Nitrogen removal from digester supernatant - comparison of chemical and biological methods. Water Science & Technology,v. 34, n. 1-2, p. 399-406, 1996. http://dx.doi.org/10.1016/0273-1223(96)00529-X

38. TCHOBANOGLOUS, G.; BURTON, F. L.; STENSEL, H. D. Wastewater engineering - treatment and reuse. 4th Ed. New York: Mc Graw Hill, 2003.

39. TORRETTA, V.; RAGAZZI, M.; TRULLI, E.;; DE FEO, G.; URBINI, G.; RABONI, M.; RADA, E. C. Assessment of biological kinetics in a conventional municipal WWTP by means of oxygen uptake rate method. Sustainability,v. 6, n. 4, p. 1833-1847, 2014. http://dx.doi.org/10.3390/su6041833

40. TSAI, C. C.; TENG, H. Regulation of the physical characteristics of titania nanotube aggregates synthesized from hydrothermal treatment. Chemistry of Materials, v. 16, n. 22, p. 4352-4358, 2004. http://dx.doi.org/10.1021/cm049643u

41. TÜNAY O.; KABDASLI, I.; ORHON, D.; KOLÇAK, S. Ammonia removal by magnesium ammonium phosphate in industrial wastewaters. Water Science & Technology,v. 36, n. 2-3, p. 225-228, 1997. http://dx.doi.org/10.1016/S0273-1223(97)00391-0

42. URBINI, G.; GAVASCI, R. VIOTTI, P.Oxygen control and improved denitrification efficiency by means of a post-anoxic reactor. Sustainability,v. 7, n. 2, p. 1201-1212, 2015. http://dx.doi.org/10.3390/su7021201

43. UNITED STATES. Environmental Protect Agency - US-EPA. Wastewater ammonia removal by ion exchange. Washington DC, 1971a. 111 p. (Water Pollution Control Research Series 17010 ECZ 02/71.)

44. UNITED STATES. Environmental Protect Agency - US-EPA. Optimization of ammonia removal by ion exchange using clinoptilolite. Washington DC, 1971b. 192 p. (Water Pollution Control Research Series 17080 DAR 09/71)

45. UNITED STATES. Environmental Protect Agency - US-EPA. Emerging Technologies for wastewater treatment

and in-plant wet weather management. Washington DC, 2013. 188 p. (EPA 832-R-12-011.)

46. VIOTTI, P.; GAVASCI, R.Scaling of ammonia stripping towers in the treatment of groundwater polluted by the leachate of municipal solid waste landfill: study of the causes and the effects on the stripping performance. Revista Ambiente & Agua,v. 10, n. 2, p. 240-252, 2015. http://dx.doi.org/10.4136/ambi-agua.1567

47. WANG, L.; CHEN, A. S. H.; WANG, A.; CONDIT, W. E. Arsenic and nitrate removal from drinking water by ion exchange. Cincinnati: US-EPA, 2011. 138 p. (EPA/600/R-11/040.)

48. WATER ENVIRONMENT FEDERATION - WEF. Design of municipal wastewater treatment plants - manual of practice 8. 5th ed. New York: McGraw-Hill Professional, 2009. 2600 p.

49. WATER QUALITY ASSOCIATION - WQA. Chloramine Fact Sheet. Lisle: WQA, 2013. 5 p.

50. WHITE, G. C. Handbook of Chlorination. 3rd ed. New York: Van Nostrand Reinhold Co., 1992.

51. ZABOCHNICKA-ŚWIĄTEK, M.; MALIŃSKA, K. Removal of ammonia by clinoptilolite. Global NEST Journal, v. 12, n. 3, p. 256-261, 2010.

52. ZDYBIEWSKA, M. W.; KULA, B. Removal of ammonia nitrogen by the precipitation method, on the example of some selected waste waters. Water Science & Technology,v. 24, n. 7, p. 229-234, 1991.

53. ZHANG, T.; DING, L.; REN, H. Pretreatment of ammonium removal from landfill leachate by chemical precipitation. Journal of Hazardous Materials,v. 166, n. 2-3, p. 911-915, 2009. http://dx.doi.org/10.1016/j.jhazmat.2008.11.101

CHAPTER 16

BIOCHEMICAL OXYGEN DEMAND AND CHEMICAL OXYGEN DEMAND FOR VARIOUS WASTEWATER TREATMENT PLANTS

Khaled Zaher Abdalla and Gina Hammam

Keywords: Biochemical Oxygen Demand; Biodegradability Index; Chemical Oxygen Demand; Wastewater Treatment Plants

INTRODUCTION

Wastewater is characterized in terms of physical, chemical, and biological composition [1]. Depending on the level of pollutants and local regulations; physical, chemical and/or biological treatment may be used [2].

In a wastewater treatment, organic matter is converted into inert mineralized products under controlled conditions and at higher rates [3]. Organic matter or compounds refer to the general class of chemicals composed of carbon and one or more of thefollowing elements: hydrogen, nitrogen, and oxygen; and are derived from material that was once alive [4,5]. Before any wastewater can be treated, it must first be characterized, because knowing the composition of the influent wastewater is essential for successful design and operation of wastewater treatment plants [6]. The impact of an effluent or wastewater discharge on the receiving water is predicted by its oxygen demand [7]. The two most common parameters used to recognize the composition of wastewater are the biochemical oxygen demand (BOD) and the chemical oxygen demand (COD). BOD_5 is a measure of how much dissolved oxygen is consumed by aerobic bacteria in 5 days at 20oC. It is the broad measure of the strength of the organic matter in a waste stream. The typical range of BOD_5 in domestic wastewater ranges from 100 to 300 mg/L. COD is chemical oxygen demand and is measured chemically by digestion with acid [8]. There exists a definite correlation between the COD and BOD under certain conditions and by determining the COD, the information about the BOD of the wastewater can be derived, but it is highly waste dependent [7, 8].

These two parameters have advantages and disadvantages, and the choice usually depends on many factors such as, the reproducibility of the determinations, time period required, location of the test [9]. Although, the BOD_5 test has been in use for more than a century, and consequently remains deeply entrenched in the practice and experience of biological wastewater treatment; the test is clearly lacking in many respects compared to the COD test [10]. On the other hand, COD analysis estimates the amount of organic matter in wastewater in only three to four hours, rather than the five days required by the BOD_5 test, and can be used as an alternative. COD results are typically higher than BOD_5 values, and the ratio between them will vary depending on the characteristics of the wastewater. This ratio has been commonly used as an indicator for biodegradation capacity [1]. It is called "Biodegradability index" (B.I.). It is generally considered the cut-off point between biodegradable and non-biodegradable waste [11]. Once an average B.I. has been established for the plant wastewater stream, COD test can be used to predict BOD_5. The BOD_5-to-COD ratio is typically 0.5:1 for raw domestic wastewater, and may drop to be as low as 0.1: 1 for a well-stabilized secondary effluent. There is no official value for BOD_5 / COD biodegradability index for different types of wastewater [12]. However, reported values for biodegradability index vary from 0.4 to 0.8 , for municipal raw wastewater. The ratio can exceed 10 for industrial wastewater [13]. In this study, eight different wastewater treatment plants (WWTPs), covered

Citation: Khaled Zaher Abdalla and Gina Hammam, Biochemical Oxygen Demand and Chemical Oxygen Demand for Various Wastewater Treatment Plants, all over Egypt and of different sizes, were investigated [10].

MATERIALS AND METHODS

The main objective of the analysis was to investigate the relation between biochemical oxygen demand (BOD) and chemical oxygen demand (COD). this includes determination of ranges and mean values of biodegradability indices in an attempt to make a kind of zoning for the BOD/COD ratio (biodegradability index). Statistical representation and analysis for the collected data were performed. The Different types and treatment capacities of investigated wastewater treatment plants are shown in Table 1.

RESULTS AND DISCUSSION

Figure 1 shows the variability in average biodegradability index for the different governorates in Egypt. These values were obtained from the available data of BOD_5 and COD for various wastewater treatment plants in each governorate.

Table 1. Investigated WWTP Design Capacities

Plant No.	Design Capacity m^3/d	Treatment Plant	Governorate
1	620.000	Arab Abu Saed	Greater
2	150.000	6 October	Giza
3	100.000	Qalioub	Greater
4	50.000	Hanoville	Alexandria
5	20.000	El Zarka	Damietta
6	20.000	El Gamalia	El Dakahlia
7	15.000	Eskan Mubarak	Alexandria
8	10.000	Bani Obeid	El Dakahlia

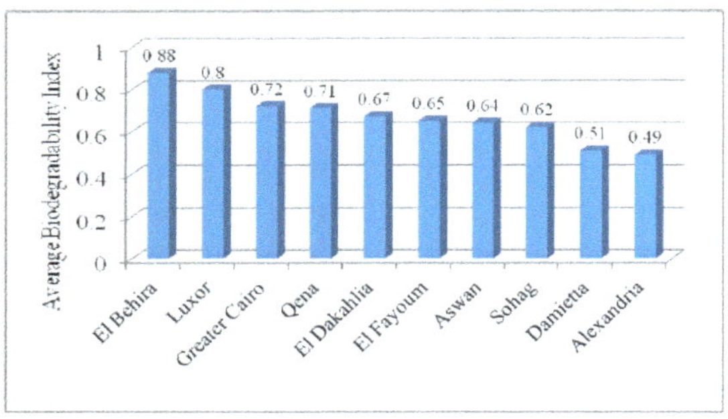

Figure 1. Average bioderadability index

Influent BOD_5 versus influent COD, of wastewater streams entering eight different wastewater treatment plants, were plotted. The data for Arab Abusaed WWTP, Qalioub WWTP, Six of October WWTP, Hanoville WWTP, El Zarka WWTP, El Gamalia WWTP, Eskan Mubarak WWTP, and Bani Obeis WWTP were shown in Figures 2, 3, 4, 5, 6,7, 8, and 9 respectively. These figures provide a clear prospect of the COD-BOD relation in different plants of variable types and treatment capacities. As shown in figures, it is clear that the correlation between BOD_5 and COD is generally linear, except in some rare cases where the relation was polynomial and sometimes doesn't have a definite pattern (dispersed). However, this rare indefinite pattern might be due to inaccuracy in determination of the BOD_5 and COD parameters in such cases. Since the BOD and COD are correlated, the estimation of BOD_5 values using the quick COD test, and plant specific biodegradability index (which is the slope of the plotted correlation) became possible and relatively reliable. Thus, it can be used as a check parameter to evaluate performance for quick action. For existing wastewater treatment plants the BOD/COD correlation should be developed, to compromise the use of these parameters. In order to establish the BOD/COD correlation for a particular existing wastewater, one should have both COD and BOD_5 values for several representative wastewater samples. Plot the BOD_5 values versus the COD values then use the regression analysis to develop the correlation. The plotted BOD_5-COD figures indicated that there is a clear linear positive correlation for most case studies, which differ from plant to another. Assessment of biodegradability index values periodically, and comparing it to the mean B.I. for the particular wastewater treatment plant can assist in monitoring the presence of toxic and non-biodegradable substances; and hence in applying the appropriate preventive actions. Before choosing the biological wastewater treatment plant technology, it is important to know the biodegradability index of the raw influent wastewater, as this choice would considerably affect the plant effluent quality. If BOD/ COD is > 0.6 then the waste is fairly biodegradable, and can be effectively treated biologically. If BOD/COD ratio is between 0.3 and 0.6, then seeding is required to treat it biologically, because the process will be relatively slow, as the acclimatization of the microorganisms that help in the degradation process takes time. If BOD/COD < 0.3, biodegradation will not proceed, thus it cannot be treated biologically, because the wastewater generated from these activities inhibits the metabolic activity of bacterial seed due to their toxicity or refractory properties. The relation between BOD_5 and COD for different WWTPs in Egypt are obtained from Figures 8 till 15. Thus, it can be used as a check parameter to evaluate performance of these WWTPS for quick action, and may also assist in monitoring the presence of toxic and non-biodegradable substances.

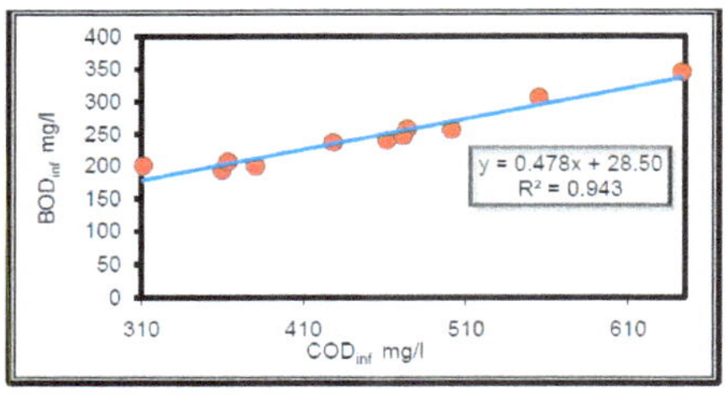

Figure 2. Arab Abu Saed WWTP, Helwan

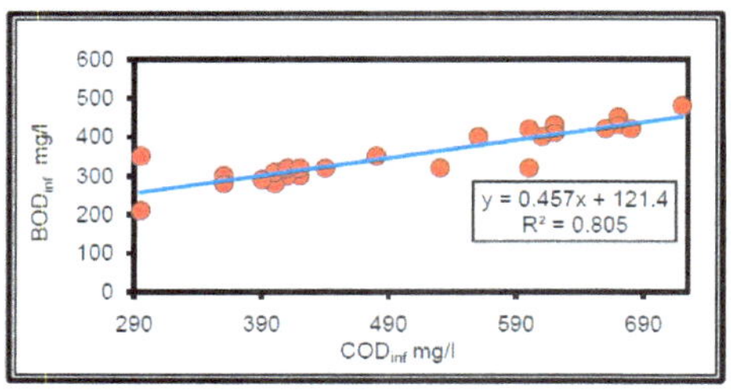

Figure 3. Qalioub WWTP, Qalioubiya

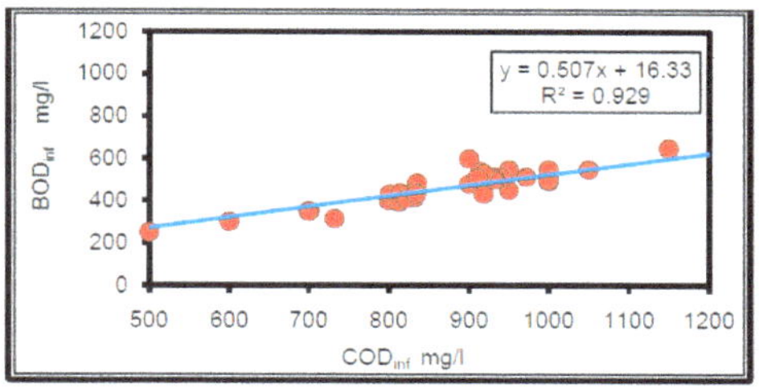

Figure 4. Six of October WWPT, Giza

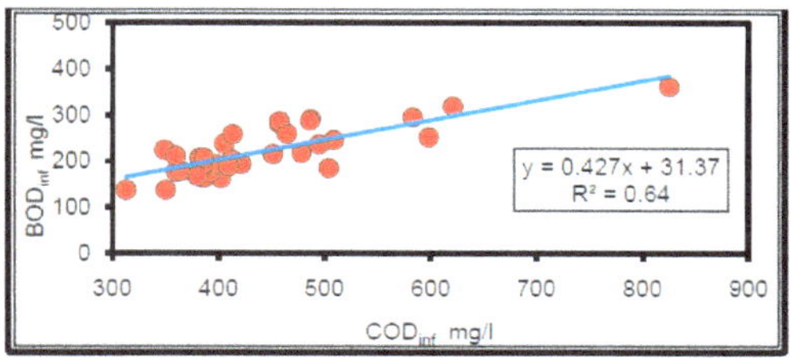

Figure 5. Hanoville WWTP, Alexandria

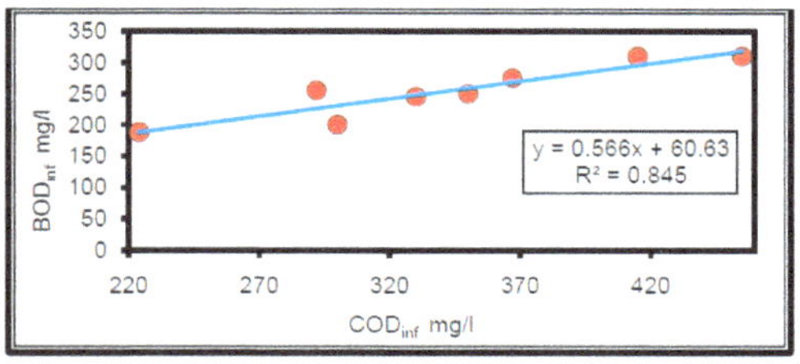

Figure 6. El Zarka WWTP, Domietta

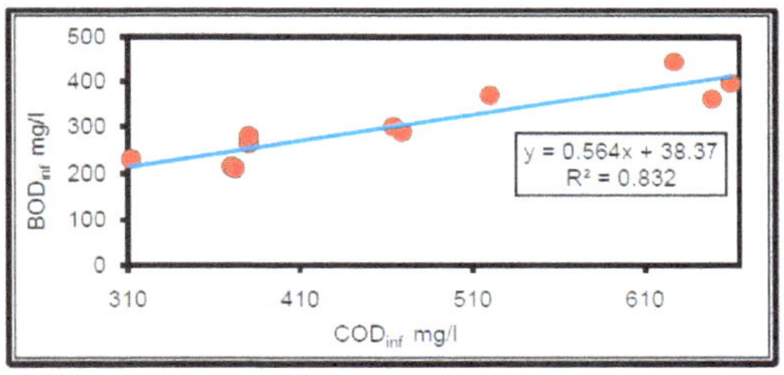

Figure 7. El Gamalia WWTP, El Dakahlia

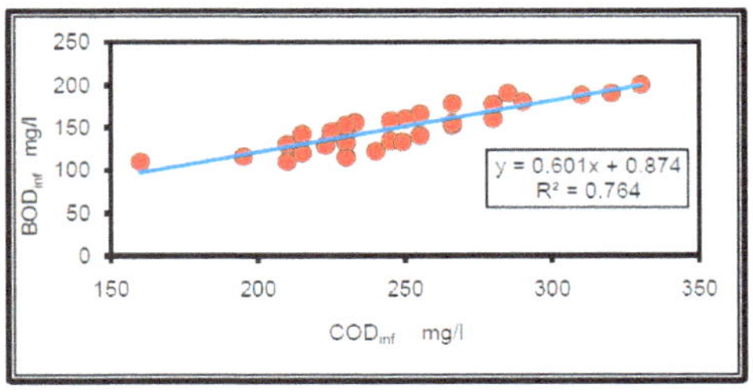

Figure 8. Eskan Mubarak WWTP, Alexandria

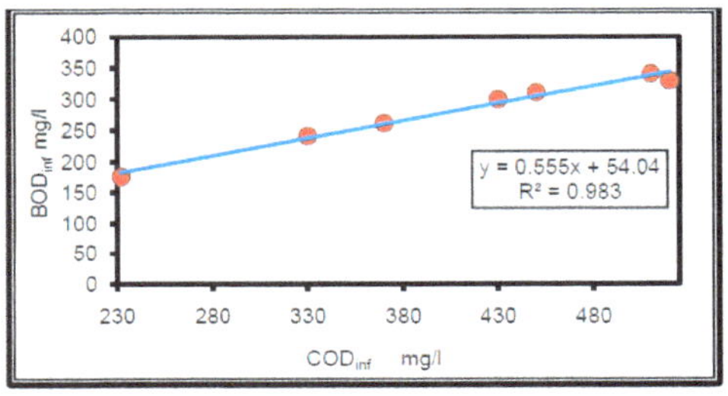

Figure 9. Bani Obeid WWTP, El Dakahlia

CONCLUSIONS

The conclusions obtained from this study are as follows;

- Biodegradability Index (B.I.) for the investigated WWTPs varied from 0.3 up to 0.96, which indicate the variability in the value of B.I. Thus, a constant ratio for B.I. cannot be assumed unless under identical environmental conditions.
- Raw wastewater BOD_5 & COD are obviously correlated and the correlation is a linear positive.
- The average B.I. of all investigated influent wastewater to wastewater treatment plants was 0.67. which indicates that generally Egyptian waste is fairly biodegradable. El Beheira governorate was found to has the highest mean B.I. of 0.88, while the lowest mean B.I. was 0.49 in Alexandria.
- Extensive observation of the COD and BOD levels in the same wastewater has shown that the COD to BOD ratio of a particular wastewater will remain constant over time. However, the correlation should be periodically rechecked due to probable seasonal variations in climatic conditions, social customs, water supply characteristics, water availability, population size, or the presence of industrial wastes.

REFERENCES

1. Metcalf & Eddy, *Wastewater Engineering: Treatment and Reuse,* 4th Ed., McGraw-Hill York, USA, 2003.
2. S. Ahmed,H. Abdelhalim, & E. Rozaik, ''Treatment of Primary Settled Wastewater Using Anaerobic Sequencing Batch Reactor Seeded with Activated EM''. *Civil and Environmental Research,* 3(11), 130-136, 2013.
3. M. Von Sperling, *Basic principles of wastewater treatment,* Vol. 2. IWA publishing, 2007.
4. F.R. Spellman, *Handbook of water and wastewater treatment plant operations*. CRC Press, 2008.
5. J.C. Crittenden, R.R. Trussell, D.W. Hand, K.J. Howe, & G. Tchobanoglous,(2012). *MWH's Water Treatment: Principles and Design*. Wiley, 2012.
6. A.P. Sincero, & G.A. Sincero, *Physical-chemical treatment of water and wastewater*. IWA Publishing, 2003
7. C.N. Sawyer, G.F. Parkin & P.L. McCarty, *Chemistry for Environmental Engineering*, 5th ed. New York: McGraw-Hill, 2003.
8. D.L. Russell, *Practical Wastewater Treatment*, John Wiley & Sons, Inc., Hoboken, NJ, USA, 2006.
9. A.A. Esener, J.A. Roels, & N.W.F. Kossen, The bioenergetic correlation of COD to BOD. *Biotechnology Letters* 3(4), 193-196, 1981.
10. G.S. Hammam, "Influence Of Raw Wastewater Biodegradability on Optimal Selection and Performance of Treatment Processes", MSc. thesis, Cairo university, Egypt, 2013.
11. U.G. Turak, & H.A. Fsar, "*Research of BOD And COD Values of Wastewaters that Contain Certain Organic Materials*", A Donor Menderes Univ., 4th AACD congress, Kusadasi – Aydin, Turkey, Proceeding Book

, p177, 2004.

12. Papadopoulous, G. Parissopoulous, F. Papadopoulos & A. Karteris, "Variations of COD/ BOD5 Ratio at Different Units of Wastewater Stabilization Pond Pilot Treatment Facility", *7th international conference on Environmental science and technology*, Syros Island, Greece, 2001.
13. G.P. Markantonatos, *Treatment and disposal of wastewater*. Athens, Greek, 1990.

CHAPTER 17

A NOVEL DESIGN CONCEPT OF MODIFIED ANAEROBIC INCLINING-BAFFLED REACTOR (MAI-BR) FOR WASTEWATER TREATMENT APPLICATION

Haider M. Zwain, Nastaein Qamaruz Zaman, Hamidi Abdul Aziz, Irvan Dahlan

Keywords: Anaerobic digestion, Anaerobic baffled reactor, Design modification, Recycled paper mill effluent, Start-up.

INTRODUCTION

With the rapid development and escalating crisis of water pollution, water resources are facing increasingly deficient. Therefore, there is a significant need to develop reliable technologies for wastewater treatment. Due to this problem, an anaerobic digestion process for wastewater treatment has gain increasing attention. It has several advantages that have attracted many researchers such as design simplicity, use of non-sophisticated equipment, low excess sludge production, high treatment efficiency and low capital and operating costs[1]. Nevertheless, anaerobic baffled bioreactor (ABR) can be described as a series of Upflow Anaerobic Sludge Blanket (UASB) for it can be divided into a few compartments. Figure 1 shows the original design of baffled bioreactor which is vertical in design. Vertical baffles arranged in series forces the wastewater to flow under and over them as it passes from the inlet to the outlet[2].

Bacteria inside the reactor gently rise and settle due to flow characteristics and gas production. The main driving force of an ABR is to enhance the solids retention capacity, however some modification have been made to treat different types of wastewater. Fannin et al.[3] introduced vertical baffle to a plug flow to treat high solids slurry in order to enhance the ability to maintain high populations of slowly growing methanogens. It proved that by applying a vertical baffle, more biomass can be retained so that slowly growing methanogens can actively produce methane gas. From the result, found that methane levels increased from 30% to over 55% with methane yield of about 0:34m3=kg VSS. In a later study, Bachmann et al.[4] comparing a baffled reactor before and after narrowing the down flow chambers and slanting the baffle edge, found that production rates and reactor efficiency were improved in the modified design.

The first several hybrid designs were introduced by Tilche and Yang[5], as shown in Fig. 2. The idea to modify the existing design was to enhance solids retention for high strength wastewater treatments. The solid settling chambers were incorporated after the last compartment and the solids washed out were collected and recycled to the first compartment. Packed Pall rings is randomly positioned at the liquid surface of the first two phase chambers and a deeper structured modular corrugated block which has a high voidage is positioned at the third chamber. Higher loading rates are possible for this structure due to minimal solids washouts during elevated gas mixing. The separation of the gas can enhance reactor stability by shielding entrophic bacteria from elevated levels of hydrogen which was found in the front compartment.

In addition, Boopathy and Sievers[6] further modified the baffled reactor to treat swine wastewater containing high content of small particulate material. The baffled reactor was modified to reduce upflow liquid velocities and to accept the whole waste, as shown in Fig. 2. The first compartment was double in size to 10L followed by 5L of second compartment. It is found that the larger compartment acted as a natural filter and provided superior solids retention for small particles. The reactor collected double the amount of solid material

Citation: Haider M. Zwain, Nastaein Qamaruz Zaman, Hamidi Abdul Aziz, Irvan Dahlan, A Novel Design Concept of Modified Anaerobic Inclining-Baffled Reactor (MAI-BR) For Wastewater Treatment Application, DOI: 10.12792/JIIAE.2.85.

(20:9g=L) than the reactor with three chambers. This was further substantiated in the solid washout data, which was lower in the two-compartment reactor despite showing lower treatment efficiency. Further analysis showed that despite losing more solids, the threecompartment reactor was more efficient at converting the trapped solids to methane.

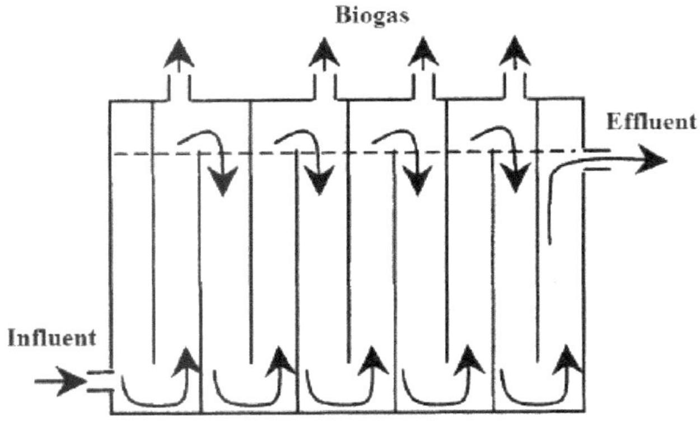

Figure 1: Original design of vertical baffled reactor.

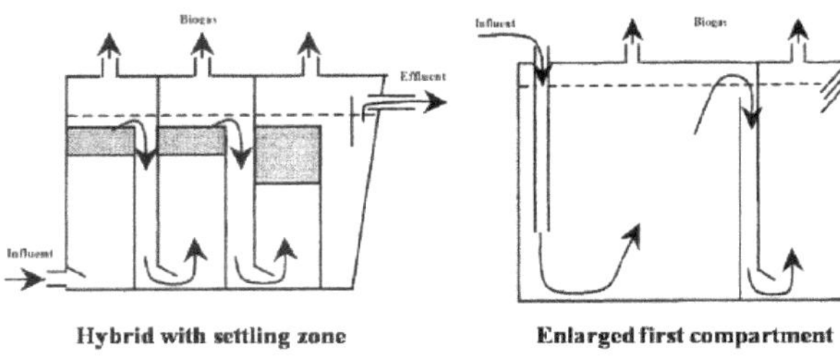

Figure 2: Modifications of the anaerobic baffled reactor.

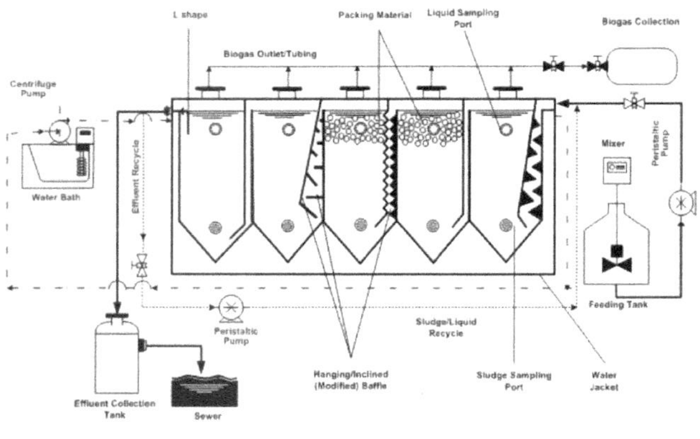

Figure 3: Schematic diagram of modified anaerobic inclining-baffled bioreactor (MAI-BR) system.

The development of effective and simple technologies for wastewater treatment is a challenging task to environmental engineers and scientists. Modification of the basic configuration of an ABR contributes to the development and improved treatment efficiency of wastewater. In this study, a novel modified anaerobic inclining-baffled reactor (MAI-BR) which is a combination of regular suspended growth and fixed biofilm systems together with the modification of baffled-reactor configurations, were developed. Furthermore, the reactor start-up and performance were investigated for the treatment of recycled paper mill effluent (RPME).

METHODOLOGY

Reactor Design and Fabrication

The schematic diagram of the laboratory-scale MAI-BR set-up is shown in Fig. 3. The MAI-BR is fabricated using polypropyleneplastic, constructed by a dimension of 80cm in length, 15cm in width and 30cm in height (without water jacket), having a total effective volume of 35L (calculated without baffles and packing materials). The MAI-BR consisted of five chambers and each chamber is separated by a modified vertical baffle, the lower portion of the hanging baffles bent to route the flow into the upflow chambers.

Each modified baffle has its own characteristics (form/ shape) to facilitate a better contact and greater mixing of wastewater and sludge at lower part of the MAI-BR. About 200g of the packing materials were added to the second and third chamber of the MAI-BR (the detail of baffle modifications and packaging materials are not shown in Fig. 3). The reactor had an attached water jacket to maintain the reactor temperature at 37°C. Peristaltic pumps are used to control the influent feed rate and effluent recycling to the first chamber of the reactor system.

Substrate and Seed inoculum

Anaerobic sludge from Palm Oil Mill Effluent (POME) (Malpom Palm Industries Bhd, Penang, Malaysia) was collected due to its ease of degradation and high COD value. The sludge was collected and kept in closed containers to avoid biological contamination by air. After collection, the sludge was immediately used and fed to the reactor. On the other hand, recycled paper mill effluent was used as substrate. The wastewater samples were collected from Muda Recycled Paper Mill, Penang, Malaysia and kept in cooling room at 4oC until used in seeding process.

Analytical Methods

For sample analysis, triplicate samples were collected for each reading and analyzed twice in order to increase the precision of the results, and only the average value was reported throughout this study. The repeatability of the experimental data was found to be sufficiently high with relative error between repeated runs less than 5%. Sludge, supernatant liquor and gas samples were taken separately for each compartment. Sample analysis including biological oxygen demand (BOD), pH, alkalinity, total solid (TS), suspended solids (SS) and volatile suspended solids (VSS); all according to standard methods [7].

The chemical oxygen demand (COD) was measured using Spectrophotometer DR-2800 according to the reactor digestion method adapted from Jirka and Carter [8]. The total volatile fatty acid (TVFA) was determined using Spectrophotometer DR-2800 according to esterification method adapted from Montgomery et al.[9]. The microbial floc size was measured using a Malvern Particle Size Analyzer model 2000. The methane gas (CH4) was determined using Shimadzu GC-FID with FTD column.

Reactor operation strategy

In order to investigate the performance of MAI-BR for the

treatment of recycled paper mill effluent (RPME), the start-up of the reactor was performed for 30 days using 1000mg=L of COD and HRT of 5 days, giving an organic loading rate of 0:2g COD/L. After the reactor was started, it was operated on continuous mood with 1000mg COD/L and 3 days HRT. An organic loading rate of 0:33g COD/L of RPM wastewater was supplied to the reactor for a period of 15 days.

Figure 4: Lab Scale MAI-BR.

Throughout the operational period, COD, biogas volume, methane content and pH were monitored every two days until steady-state condition achieved. At the steady-state condition, samples from each compartment, influent and effluent were taken and COD, BOD, TSS, VSS, Alkalinity, VFA, pH, lignin and floc size were analyzed.

RESULTS

The Modified Anaerobic Inclining-Baffled Reactor (MAI-BR)

The lab scale set up of MAI-BR is shown in Fig. 4. The compartmentalization MAI-BR results in a buffering zone between the primary acidification zone (downflow chamber) and active methanogenesis zone (upflow chamber)[1]. MAI-BR behaves partly as a fluidized bed reactor similar to up-flow anaerobic sludge blanket (UASB) reactor and it also acts as an activated sludge reactor as well as trickling submerged fixed film reactor.

This bioreactor combines suspended- growth and attachedgrowth processes in a single reactor to take advantage of both biomass types which theoretically advantageous to reactor kinetics and process optimization. The most significant advantages of this design was its ability to nearly perfectly realize the staged multi-phase anaerobic theory, allowing different bacterial groups to develop under more favorable conditions, low costs and without the associated control problems. Other advantages include reduced sludge bed expansion, no special gas or sludge separation required and high stability to organic and hydraulic and toxic shock loads.

The start-up of modified anaerobic inclining-baffled reactor (MAI-BR)

Initially, the reactor was started-up on continuous feeding with 1000mg COD/L giving an organic loading rate of 0:2g COD/day of RPME [10]. Over 30 days of continuous operation, the system demonstrated a remarkable performance of high COD removal efficiency (up to 71%). Similarly, a maximum COD reduction of 60% was recorded using unmodified anaerobic baffled reactor for the treatment of pulp and paper mill black liquors[11]. Nevertheless, the effluent pH was slightly decreased (from 7:3 to 6:2), while the methane production in- creased from 0:018 to 0:808L CH4/day in 30 days. It was believed that during anaerobic reactor start-up, the biomass is acclimatized to new environmental conditions, such as substrate, operating strategies, temperature and reactor configuration. Moreover, the methanogens and certain acetogens may be greatly outnumbered by the fast growing acidogens. Consequently, an accumulation of volatile fatty acids (VFAs) and dissolved H2 were occurred. Therefore, the pH was found slightly decrease since day 1 until day 30.

The Performance of Modified Anaerobic Inclining Baffled Reactor (MAI-BR)

The performance of the MAI-BR was investigated for the treatment of recycled paper mill effluent (RPME). The continuous operation of the MAI-BR was started using an initial COD concentration of 1000mg=L at HRT of 3 days. For 15 days, MAI-BR was run continuously and observations were made at this particular OLR and HRT. Biogas production, methane composition, pH profile and COD removal efficiency were monitored parameters during the continuous feeding technique. Fig.5 shows temporal changes in the total RPM COD removal RPME using MAI-BR.

As COD removal rate increased from 81% to 94%, the quick steady state was achieved. The COD removal rate in the reactor was comparatively high which indicate that MAI-BR has successfully performed. The above results on COD removal are comparable with other studies on treatment of municipal wastewater using modified anaerobic baffled reactor, when a total of 84% COD was removed [12].

One of the important observations in anaerobic process is pH levels. Fig. 6 shows the pH profile during 15 days of operation. The pH slightly decreased (from 7:1 to 6:7), indicating the adjustment of the microorganisms to maintain the desired pH levels (6:6 to 7:7) for an anaerobic reactor [13]. Methane production during the MAI-BR operation was analyzed and the results are shown in Fig. 7. Through 15 day, the biogas volume was increased from 0:88L to 1:05L per day, while the methane content increased from 28% to 55%. At the end of this phase, a total of 0:54L CH4/day was yielded. The increase in methane content may be attributed to the increase in the growth of anaerobes particularly methanogens in the newly synthesized biomass.

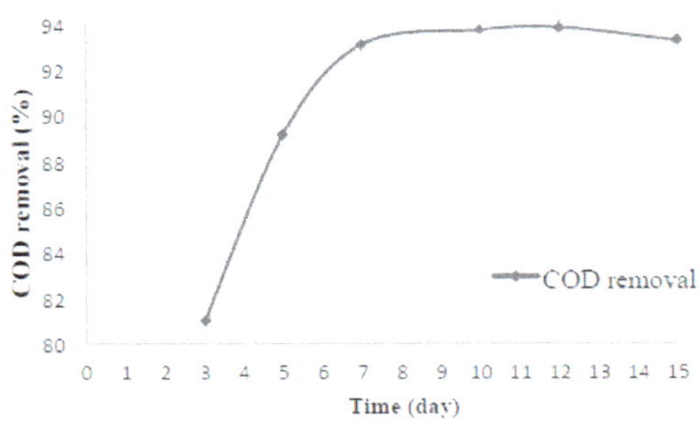

Figure 5: COD removal efficiency profile (%) at OLR of 0.33 g COD/day and HRT of 3 days.

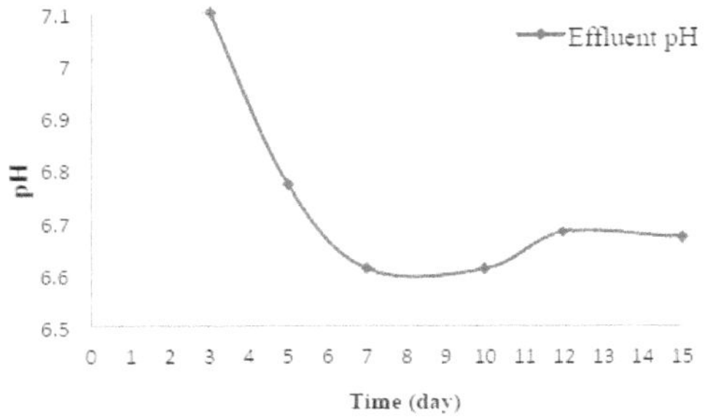

Figure 6: Effluent pH profile at OLR of 0.33 g COD/day and HRT of 3 days.

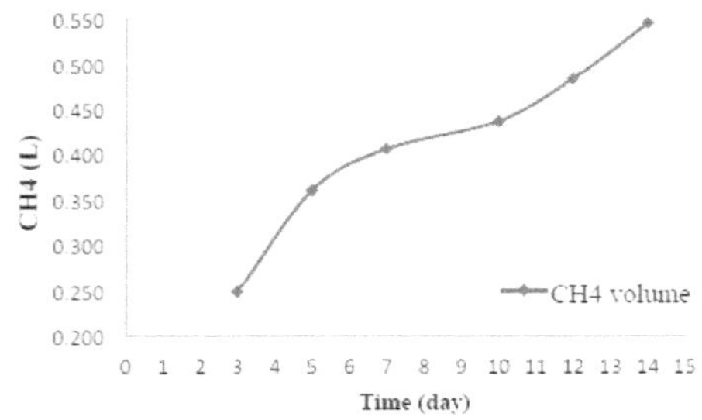

Figure 7: Methane production volume profile (L) at OLR of 0.33 g COD/day and HRT of 3 days.

The Performance of MAI-BR during Steady State

The performance of MAI-BR at steady state is shown in Table 1. Eleven parameters were tested for each compartment, influent and effluent. The contribution to total COD and BOD removals by each compartment was digressive longitudinally down the reactor, thus organic matter significantly removed in the 1st compartment. As illustrated in Table 1, the pH values of each compartment increased in the sequence of flow pattern from the influent to the effluent under all conditions tested. The pH levels in the individual compartments varied in a narrow range with no statistically difference. In addition, a low VFAs accumulation was observed in compartment, whilst the rest compartments maintained the acidity in terms of pH and VFA concentration at favorable levels that allowed effective anaerobic digestion to proceed [14]. sharply in the following compartments was consistent with COD concentration profiles. In term of alkalinity, it was observed the system has enough bicarbonate with slightly constant value of 214mg=L throughout all compartments. It's believed that effluent recycling has enhanced the alkalinity content in the system, system, was also found to alleviate the problems of low pH. Nevertheless, neutral pH level, high alkalinity and low VFAs suggested that methanogenesis appears to be dominant in all compartments.

Table 1 shows the concentrations of TS, TSS and VSS for each compartment influent and effluent. The influent solids (TS) concentration is 1382mg=L with 83% of it as total suspended solids (TSS), whilst the total suspended solids (TSS) concentration is 1142mg=L with 89% of it as a volatile fraction (VSS). Up to 73%, 83% and 89% removals were observed for TS, TSS and VSS, respectively. The TS, TSS and VSS concentrations in the treated effluent were 368mg=L; 192mg=L and 116mg=L, respectively.

Table 1: The performance of MAI-BR at the steady state condition

	COD mg/L	BOD mg/L	TSS mg/L	TDS mg/L	TS mg/L	VSS mg/L	PH	Fatty acids mg/L	Alkalinity mg/L	Lignin mg/L	Floc Size μm
In	949	453	1142	240	1382	1024	6.75	75	180	13.75	179
C1	76	21	29430	730	30160	13373	6.57	30	220	7.13	104
C2	64	20	20410	700	21110	10660	6.56	9	215	9	115
C3	74	28	2420	760	3180	2240	6.61	6	210	7.25	88
C4	60	26	1520	720	2240	1293	6.62	5	220	7.88	97
C5	60	29	1030	660	1690	447	6.61	3	210	7	64
Out	64	32	192	176	368	116	6.67	4	210	6.5	207

A floc samples were taken from the bottom of each compartment and analyzed for its size. Compartments 1 and 2 had the largest floc size as it's contain the highest COD levels which encourage the growth of microbes in the front of the reactor. The floc size decreased down of the reactor due to the low substrate. Lignin content in influent and effluent were 13:75mg=L and 6:5mg=L, respectively. About 53% of the lignin was removed even though there is no clear understanding on the biodegradability of lignin in anaerobic system [15].

CONCLUSION

The design of MAI-BR shows that it is capable to treat complex type of wastewater. The physical structure of the reactors allows various modifications to be made that providing the capability to treat wastewaters that currently require at least two separate units, therefore substantially reducing capital costs. During 30 days of the reactor start-up, up to 71% of COD removal was observed, while methane yield increased from 0:018 to 0:808L CH4/day. Even though slightly low pH levels were noted, high COD removal efficiencies confirm the ability of MABR configuration to overcome the adverse effect of pH. The performance of MAI-BR treating RPME has showed that COD and BOD removal efficiencies were 94% and 93%, respectively. A total of 73%, 83% and 89% removals were observed for TS, TSS and VSS. In the short period of 15 days, the biogas volume increased up to 1:05L per day, while the methane content reached 55%. Finally, a total of 0:55L CH4/day was yielded.

These results clearly indicate that the start-up and operation of MAI-BR led to a more complete biological degradation of the organic matter and a better adaptation of the biomass for the degradation of the substrates.

ACKNOWLEDGMENT

The authors wish to acknowledge the financial support from the Universiti Sains Malaysia (RU-I Grant A/C. 1001/PJKIMIA/814148).

REFERENCES

1. S.R. Hassan, H.M. Zwain and I. Dahlan, "Development of Anaerobic Reactor for Industrial Wastewater Treatment: An Overview, Present Stage and Future Prospects", Journal of Advanced Scientific Research, Vol.4, No.1, pp.07-12, 2013.
2. P.L. McCarty, "One Hundred Years of Anaerobic Treatment Digestion", Hughes (Ed.) Anaerobic Digestion, Elsevier Biomedical Press, 1981.
3. K.F. Fannin, J.R. Conrad and D.P. Chynoweth,

"Marine Biomass Program: Anaerobic Digester System Development" Annual Report for General Electric Company, IGT Project 65044, Institute of Gas Technology, 1981.

4. A. Bachmann and P.L. McCarty, "Comparison of Fixed Film Reactors with a Modifed Sludge Blanket Reactor", Y.C. Wu, E.D. Smith (Eds.) Fixed Film Biological Processes for Wastewater Treatment, Noyes Data, 1983.

5. A. Tilche and X. Yang, "Light and Scanning Electron Microscope Observations on The Granular Biomass of Experimental SBAF and HABR Reactors", In Proceedings of Gasmat Workshop, Netherlands, 1987.

6. R. Boopathy and D.M. Siever, "Performance of a Modified Anaerobic Baffled Reactor To Treat Swine Waste", Transactions of the ASAE, Vol34, No.6, pp.2573-2578, 1991.

7. APHA, AWWA and WEF, "Standard Methods for the Examination ofWater &Wastewater", American Public Health Association (APHA), American Water Works Association (AWWA) & Water Environment Federation (WEF), Washington, DC, USA, 2012.

8. A.M. Jirka and M.J. Carter, "Micro semiautomated analysis of surface and waste waters for chemical oxygen demand", Analytical Chemistry, Vol.47, No.8, pp.1397-1402, 1975.

9. H.A.C. Montgomery, N.S. Thom and J.F. Dymock, "The rapid colorimetric determination of organic acids and their salts in sewage-sludge liquor", The Analyst, Vol.87, No.1041, pp.949-955, 1962, DOI: 10.1039/AN9628700949.

10. H.M. Zwain, S.R. Hassan, N.Q. Zaman, H.A. Aziz and I. Dahlan, "The start-up performance of modified anaerobic baffled reactor (MABR) for the treatment of recycled paper mill wastewater", Journal of Environmental Chemical Engineering, Vol.1, No,1-2, pp.61-64, 2013.

11. R. Grover, S.S. Marwaha and J.F. Kennedy, "Studies on the use of an anaerobic baffled reactor for the continuous anaerobic digestion of pulp and paper mill black liquors", Process Biochemistry, Vol.34, No.6-7, pp.653-657, 1999, DOI: 10.1016/S0032-9592(98)00138-1 .

12. S.Y. Bodkhe "A modified anaerobic baffled reactor for municipal wastewater treatment", Journal of Environmental Management, Vol.90, No.8, pp.2488-2493, 2009, DOI: 10.1016/j.jenvman.2009.01.007.

13. H.M. Zwain and I. Dahlan, "Biological Treatment of Recycled Paper Mill Wastewater Using Modified Anaerobic Inclining-Baffled Bioreactor (MAIB-R)", Wastewater Engineering: Advanced Wastewater Treatment Systems, IJSR Publications, pp.72-85, 2014.

14. S.R. Hassan, H.M. Zwain, N.Q. Zaman and I. Dahlan, "Recycled paper mill effluent treatment in a modified anaerobic baffled reactor: start-up and steady-state performance", Environmental Technology, Vol.35, No.3, pp.1-6, 2014, DOI:10.1080/09593330.2013.827222.

15. H.M. Zwain, N.Q. Zaman, H.A. Aziz and I. Dahlan, "Evaluation of Lignin Removal from Recycled Paper Mill Effluent (RPME) under Different Start-Up's Steady State Condition of Modified Anaerobic Inclining-Baffled Reactor (MAIBR)", Current Environmental Engineering, Vol.1, No,1, pp.60-63 (2014).

CHAPTER 18

DISSOLVED ORGANIC NITROGEN IN WASTEWATER TREATMENT PLANTS

Mohammed Wedyan & Ahmed Al Harahsheh & Esam Qnaisb

Keywords: Total dissoloved nitrogen, dissolved organic nitrogen (DON), bioavailable/biodegradable dissolved organic nitrogen, wastewater treatment plants, Jordan.

INTRODUCTION

The wastewater effluent can be an important source of anthropogenic nitrogen to aquatic environments and possesses undesirable influences over the water quality, particularly in effluent-dominated waters (Pagillaet al 2008, Bronk et al 2010, Liu et al 2012).In order to minimize the eutrophication with the aquatic environment, biological enhanced nitrogen removal (BENR) operations that eliminate the majority of the dissolved inorganic nitrogen tend to be regularly used within wastewater treatment (Czerwionka et al 2012). Because of successful of BENR processing, the most important part of the remaining nitrogen within effluent released to the responsive environments is made of dissolved organic nitrogen (DON) (Pagilla et al 2006, Liu et al 2012). Earlier reports revealed that effluent DON is usually bioavailable to algae and also plankton (Pehlivanoglu-Mantas and Sedlak2004, Sattayatewa et al 2009, Simsek et al 2013). Accordingly, DON is actually drawing a greater number of interesting in waste water treatment plants (WWTPs) recently as a result of increasing worries including the revitalizing algal continuing growth of aquatic environments (Pehlivanoglu-Mantas and Sedlak 2004) and also the forming nitrogenous disinfection by-product N-nitrosodimethylamine (Pehlivanoglu-Mantas and Sedlak 2006, 2008, Lee et al2007). The rising issues regarding wastewater-derived DONhave raised the necessity to know more about their levels, compositions and properties.

DON is usually driven by subtracting dissolved inorganic nitrogen (DIN, the sum of ammonium, nitrate and nitrite) concentrations out of the total dissolved nitrogen (TDN) concentrations. Low DON concentration within waters abundant with proportion of DIN to TDN using the available techniques is commonly imprecise and DON estimating frequently possess high standard deviations (Lee and Westerhoff 2005, Vandenbruwane et al 2007). In order to improve the precision and accuracy of DON estimating, many pretreatment techniques were applied to eliminate DIN species in waters, for example dialysis (Lee and Westerhoff 2005) and nano filtration (NF) (Xu et al 2010). The restricted accessible techniques frustrated researchers looking into the characteristics and behavior of DON in wastewater treatment plants. As a substitute for determining DON as being a big issue in wastewater, research study has quantified certain organic nitrogen made up of substances like dissolvedfree and combined amino acids (DFAA and DCAA), dimethyl amine (DMA), and ethylenediaminetetraacetic acid (EDTA) (Pehlivanoglu-Mantas and Sedlak 2008).

Previous reports demonstrated that about 70% of waste water-derived DONalso cannot be described with available techniques (Pehlivanoglu-Mantas and Sedlak 2006, 2008, Simsek et al2012). Earlier investigations on wastewater-derived DON focused on two questionable problems; where DON is eliminated or even created in BENR processes and what exactly is the influence of BENR process on DON (Czerwionka et al 2012,Simsek et al 2013, Huo, et al., 2013). Sattayatewa et al (2009) reported that around 28-57% of the effluent DON was bioavailable or biodegradable, Simsek

Citation: Mohammed Wedyan & Ahmed Al Harahsheh & Esam Qnaisb, Dissolved Organic Nitrogen in Wastewater Treatment Plants, doi: 10.12973/ijese.2016.344a.

and his colleague (2013) determined biodegradable dissolved organic nitrogen (BDON) and bioavailable dissolved organic nitrogen(ABDON) in activated sludge (AS) and trickling filter (TF)wastewater treatment processes. They recorded that 65%BDON and 63% ABDON had been eliminated in the TF facility and 68%, 56%, respectively within the AS facility. Understanding the behavior of DON across the biological processes inside the WWTPs plays important role in realizing the function of treatment units in the particular elimination of DON. Even so, present data around the qualities of DON in BNR WWTPs continue to be constrained and inadequate to know the future or alterations of DON over the activated sludge process (Czerwionka et al 2012).

In Jordan many farmers are turned to treated wastewater as a low-cost alternative to conventional irrigation freshwater. However, wastewater is rich in essential nutrients such as nitrate, nitrite, ammonia, organic nitrogen and phosphate which lowers the cost of fertilizers additives. Meanwhile, wastewater used for irrigation may possess a health risk to peoples as it may contain excessive nutrients. Therefore, this study comes to highlight the risk of high organic nitrogen contents in three main waste water treatment plants that were investigated; Al Zarqa, Mafraq, and Irbid to achieve more information on wastewater-derived DON in WWTPs in Jordan. And to assess the effect of biological treatment on DON and its components. Finally, provide a better understanding of the fate of the unidentified DON WWTPs.

MATERIALS AND METHODS

Sample Sources

Samples were obtained from three different treatment plants effluent, which are Al-Zarqa (ZTP), Irbid (ITP), and al Mafraq (MTP) (Figure 1). The selected plants treat about 90% of the domestic wastewater in Jordan and serve more than 2 million inhabitants (Bataineh et al., 2002), the treatment process shown in Table 1.

All plants have to comply with the discharge limits for biochemical oxygen demand (BOD) and ammonia (based on the receiving river flow rate) but are not subject to any total nitrogen or total phosphorus limits.

Sample Collection and Preparation

The samples were collected from all of the WWTPs during different seasons, specifically February, April , and June. All samples were collected in polyethylene containers (acid-washed and rinsed with ultrapure water) (Mill-Q, Millipore Corp. USA) before used, then were delivered to the laboratory on ice, filtered through 0.45 μm cellulose acetate membranes upon arrival, and then were stored at 4°C in the dark.

Dissolved inorganic Nitrogen (DIN) species

Total dissolved nitrogen was measured with the standard persulfate digestion method (APHA, 1998). Nitrate concentrations were measured with a ion chromatograph (equipped with Ion Pac AS-14 4x250mm column), after conversion of all the nitrogen forms to nitrate instead of the standard Cd-column reduction method (APHA, 1998). Nitrate and nitrite in undigested samples also were measured with the aforementioned ion chromatographic method. Ammonium (NH^{4+}) was measured with the standard phenate method (APHA, 1998).

Dissolved Organic Nitrogen (DON) Calculation and Biodegradable Dissolved Organic Nitrogen (BDON) determination procedures

DON was calculated as the difference between the total dissolved nitrogen and the sum of inorganic nitrogen species (i.e., NO^{3+}, NO^{2+}, NH^{4+}) using equation (1). The BDON procedure is as follows. All the samples were filtered through a 0.2 mm pore-size cellulose acetate membrane filter (WhatmanInc) within an hour after collection. A portion of the filtered samples were used for immediate analysis of total nitrogen and inorganic nitrogen species (ammonia, nitrite, and nitrate). The value was recorded as initial DON (DONi). Two hundred milliliters of the remaining filtered sample were mixed with 2 mL of acclimated inoculum bacteria in a 250 mL amber bottle. Raw wastewaters (collected from locations) were used as the inoculum.

Table 1. The treatment plants concerned in the study (Bataineh et al., 2002)

Plant	Treatment process	Remarks
Al-Zarqa	Stabilization ponds (natural aeration, facultative, anaerobic lagoons)	
Irbid	Screen, grit removal, primary sedimentation, biological process, secondary sedimentation, disinfections	Trickling filter & activated sludge
Mafraq	Screen, grit removal, biological process, secondary sedimentation, Polishing pond, infiltration, disinfections	Activated sludge with nitrogen removal technique

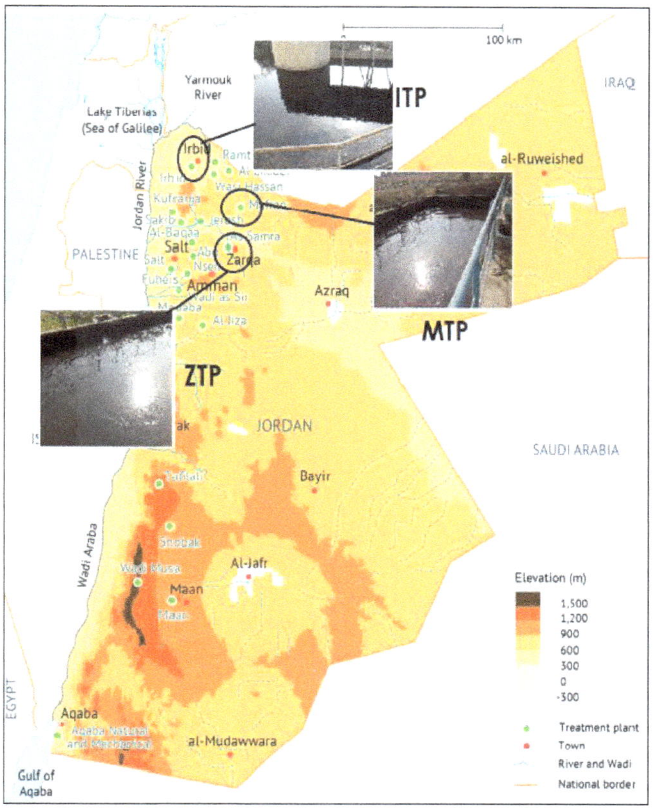

Figure 1. Jordan map indicating the sampling site

The solution in the bottle was shaken thoroughly to aerate and placed in an incubator in the dark at 20 °C for 16 days. During the incubation period, the solution in the bottle was manually shaken to aerate at least once every day to maintain aerobic conditions. A seed control (sample b), which was treated the same way as the samples, were prepared by adding the inoculum to 200 mL of de-ionized distilled water. After 16 days of incubation, all nitrogen species in the supernatant were measured to determine final DON (DONf). BDON was calculated according to equation (2).

$$\text{DON (mg/L as N)} = \text{TDN} - \text{DNH}_3 - \text{DNO}_2 - \text{DNO}_3 \quad (1)$$

$$\text{DON (mg/L as N)} = (\text{DON}_i - \text{DON}_f) - (\text{DON}_{bi} - \text{DON}_{bf}) \quad (2)$$

where DNH_3, DNO_2, and DNO_3 are dissolved ammonia, nitrite and nitrate, respectively; DONi and DONf are DON before and after incubation for samples; and DONbi and DONbf are DON before and after incubation for control.The rate constant for organic nitrogen degradation was evaluated and computed by measured rate of disappearance of the DON using the equation: $dN/dT=-Kn .N$ (Avnimelech et al 1995).

RESULTS AND DISCUSSIONS

The profile of different dissolved nitrogen species (ammonia, nitrite, nitrate, total nitrogen and organic nitrogen) in the three different wastewater treatment plants across the northern part of Jordan (Table 2).

Due to the environmental and weather conditions the concentrations of the nitrogen were fluctuation.

Dissolved nitrogen species

Dissolved total nitrogen (DTN)

The results show that the average concentration was ranged from 19.03 to 41.80 (mg - N/L). The highest concentration of DTN was found in ITP sample with average was 41.8±3.39 (mg/L ± SD) and the lowest concentration was detected in ZTP with average 19.30±6.60 (mg/L ± SD) (Table 2).

Dissolved Inorganic Nitrogen (DIN) species

All the nitrogen species in the dissolved fraction from effluent were detected the collected samples during this study (Table 2). The data show that the average concentrations of NH4-N 0.039 ± 0.003, 0.06 ±0.019 and 0.70 ± 0.048 (mg-N/L ± SD) (ZTP, MTP and ITP, respectively). A small variation in ammonia concentrations (3%) was typically observed in the samples collected from ZTP and MTP. But highly significant difference with ITP samples was (85%). The plants process achieved almost complete ammonia removal during the wastewater treatment processes in three different plants, process in the ITP does not remove ammonia may be because of the toxicity of high oxygen concentration to the nitrifying microorganisms (Uemoto et al., 2000). Similar to the Fargo WWTF results, ammonia inall the samples were totally nitrified during the incubation (Pehlivanoglu-Mantas and Sedlak, 2008; Simsek et al, 2012, Liu, et al., 2013).

The results show that the highest concentration of NO3 was in MTP sample with average was 7.83±6.49 (mg/L ± SD) and the lowest concentration was detected in ITP with average was 6.33±3.80 (mg/L ± SD) (Table 2).Nitrite was present at very low concentration in all WWTP. Generally, nitrite concentration was always less than 0.8 mg N/L in all collected samples. The highest concentration of NO_2 was in ITP samples with average 0.7±0.48 and the lowest concentration was detected in MTP 0.013±0.002 (mg/L ± SD). Nitrite concentrations in all samples were low (<1 mg/L, Table 2). For the samples from WWT"Ps which have high ammonia concentrations, partial nitrification to nitrite was expected. Final effluent samples were low because of full nitrification in the during the treatment processes. Corresponding to nitrite concentrations, nitrate concentrations were high for final effluent samples. They show the same trend as the data for the Fargo WWTF (Simsek et al, 2012; Liu, et al., 2013).

Table 2. Concentration (mg-N/L) of Dissolved Nitrogen in Different locations, (average ±SD)

	ZTP	ITP	MTP
DNO_3-N	6.81 ± 2.40	6.33 ± 3.80	7.83 ± 6.49
DNO_2-N	0.48±0.040	0.70±0.048	0.013±0.002
DNH_4-N	0.039±0.0030	0.70±0.48	0.06±0.019
DTN	19.30±6.60	41.8±3.39	33.21±5.01

Dissolved organic Nitrogen (DON)

Due to the difficulties associated with the measurement of DON, there is a shortage of information about the structure and behavior of wastewater- derived DON. By measuring DON and nitrogen containing compounds combined with DON characterization, we have gained new insight into the nature and properties of wastewater-derived DON. The results show that the average concentration was ranged from 11.71 to 34.82 (mg - N/L). The highest concentration of DON was noticed in ITP sample with average 34.82±3.68 (mg/L ± SD) and the lowest concentration was detected in ZTP sample with average 11.71±1.16 (mg/L ± SD) (Figure 2).

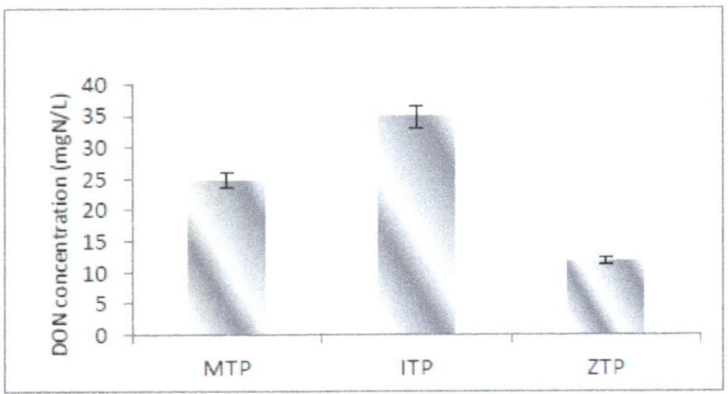

Figure 2. Average DON profile from three WTTPs.

The differences in the concentrations of DON were suggested to be nitrogenous organic compounds could be removed during transportation through the ditch canals by biological processes. The nitrogen present in wastewater is primarily in proteins and urea. Decomposition readily changes the organic form to ammonia. The age of wastewater is indicated by the relative amount of ammonia that is present (Metcalf and Eddy, 2003).

The Figure 3 represented the proportion of DON to TDN in different three WWTPs with different time. The data revealed that the percentage of DON varies with the time of collection and the highest percentage found in the spring in the ITP and MTP while in ZTP in winter was significantly appear (Figure 3). The results shown that the averages of the proportion of the DON were 76.3, 75.2 and 56.75 in the ITP, MTP and ZTP respectively.

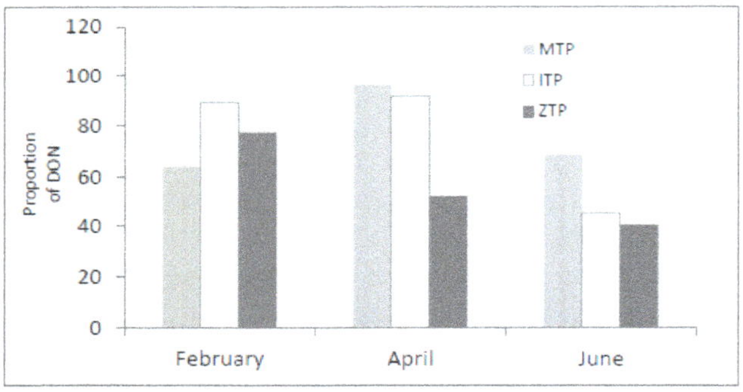

Figure 3. Represent the proportion of DON to TDN in different three WWTPs with different time.

The data revealed that the treatment plant removed of the DON. Similar to inorganic nitrogen removal, major removal of DON around 43.3% was observed in the ZTP which mainly operated by the biological processes of the plant. DON fractions of TDN were 24% and 25% in the plant effluent of ITP and MTP, respectively. The results agreed with the previous observation (Huo, et al., 2013)

Figure 4 shown that the concentrations of DIN were lower than the DON with average 7.01±4.04, 8.43±6.56 and 7.58±1.75 mg/l, (ZTP, MTP and ITP, respectively). To compare the concentrations of DIN in different plants with the concentrations of DON it found that the mechanisms in the plants tried to remove a large amount of DIN fraction compare to DON fraction.

DIN and DON from three WTTPs.

In addition, all dissolved nitrogen species were monitored in the three WWTPs and the results shown in the Figure 5. It was found that there was very little or no nitrite in the samples. the dissolved nitrogen concentration ranged from 0.3 to 7.8 mg N/L. Nitrate was the dominant species and was approximately 8-21% of dissolved nitrogen.

The most variable plant is the MTP, which comprised of up to 24 percent of influent N as ON. Similar results were reported by Pagilla et al. (2008) in the samples from 3 fully nitrified WWTPs in Illinois and 4 BNR plants in Poland. The fraction of influent ON ranged from 2 to 21 % and 13 to 27% in Illinois and Polish plants, respectively. In high influent ON concentration, the results show that ON concentration decreased during the primary treatment process and along the treatment processes. This could be due to dilution from recycle flows or biological processes when wastewater was exposed to treatment conditions in preliminary and primary treatment processes.

Ammonium removal was mainly observed in biological process. The results showed that all WWTPs efficiently removed ammonium and nitrate was present in very low concentration in the effluent. It was found that complete nitrification was achieved. The similarity was found in all 3 plants.

Biodegradable dissolved organic nitrogen BDON

Out of the initial DON of 3.18 mg N/L, nearly 1.95 mg N/L (30% DON bioavailable) was bioavailable to algae at the end of the 14-day bioassay as seen by the net DON concentration reduced over the 14-day period (Fig 6). The biodegradability/bioavailability of DON drops significantly. The removal percentage of the DON was increased during the biodegradation process so the removal BDON was about 30% within 14 days (Figure 6). The BDON degradation rate constant was found to be 0.03/day.

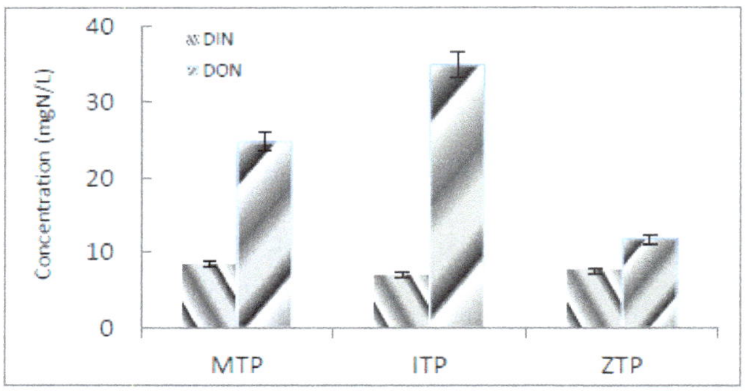

Figure 4. represented the comparison between the concentrations of DIN and DON from three WTTPs.

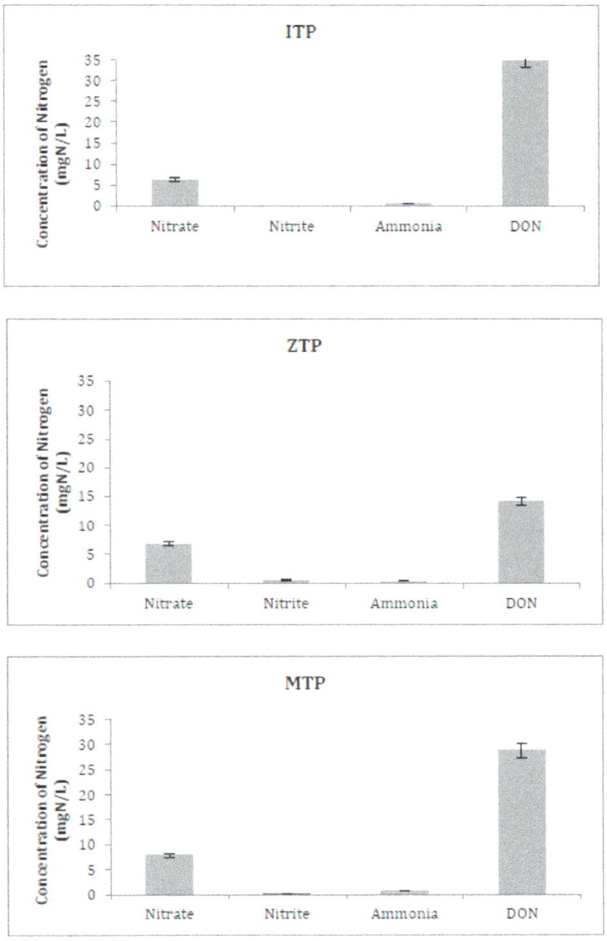

Figure 5. Summary nitrogen species concentrations in the studied WWTPs.

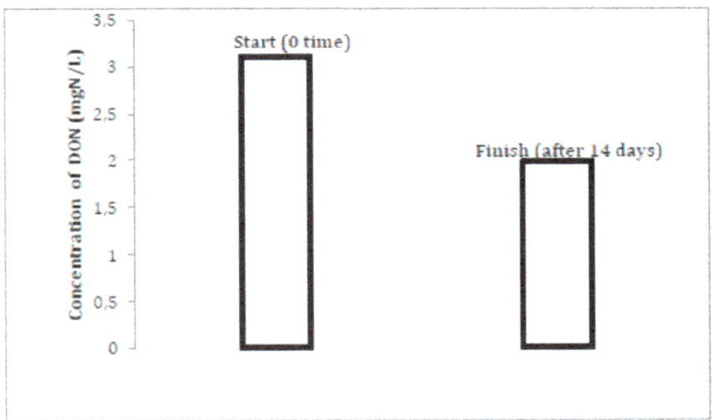

Figure 6. Represent the concentration of BDON from the first day to 14 days of processing

The results from DON bioavailability experiments showed that DON derived from WTPs is bioavailable either directly or indirectly to algae as seen from decrease in DON concentration, at the end of the 14-day incubation period.

Similar trends of BDON concentrations along treatment train were also reported for e US WWTPs (Czerwionkaet al 2012).These studies showed that a portion of effluent DON was biodegradable and/or bioavailable regardless of the type of wastewater treatment processes (Bronk, et al., 2010), and a significant portion of effluent DON was difficult to remove during biological treatment (Pehlivanoglu-Mantas and Sedlak 2008). The information of DON profile along treatment train would help to understand the roles of wastewater treatment process in the removal of this fraction of DON (Simsek et al, 2012).

As a conclusion the results were provide insight into the nature of wastewater-derived DON in some Jordanian WWTPs. The total DON account approximately for more than 70% of the TDN from the effluent leaving WWTP. Beside the dissolved amino acids 70% of the DON unidentified. It is likely that these compounds consist of a complex suite of partially metabolized compounds of biogenic origin. Although it may not be possible to chemically characterize, and identify all the individual nitrogen-containing compounds.

ACKNOWLEDGMENT

This research was supported by Deanship of Research/ The Hashemite University. We thank the technician in our lab for assistance for analyses, and for comments that greatly improved the manuscript.

REFERENCES

1. American Public Health Association. Water Environment Federation (APHA). (1998) Standard Methods for the Examination of Waters and Waste waters,.
2. Avnimelech, Y., Mozes, N., Diab, S., & Kochba, M. (1995). Rates of organic carbon and nitrogen degradation in intensive fish ponds. *Aquaculture, 134*(3), 211-216.
3. Bataineh, F., Najjar, M., &Malkawi, S. (2002) *Wastewater Reuse, in water demand management forum.* Amman, Jordan
4. Bronk D A, Roberts Q N, Canuel E A, Mesfioui R, Filippino K C, Mulholland M R and Love N G (2010). Effluent organic nitrogen bioavailability and photochemical and salinity-mediated release. *Environ. Sci. Technol. 44.* 5830–5.
5. Bronk, D. A., See, J. H., Bradley, P., &Killberg, L. (2007). DON as a source of bioavailable nitrogen for phytoplankton. *Biogeosciences, 4*(3), 283-296.
6. Czerwionka K, Makinia J, Pagilla K R and Stensel H D (2012). Characteristics and fate of organic nitorgen in municipal biological nutrient removal wastewater treatment plants. *Water Res. 46,* 2057–66.
7. Huo, S., Xi, B., Yu, H., Qin, Y., Zan, F., & Zhang, J. (2013). Characteristics and transformations of dissolved organic nitrogen in municipal biological nitrogen removal wastewater treatment plants. *Environmental Research Letters, 8*(4), 044005
8. Lee W and Westerhoff P (2005). Dissolved organic nitrogen measurement using dialysis pretreatment. *Environ. Sci. Technol. 39,* 879–84.
9. Lee W, Westerhoff P and Croues J P (2007). Dissolved organic nitrogen as a precursor for chloroform, dichloroacetonitrile, N-nitrosodimethylamine, and trichloronitrometha. *Environ. Sci. Technol. 41,* 5485–90.
10. Liu H Z, Jeong J, Gray H, Smith S and Sedlak D L (2012). Algal uptake of hydrophobic and hydrophilic dissolved organic nitrogen in effluent from biological nutrient removal municipal wastewater treatment systems. *Environ. Sci. Technol. 46*, 713–21.
11. Metcalf, E. (2003). Inc., *Wastewater Engineering,*

Treatment and Reuse. New York: McGraw-Hill

12. Pagilla K R, Czerwionka K, Urgun-Demirtas M and Makinia J (2008). Nitrogen species in wastewater treatment plant influents and effluents-the US and Polish case studies. *Water Sci. Technol. 57,* 1511–7.
13. Pagilla K R, Urgun-Demirtas M and Ramani R (2006). Low effluent nutrient technologies for wastewater treatment. *Water Sci. Technol. 53*, 165–72.
14. Pehlivanoglu-Mantas E and Sedlak D L (2004). Bioavailability of wastewater-derived organic nitrogen to the alga Selenastrum Capricornutum. *Water Res. 38*, 3189–96.
15. Pehlivanoglu-Mantas E and Sedlak D L (2006). The fate of wastewater-derived NDMA precursors in the aquatic environment. *Water Res. 40,* 1287–93.
16. Pehlivanoglu-Mantas E and Sedlak D L (2008). Measurement of dissolved organic nitrogen forms in wastewater effluents: concentrations, size distribution and NDMA formation potential. *Water Res. 42,* 3890–8.
17. Sattayatewa C, Pagilla K, Pitt P, Selock K and Bruton T (2009). Organic nitrogen transformations in a four-stage Bardenpho nitrogen removal plant and bioavailability/biodegradability of effluent DON. *Water Res. 43,* 4507–16.
18. Simsek H, Kasi M, Ohm J B, Blonigen M and Khan E (2013). Bioavailable and biodegradable dissolved organic nitrogen in activated sludge and trickling filter wastewater treatment plants. *Water Res. 47,* 3201–10.
19. Simsek H, Kasi M, Ohm J B, Wadhawan T, Bye C, Blonigen M and Khan E (2012). Fate of dissolved organic nitrogen in two stage trickling filter process. *Water Res. 46*, 5115–26
20. Vandenbruwane J, Neve S D, Qualls R G, Salomez J and Hofman G (2007). Optimization of dissolved organic nitrogen (DON) measurements in aqueous samples with high inorganic nitrogen concentrations. *Sci. Total Environ. 386,* 103–13.
21. Xu B, Li D P, Li W, Xia S J, Lin Y L, Hu C Y, Zhang C J and Gao N Y (2010). Measurements of dissolved organic nitrogen (DON) in water samples with nanofiltration pretreatment *Water Res. 44,* 5376–84.

CHAPTER 19

EXPERIMENTAL DESIGN OF ELECTROCOAGULATION AND MAGNETIC TECHNOLOGY FOR ENHANCING SUSPENDED SOLIDS REMOVAL FROM SYNTHETIC WASTEWATER

Moh Faiqun Ni'am, and Fadil Othman

Keywords: Electrocoagulation; Magnetic field; Wastewater treatment; Experimental design; Response surface methodology

INTRODUCTION

Colloidal or fine particles, commonly called suspendedsolids, are extremely small and light, which will not settleout by gravity. Colloidal particle size is approximately0.001 microns to 1 microns (1 nm to 1 μm) in dimension.One of the most common example of colloids is milk(Shaw, 1970; Hiemenz and Rajagopalan, 1997). Becauseof difficult to settle under gravity, hence coagulation areneeded so that the particle size becomes relatively larger.An alternative separation treatment for ultrafine particlesis electrocoagulation (EC), which involves the in situformation of ions by electrolysis. The key foundationsciences for electrocoagulation are electrochemistry,coagulation, and flotation.

Electrocoagulation technology, has reached profitablecommercialization, and existing ones improved that donot require chemical additions (Mollah et al., 2001).These include electrocoagulation (EC), electroflotation(EF), electrodeposition (ED), electrooxidation (EO), andothers (Chen, 2004). The main based ofelectrocoagulation technology is electrochemistry,coagulation, and flotation, and the relevant mechanism oftheir interaction, which are extracted and applied in theelectrocoagulation context.

EC has been used by various authors for the treatmentof wastewater for different wastes, e.g. soluble oils, liquidfrom the food, textile industries, or cellulose and effluentsfrom the paper industry, landfill leachate, restaurantwastewater, saline wastewater, tar sand and oil shalewastewater, urban wastewater, laundry wastewater,nitrate and arsenic bearing wastewater, and chemicalmechanical polishing wastewater (Holt et al., 2002; Calvoet al., 2003; Kobya et al., 2003; Larue et al., 2003;Daneshvar et al., 2004; Kumar et al., 2004; Can et al.,2006; Carmona et al., 2006; Ilhan et al., 2008). Severaldifferences were found in comparison to the chemicalcoagulation process (Zhu et al., 2005; Meunier et al.,2006) and economical evaluation (Kobya et al., 2007 and2009).

Aluminium or iron are usually used as electrodes andtheir cations are generated by dissolution of sacrificialanodes upon the application of a direct current (Carmonaet al., 2006; Ilhan et al., 2008; Hernandez et al., 2009;Zongo et al., 2009; Tchamango et al., 2010). The resultsobtained that the aluminium and iron are suitableelectrode materials for the treatment of textilewastewaters by electrocoagulation.

Citation: Ni'am, M.F. and Othman, F. (2014 Experimental Design of Electrocoagulation and Magnetic Technology for Enhancing Suspended Solids Removal from Synthetic Wastewater. International Journal of Science and Engineering, 7(2),178-192. Doi: 10.12777/ijse.7.2.178-192]

Whereas, magnetic treatments for water andwastewater attract a special attention due to their safety,ecological purity, environmental friendly and simplicity.Magnetic treatment of water was first patented byVermeiren in Belgium in 1945, and he is recognized as thediscoverer of the fact that magnetic fields affect water.Magnetic treatment of water is an attractively simpleapproach by which the water to be treated flows througha magnetic field, and consequently changes some of itsphysicochemical properties (Othman et al., 2001; Sohailiet al., 2004; Ni'am et al., 2006).

Magnetic field is known to create the asymmetry ofhydrated shells due to its effect on water moleculessituated around the charged particles (colloid). Exposureto magnetic field would lead to higher electro-kineticmovement among the colloid. This definitely will help inattributing to a higher probability of attracting particlesto cloak with one another. The theory of magnetic fieldimpact on technological processes for water treatmentfalls into two main categories; crystallization at magneticwater preparation and impurity coagulation in watersystems (Othman et al., 2001).

In fact, from the literature study of previous research,very little available literature on water and wastewatertreatment using a combination of EC and magnetic fields.Electrocoagulation (EC) has investigated by Tsouris et al.(2001), as a magnetic seeding method to be used prior toa high-gradient magnetic separation (HGMS) process.Ihara et al. (2004) have treated landfill leachate in abench scale plant by high gradient magnetic separationcombined with electrocoagulation using iron electrodesand electrochemical oxidation. Meanwhile, Ghernaout etal. (2009), have studied the reduction of humic acids (HA)by continuous electromagnetic (EM) treatment followedby electrocoagulation (EC) in batch of synthetic HAsolution using two aluminium electrodes at neutral pH.

In the preliminary research, the authors haveinvestigated the effect of EC with iron (Fe) bar and plateelectrodes in the batch experiment (Ni'am et al., 2005,2007a,b and 2008; Othman et al., 2006). The results of acombined magnetic field and EC with Fe bar electrodesshown that the SS and turbidity removal are as high as91.4 % and 85.5 % with the combined process, while forEC process is as high as 88 % and 72.1 % (Ni'am et al.,2005).

In another research, Othman et al. (2005) haveinvestigated that the SS and turbidity removal are as highas 92.3 % and 81.25 % with the combined process, whilefor EC process is as high as 89.3 % and 75.16 %. Theseresults obtained from batch experiment by a combinedmagnetic field and EC with iron (Fe) plate electrodes.The objectives of this study were to investigate thesuitability of magnetic fields and electrocoagulationtechnologies for the treatment of wastewater inenhancing removal of suspended solids and itssedimentation.

MATERIALS AND METHODS

This research is mainly focused on the capability ofmagnetic field and EC technology to removal and increasethe sedimentation of suspended solid through single flowprocessing methods. Response surface methodology (RSM) is employed to optimize and to analyzeinteractions of influencing factors on the treatmentefficiency of synthetic wastewater with some plannedexperiments.

Synthetic Wastewater

The synthetic wastewater were prepared from milkpowder with concentrations of 700 mg/L and treated byusing a buffer solution. The buffer solution consisted of 5mL/L HCl (0.5 %) and 125 mg/L NaCl as pH adjusmentand electrolyte. From these concentrations (after adjustedby buffer solution), the initial pH of the wastewater wasfound to be 2.91 and maximum current was 1.4 A at avoltage of 30 V. The current was adjusted to a desiredvalue between 0.50 A and 1.1 A before the coagulationprocess started.

Experimental Setup

The experiments were carried out in this work as asingle flow method. EC and the synthetic wastewaterwere performed in the reactor glass cell (volume 2000mL). Fifteen monopolar plate electrodes (100 x 50 x 4mm) from iron (Fe) plates were set up as a baffle atdistances of 14 mm and placed in the reactor. The wastewater flows from reservoir (volume 8000mL) throughout the system is provided by means of apump controlled and can be adjusted from zero tomaximum 15 mL/s. In this research, the wastewater wascirculated with a constant flowrate, varied between 1.00mL/s and 3.50 mL/s. The experimental setup was shownin Figure 1.

One permanent magnets namely AlNiCo with amagnetic strength of 0.16T was used. Three pairs ofmagnets have been placed under the reactor. Each pairconsists of two-piece magnets of cubic-shaped rare earthpermanent magnets (size 50 mm x 50 mm x 20 mm).Furthermore, the treated wastewater was placed ingraduated sedimentation columns of 5 cm internaldiameter and 1 m height for 6 to 24 hours settling time. Samples were collected at 5 cm from the surface ofwastewater using pipette, at 30, 60, 90 minutes and 2, 3, 4,6, 8, 12 hours of settling time. The effects of relevantwastewater characteristic such as turbidity, COD, and SSremoval efficiencies have been explored.

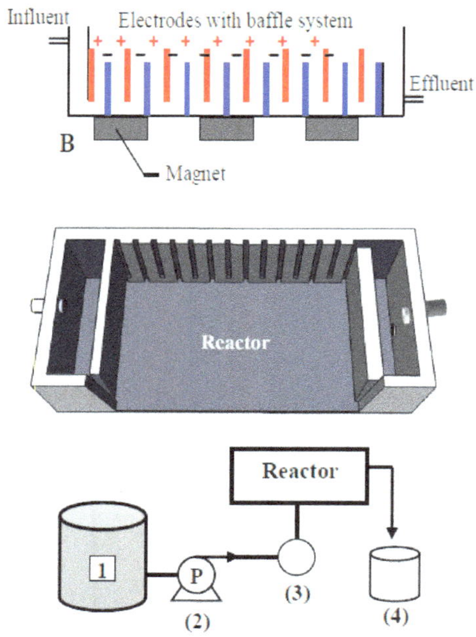

Figure 1. Detail of EC reactor and Experimental setup: (1) reservoir with mixing, (2) circulation pump, (3) flowrate meter, reactor, and (4) effluent & sample collecting

Experimental Design and Analytical Method

Optimization of suspended solids removal fromwastewater was carried out using RSM. According toMontgomery (2001), RSM is a collection of mathematicalmethods and statistical techniques that aims to createmodels and perform analysis of the responses that areinfluenced by several variables and the objective is tooptimize this response.

For statistical analysis, analysis of variance, betterknown by the name of ANOVA from MINITABTM (Version14.0) was used. Analysis of Variance (ANOVA) is a methodof analysis used to investigate the relationship betweenthe response variable (dependent) with one or morepredictor (independent). ANOVA is similar to regression,but the data's scale of independent variable is data ofcategories, i.e.: ordinal or nominal scale. ANOVA also didnot have the coefficient (parameter) of the model.

The turbidity removal was measured from wastewatersamples by HACH DR/4000 (HACH Method 10047). CODmeasurements were determined according to theStandard Methods for Examination of Water andWastewater (APHA, 2005). The COD samples wereanalyzed using UV-Vis HACH DR/4000spectrophotometer (HACH Method 8000).To measure total suspended solid (TSS), thewastewater samples were filtered through a standardGF/F glass fiber filter. The residue retained on the filterwas dried in an oven at 1030C to 1050C until the weight ofthe filter no changes. The increase in weight of the filterrepresents the total suspended solids (APHA Method2540 D).

The calculation of turbidity, COD and suspended solid removal efficiencies after electrocoagulation treatment were performed using this formula (Daneshvar et al, 2006):

$$CR(\%) = \frac{C_0 - C}{C_0} x\ 100$$

Where C_0 and C are concentrations of wastewater before and after electrocoagulation in NTU or mg/L, respectively.

According to Matteson et al. (1995), the rate of change of wastewater concentration, such as turbidity, COD and suspended solid removal can be expressed as a first order kinetic model, as follows:

$$\frac{C}{C_o} = e^{-k_2 t}$$

Where C, C_0, and k_2 are wastewater concentrations after EC, initial, and kinetic constant, respectively.

Hence, the loss of particles due to coagulation after treatment process (Sohaili, 2003), as follows:

$$\frac{C}{C_o} = a(b - e^{-kt})$$

Where k is kinetic constant; a and b are constant values.

RESULT AND DISCUSSION

Experimental Design

For obtain the optimum values, this study was carried out in the preliminary tests by experimental design using Response Surface Methodology (RSM) in the batch experiment. Montgomery (2001) and Antony (2003) explained that the experimental design is an important role in the development process and can be used to solve the problems in the process for improved of process performance. Meanwhile in the second stage, the main experiment was performed with a single flow system using the optimum values have been obtained from simulation of the preliminary tests. To create the experimental design and analyze statistical data, the MINITAB^{TM} (Version 14.0) statistical application program was used.

Table 1: Values of variables will be tested in the batch experiment to obtain optimum value

Variables/Factors	Levels	Values		
Distance of electrodes (d)	3	14 mm	32 mm	50 mm
Current (i)	3	0.5 A	0.8 A	1.1 A
Treatment Time (t_1)	3	20'	30'	40'

Three variables and three levels Central Composite Design (CCD) consisting of twenty experimental runs order were employed. The design variables were distance between electrodes (d), direct current (i), and treatment time (t1) while three important response variables were Suspended Solids (SS), turbidity, and COD removal efficiencies. Values of each variable that will be tested in this experiment conducted in one replicate, as indicated in Table 1.

Table 2: Values of actual level for RSM design by using MINITAB

Factors	Values				
	-1.68	-1.00	0.00	1.00	1.68
X_1= distance (mm)	1.72	14	32	50	62.28
X_2= current (A)	0.3	0.5	0.8	1.1	1.3
X_3= time (minute)	13.18	20	30	40	46.82

Table 3: Experimental design and results information by RSM using CCD type

Std Order	Run Order	Pt Type	Blocks	Treatment (t) (min)	Distance (d) (mm)	Current (i) (A)	% SS Removal	% Turbidity Removal	% COD Removal
1	1	1	1	20	14	0.5	66.643	81.874	49.035
2	2	1	1	40	14	0.5	85.643	89.817	75.702
3	3	1	1	20	50	0.5	54.071	50.509	43.596
4	4	1	1	40	50	0.5	72.357	78.615	66.053
5	5	1	1	20	14	1.1	78.786	90.428	77.193
6	6	1	1	40	14	1.1	74.714	88.595	55.439
7	7	1	1	20	50	1.1	74.000	66.599	68.070
8	8	1	1	40	50	1.1	83.214	97.963	79.386
9	9	-1	1	13.18	32	0.8	66.000	34.000	46.500
10	10	-1	1	46.82	32	0.8	80.000	96.000	76.000
11	11	-1	1	30	1.72	0.8	80.643	88.595	76.228
12	12	-1	1	30	62.28	0.8	74.143	72.301	68.246
13	13	-1	1	30	32	0.3	44.071	74.134	31.316
14	14	-1	1	30	32	1.3	85.500	97.352	82.193
15	15	0	1	30	32	0.8	84.214	96.130	69.386
16	16	0	1	30	32	0.8	84.214	96.130	69.386
17	17	0	1	30	32	0.8	84.214	96.130	69.386
18	18	0	1	30	32	0.8	84.214	96.130	69.386
19	19	0	1	30	32	0.8	84.214	96.130	69.386
20	20	0	1	30	32	0.8	84.214	96.130	69.386

By using RSM, the experimental work in this study was designed statistically using Selections of the values for each independent variable were decided upon based on the variable range input to the MINITAB Statistical Software. By selecting the design type of CCD, values of actual levels of each factor in Table 1 are recalculated using the following equation:

$$X_1 = \frac{\text{distance} - 32}{18}$$

$$X_2 = \frac{\text{current} - 0.8}{0.3}$$

$$X_3 = \frac{\text{treaatment time} - 30}{10}$$

The calculation result of the actual level values using the formulas above, as shown in Table 2 and then performed of design analysis by using MINITAB software. The experiment was done with the values of the design information that has been combined by MINITAB (Table 3). The experimental results of SS, turbidity and COD removal as shown in Table 3, then used as input data in MINITAB program to be modeled and analyzed the design and statistics. Output of response surface regression of SS removal efficiency as shown as follows:

Term	Coef	SE Coef	T	P
Constant	84.127	2.815	29.881	0.000
Treatment (t1)	**4.831**	**1.868**	**2.586**	**0.027**
Distance (d)	-2.422	1.868	-1.297	0.224
Current (i)	**7.445**	**1.868**	**3.986**	**0.003**
Treatment (t1)*Treatment (t1)	**-3.393**	**1.818**	**-1.866**	**0.092**
Distance (d)*Distance (d)	-1.840	1.818	-1.012	0.335
Current (i)*Current (i)	**-6.297**	**1.818**	**-3.463**	**0.006**
Treatment (t1)*Distance (d)	1.572	2.441	0.644	0.534
Treatment (t1)*Current (i)	-4.018	2.441	-1.646	0.131
Distance (d)*Current (i)	3.697	2.441	1.515	0.161

S = 6.903 R-Sq = 81.6% R-Sq(adj) = 65.0%

Analysis of Variance for % SS Removal

Source	DF	Seq SS	Adj SS	Adj MS	F	P
Regression	9	2110.6	2110.6	234.52	4.92	0.010
Linear	**3**	**1155.8**	**1155.8**	**385.26**	**8.08**	**0.005**
Square	**3**	**696.6**	**696.6**	**232.21**	**4.87**	**0.024**
Interaction	**3**	**258.2**	**258.2**	**86.07**	**1.81**	**0.210**
Residual Error	10	476.5	476.5	47.65		
Lack-of-Fit	**5**	**283.8**	**283.8 5**	**6.76**	**1.47**	**0.341**
Pure Error	5	192.7	192.7	38.55		
Total	19	2587.2				

Estimated Regression Coefficients for % SS Removal using data in uncoded units

Term	Coef
Constant	84.1268
Treatment (t1)	4.83077
Distance (d)	-2.42191
Current (i)	7.44498
Treatment (t1)*Treatment (t1)	-3.39301
Distance (d)*Distance (d)	-1.83985
Current (i)*Current (i)	-6.29727
Treatment (t1)*Distance (d)	1.57150
Treatment (t1)*Current (i)	-4.01800
Distance (d)*Current (i)	3.69650

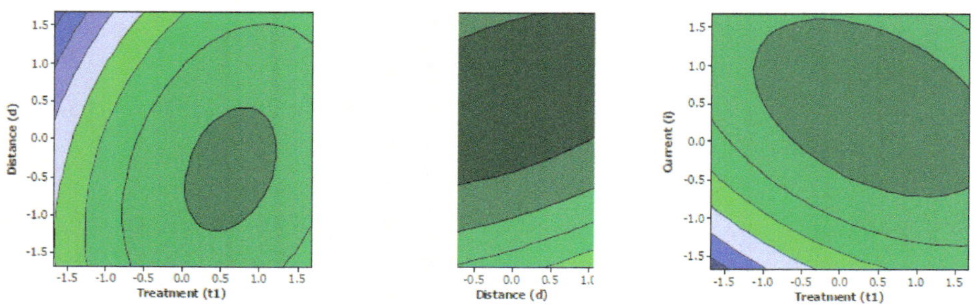

(a)

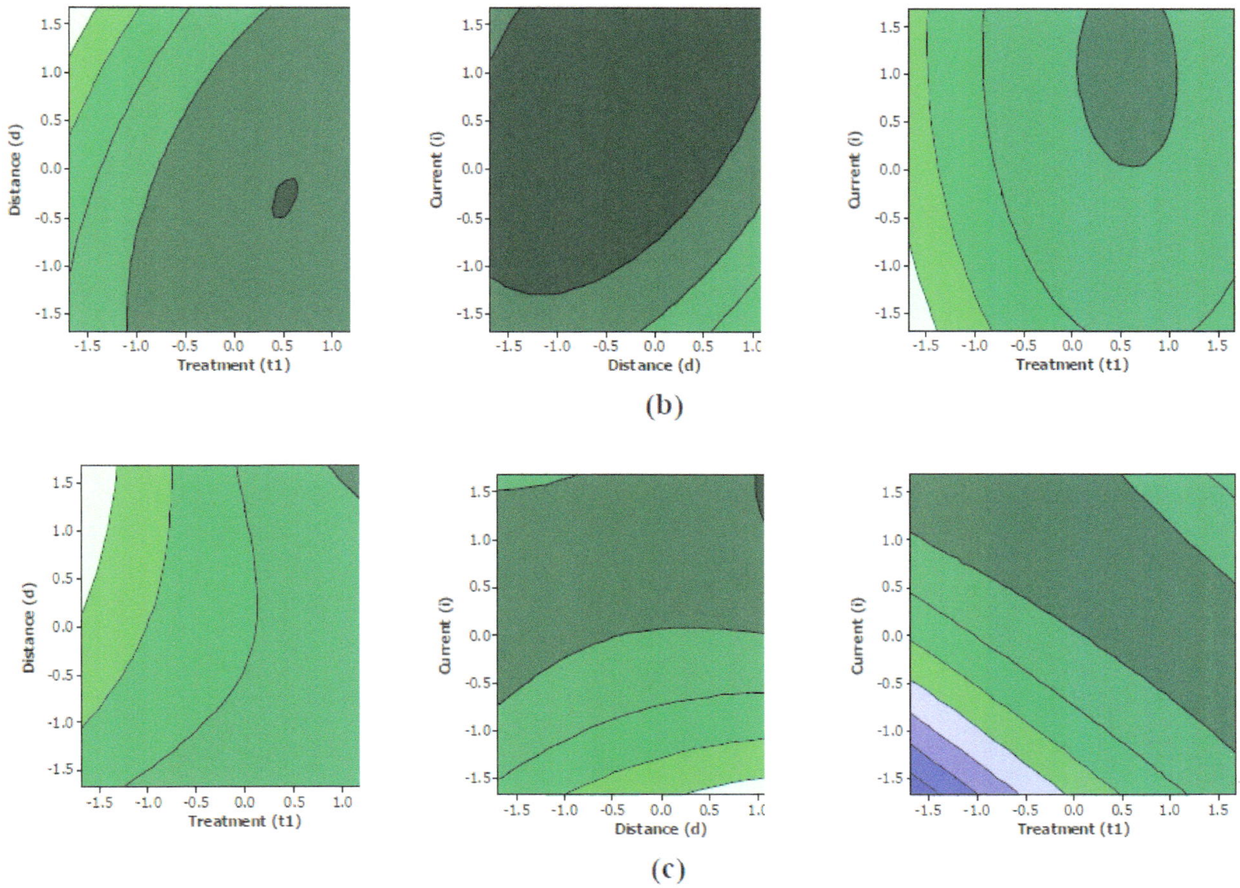

Figure 2: Contour plot for combine effect of treatment time-distance, distance-current, and treatment time-current on percentage of (a) SS, (b) turbidity, and (c) COD removal.

The output above shows of several sections, including: the results of the estimated parameters of the model, ANOVA table used to test the suitability of the model with the data, and unusual observation. To perform the nalysis of output, the first step is to detect the significances of the model. The significance of the model can be observed in the ANOVA table, which shows that the linear model (p-value = 0.005) and quadratic model (p-value = 0.024) is significant because the p-value of both less than a= 0.1. Conversely, nonlinear model that includes interactions between factors (p-value= 0.210) were not significant. This means, the right model is a quadratic model. On the other hand, the parameters test of model also shows that the variable "treatment time","current", "quadrate of treatment" and "quadrate of current" have significances because those terms have a small p-value or less than 10%.

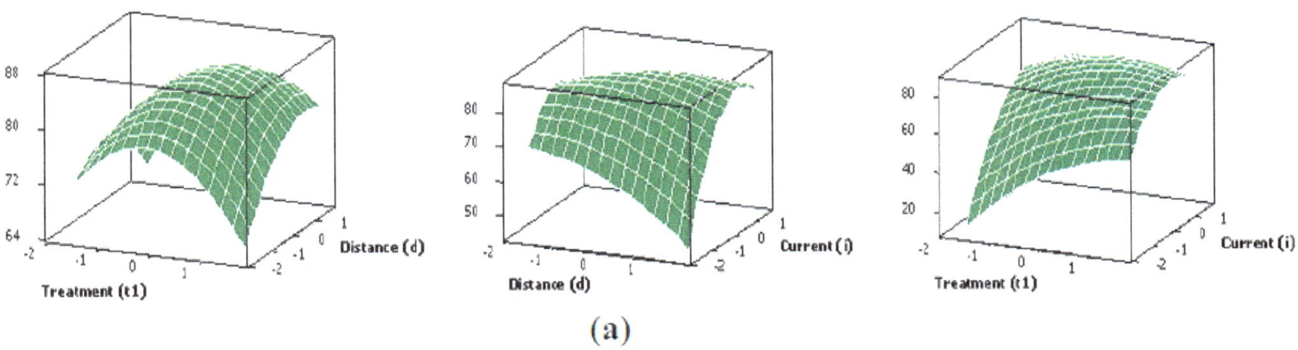

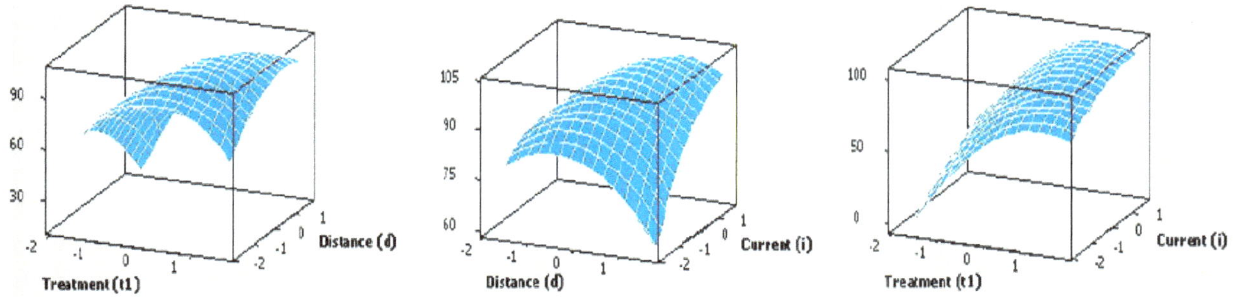

(b)

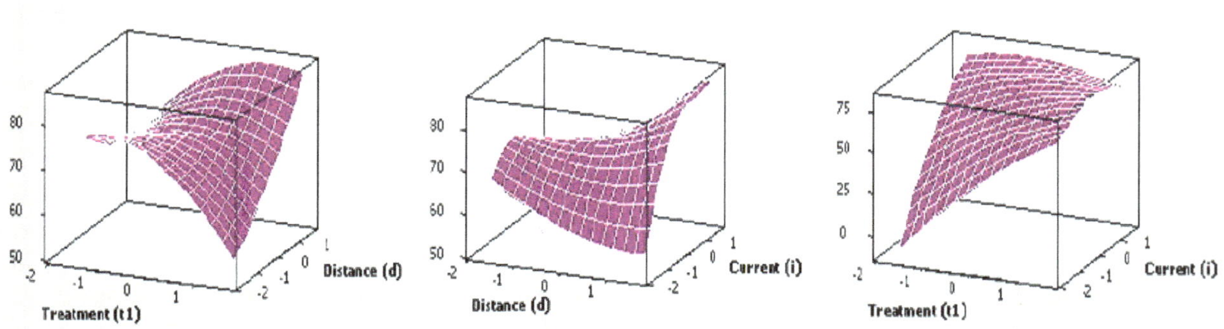

(c)

Figure 3: 3D graphs of response surface plot for combine effect of treatment time-distance, distance-current, and treatment timecurrent on percentage of (a) SS, (b) turbidity, and (c) COD removal.

For the optimization model, the data are analyzed by making a contour plot of the response variable. To visualize the response in three-dimensional images, it would require two independent variables and one variable is set its value at normal conditions. In this analysis, the value of a specified variables are the variable "current", "treatment time", and "distance" with the value at normal conditions are 0.8 A, 30 minutes, and 32 mm, respectively. Based on the output shown in Figure 2 and Figure 3. it can be concluded that the efficiency will be optimum when the treatment time is at a level between 0.0 to 1.0 (between 30 to 40 minutes) and distance between electrodes at a level between -1.0 to 0.0 (14 mm to 32 mm). Based on statistical analysis of the estimated regression coefficient using data in uncoded units for the experimental condition of this study, obtained following equation:

$$y = 84.13 + (4.83*t_1)-(2.42*d)+(7.45*i)-(3.4*t^2)-(1.84*d^2)-(6.3*i^2)+(1.57*t*d)-(4.02*t*i)+(3.7*d*i)$$

Where:

y = percentage of SS removal.

t_1 = treatment time.

d = distance between electrodes.

i = direct current from power supply

Furthermore, based on one way method of statistical analysis of ANOVA on SS removal efficiency for interaction between three variables (Figure 4), obtained results that the optimum value will be reached at distance= 14 mm and current= 0.8 A at treatment time= 30 minutes.

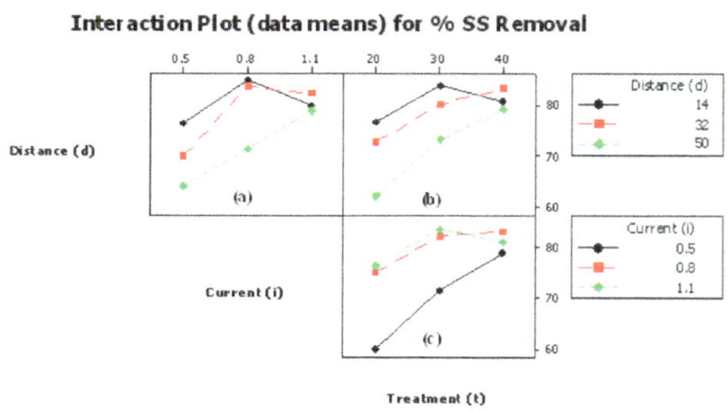

Figure 4: Output of one way statistical analysis of ANOVA to predict of optimum value for three variables

Effect of Flowrate and Magnetic Fields

Studies of influence of flowrate have been done by using single flow experiment. Wastewater were circulated through the reactor with three variations of flowrate are 1.1 mL/s, 1.7 mL/s and 3.3 mL/s. Determination of optimum flowrate is based on the reactor volume (2000 mL) and optimum treatment time (30 minutes). To determine the optimum flowrate in the reactor, calculated using the following equation:

$$Q = \frac{Volume}{time} = \frac{2000\ mL}{30 x\ 60\ s} = 1.111\ mL/s$$

By the same method, obtained values of other flowrates are 1.7 mL/s for t= 20 minutes and 3.3 mL/s (t= 10 min). As for the parameters that have been tested in this experiment, can be seen in Table 4. Furthermore, those parameters as shown in Table 4 are simulated on a single flow reactor that has been designed. From the data have been obtained as shown inthe curves of Figure 5 and Figure 6, it appears that Suspende d solids removal in the single flow reactor are also influenced by the contact time and flowrate. The longer of contact time or the slower of flowrate, will get a better result removal. However, the length of contact time should not exceed of the optimum time.

Table 4: Values of variables have been tested in the single flow experiment

Variables/Factors	Levels	Values		
Magnetic fields (B)	2	Without Magnet		0.16 T
Flowrates (Q)	3	1.1 mL/s	1.7 mL/s	3.3 mL/s
Cell current (i)	1	0.80 A		

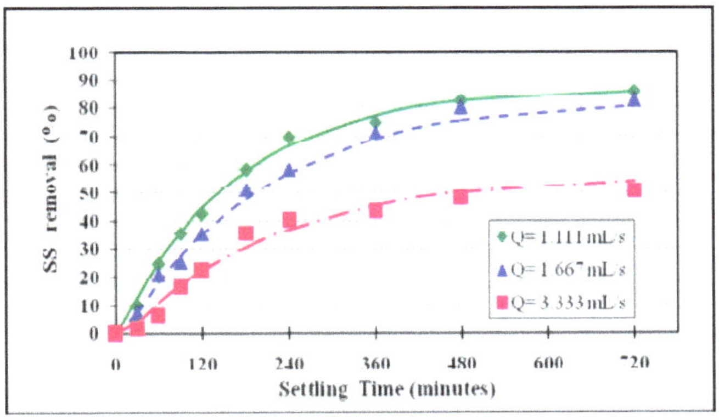

Figure 5: Percentages removal of suspended solids by Electrocoagulation using single flow method at cell current of 0.8A.

As seen from Figure 5, the optimum result of SS removal efficiencies by using treatment process without magnetic fields (B= 0.00T), for the flowrate of 1.1 mL/s is reached more than 85% (exactly is 85.9%). Whereas for the flowrate of 1.7 mL/s and 3.3 mL/s are reached more than 82% and 54%, respectively. On the other hand, Figure 6 illustrates of SS removal efficiencies by using combined process of EC and magnetic fields (B= 0.16T). In this combined process, besides of the influences of flowrate are also seen a significant influence of the magnetic fields.

Shown in the curves of Figure 6, the faster of the flowrate will have unsatisfactory results. Conversely, if the flowrate are set close to the optimum time, then the results become more increased. Moreover, if compared with Figure 5 then the curves in Figure 6 also showed improved results due to the influence of magnetic fields. With this combined processing techniques, SS removal efficiencies are increased reached more than 92%, 86% and 62% at flowrate of 1.1 mL/s, 1.7 mL/s and 3.3 mL/s, respectively.

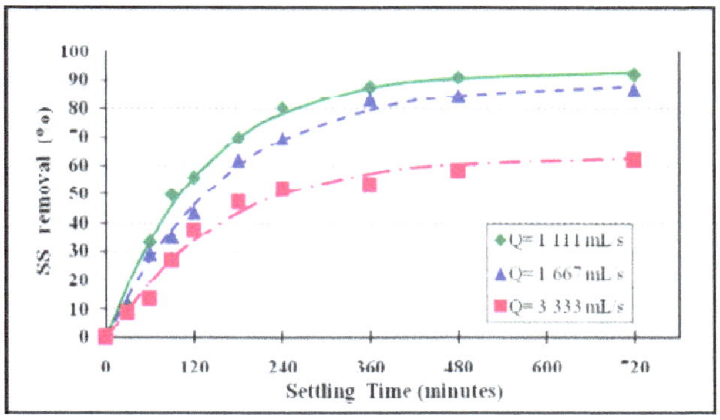

Figure 6: Percentages removal of suspended solids by combined of EC and magnetic fields (B= 0.16T) using single flow method at cell current of 0.8A.

A clearer comparison of the influence of these magnetic fields can be seen on the curve in Figure 7. Wherein shown in Fig. 7, SS removal by electrocoagulation technique at flowrate of 1.1 mL/s are obtain optimum results of 86%, when combined with the technique of EC-magnetic fields (B = 0.16T) are obtain better results more than 92%.

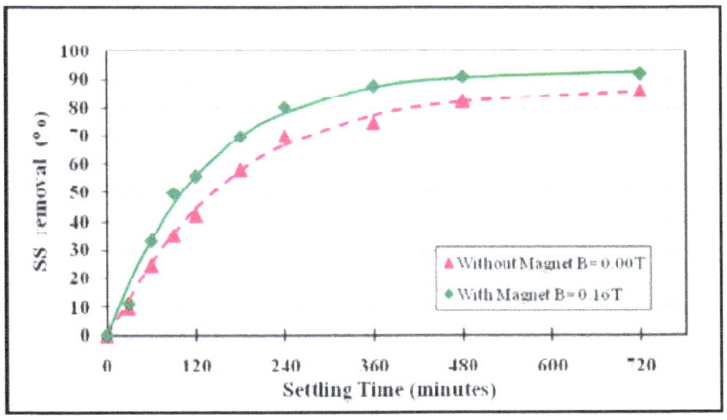

Figure 7: Comparison of magnetic fields ability on SS removal percentage by using single flow method at Q= 1.1 mL/s and i= 0.80A.

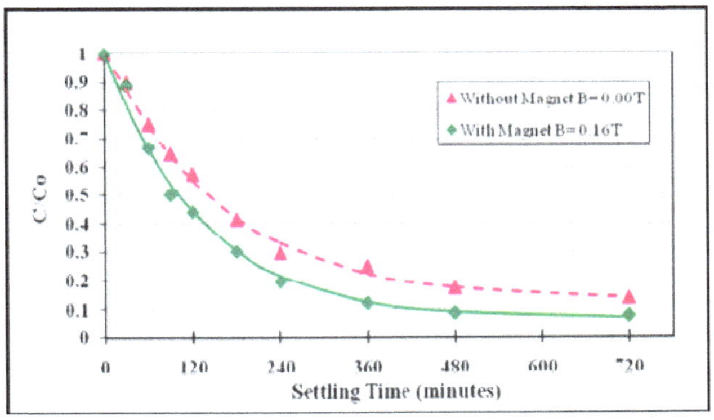

$$C/Co_{(B=0.00T)} = -0.8949(-0.1459-e^{(-0.00624t)})$$
$$C/Co_{(B=0.16T)} = -0.9448(-0.0713-e^{(-0.00770t)})$$

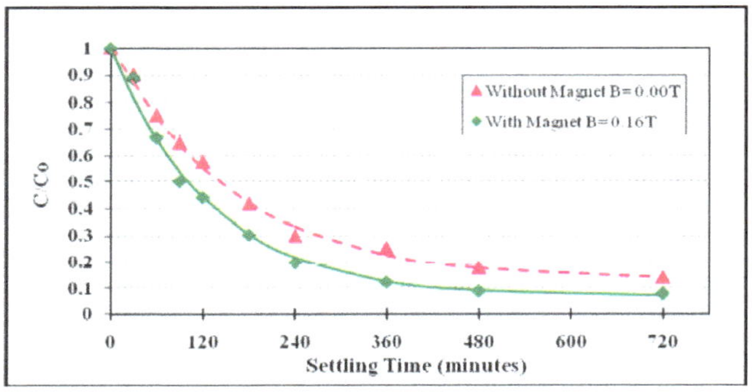

$$C/Co_{(B=0.00T)} = -0.8949(-0.1459-e^{(-0.00624t)})$$
$$C/Co_{(B=0.16T)} = -0.9448(-0.0713-e^{(-0.00770t)})$$

Figure 8: Comparison of magnetic fields ability on rate of change of SS removal by using single flow method at Q= 1.1 mL/s and i= 0.8A

On the other hand, if the results of these experimentswere analyzed using the curve of rate of change (C / Co),then the result will be presented as shown in Figure 8.From this figure, the study results indicate that theremoval of suspended solids by using EC technologycombined magnetic fields, also in accordance withequation(3).As Shown in Figure 8 that the k values willincrease with applied of the combined technique, whencompared with the electrocoagulation process alone. Thismeans that the ability of the magnetic fields has asignificant influence on the process of suspended solidsremoval from wastewater. Furthermore, this study also analyzed using factorialdesign graph, which the output of this analysis are twographs plot, i.e.: main effect plot and interaction plot,Based on factorial design graphs as shown in Figure 9 andFigure 10, it were clear that wastewater treatment byusing combined of EC and magnetic fields gives betterresults than without magnetic fields.

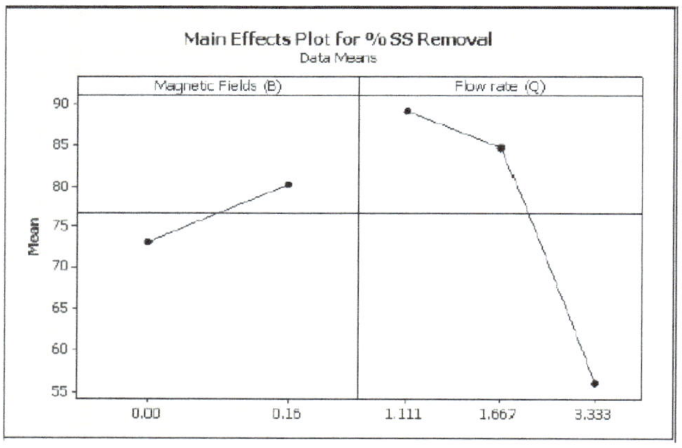

Figure 9: Main Effect Plot for SS removal using Factorial Design graphs.

As depicted in Figure 9, that the magnetic field and the flowrate have major influences on suspended solids removal. These two factors have the opposite effect. The stronger of magnetic field, it will give a greater effect on suspended solids removal. Conversely, the faster of flowrate, it will give a smaller effect on suspended solids removal. While in Figure 10, illustrates the interaction between flowrate and the two technologies are simulated for SS removal. It appears that both technologies have the same effect in each of flowrate, but the combined technology with magnetic fields are gives better results.

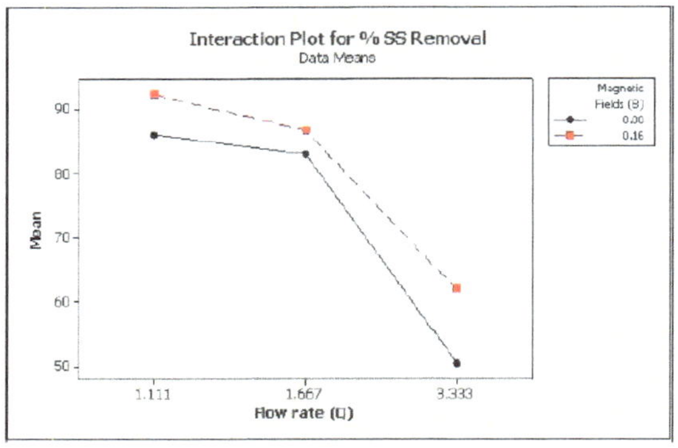

Figure 10: Interaction Plot for SS removal using Factorial Design graphs.

Iron (Fe) Content in the Sludge

Suspended solids that have been coagulating will be settling to the bottom, as shown in Figure 11. To determine the presence of Fe content and concentration, make do with the SEM (Scanning Electron Microscope) analysis on standard glass-fiber filter that have been used to filter for suspended solids (see in Fig. 11 and Fig. 12).

Figure 11: Sludge of settled SS on the bottom of settling column and the glass-fiber filter that have been filtered the SS at 0 min (before treatment), 10 min, 15 min, 20 min, 25 min and 30 min, respectively.

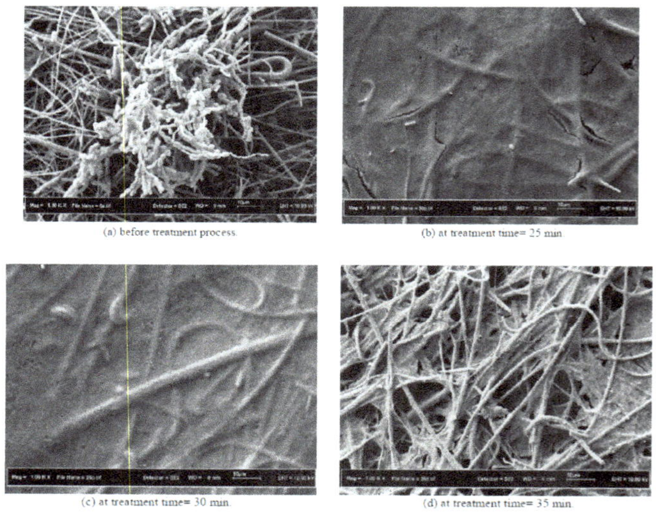

Figure 12: SEM images of coagulation process of suspended solids.

In the SEM image as shown in Figure 12 can explain more details about the optimum time. As shown in Figure 12(a) at the beginning before the wastewater treatment is seen that there is no coagulation of the suspended solids. Fig. 12(b) and Fig. 12(c) explained that coagulation process begins when the treatment processing has been going on for 15 minutes, and complete coagulation occurs between 25 to 30 minutes. In Figure 12(d) when the processing time more than 35 minutes there was an anti-climax of coagulation process, where the coagulation has been experiencing breakdown again or rupture. Thus, based on Figure 12 can determined that the optimum time of the EC process is 30 minutes by using the current (i) = 0.80 A and the distance between electrode (d) = 14 mm. This complements the results of the ANOVA statistical analysis shown in Figure 4.

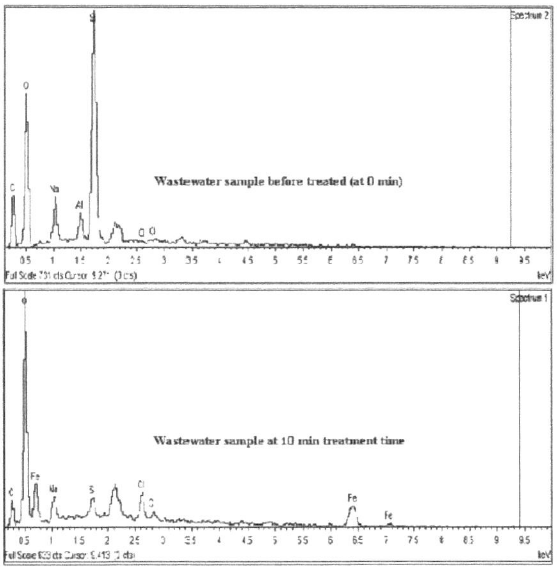

Figure 13: The spectrum image of chemical elements quantity of suspended solids sample on wastewater before treated and at 30 min treatment time.

Table 5: Weight percentages of Fe contents in the sludge

No.	Treatment Time (minutes)	Weight Percentages (%)
1	0	0
2	25	38.34
3	30	37.42
4	35	29.25

The results of SEM analysis can be used to determine the quantities of chemical elements, especially Fe contained in the suspended solids are filtered by glassfiber filters. The results of chemical analysis of Fe content in the SEM examination have been done, it is known that the Fe content in the filter samples before treatment was 0%. Furthermore, the quantity of Fe content increased to 38.34% and 37.42% in the filter samples were treated for 25 minutes and 30 minutes, respectively. More detail can be seen in the spectrum image is shown in Figure 13. From the data, the weight percentage of Fe at treatment time of 25, 30 and 35 minutes are 38.34%, 37.42% and 29.25%, respectively. More specifically, the weight percentage of Fe in the sludge during the treatment time, can be seen in Table 5 and Figure 14.

As shown in Figure 14, at the beginning before the wastewater treatment is seen that there is no coagulation of the suspended solids. The coagulation process begins when the treatment processing has been going on for 15 minutes, and complete coagulation occurs between 25 to 30 minutes. As explained by Kobya et al. (2009 and 2011), in this processes, Fe^{2+} (ferrous) ions are released. And when the processing time more than 35 minutes there was an anti-climax of coagulation process, where Fe^{3+} (ferric) ions are released and the coagulation has been experiencing breakdown again or rupture. Ferrous (Fe^{2+}), in chemistry, indicates a divalent iron compound (+2 oxidation state), as opposed to ferric, which indicates a trivalent iron compound (+3 oxidation state).

CONCLUSION

From the experiment and statistical analysis using RSM and ANOVA obtained optimum values, are: optimum time is 30 minutes, optimum distance is 14 mm and optimum cell current is 0.8 A. From the experimental results also found that after passing through the optimum time, coagulation will be break down again or experiencing rupture. And if this process is continued, wastewater will turn brownish green color because of the increasing Fe^{3+} ions are released.

As shown in experimental results and the statistically analysis using Factorial Design from MINITAB, that the magnetic field and the flowrate has a major influences on suspended solids removal. These two factors have the opposite effect. The stronger of magnetic field, it will give a greater effect on suspended solids removal. Conversely, the faster of flowrate, it will give a smaller effect on suspended solids removal.

ACKNOWLEDGMENT

The authors are grateful for the financial support IRPA Grant: VOT 74146, by the Malaysian Ministry of Science, Technology and Innovation (MOSTI). We also wish to thank the Universiti Teknologi Malaysia (UTM) and Universitas Islam Sultan Agung (UNISSULA) for financial and other support provided.

REFERENCES

1. Antony, J., (2003). Design of experiments for engineers and scientists, 1st Edition, Butterworth Heinemann, Oxford.
2. APHA, (2005). Standard methods for the examination of water & wastewater, 21st Edition The American Public Health Association.
3. Calvo, L.S., Leclerc, J.P., Tnguy, G., Cames, M.C., Paternotte, G., Valentin, G., Rostan, A. and Lapicque, F., (2003). An electrocoagulation unit for the purification of soluble oil wastes of high COD, Environmental Progress, 22: 1, 57–65.
4. Can, O.T., Kobya, M., Demirbas, E. and Bayramoglu, M., (2006). Treatment of the textile wastewater by combined electrocoagulation, Chemosphere, 62, 181–187.
5. Carmona, M., Khemis, M., Leclerc, J.P. and Lapicque,

F., (2006). A simple model to predict the removal of oil suspensions from water using the electrocoagulation technique, Chemical Engineering Science, 61, 1237–1246.

6. Chen, G., (2004). Electrochemical technologies in wastewater treatment, Separation and Purification Technology, 38, 11–41.
7. Daneshvar, N., Sorkhabi, H.A. and Kasiri, M.B., (2004). Decolorization of dye solution containing acid red 14 by
8. electrocoagulation with a comparative investigation of different electrode connections, Journal of Hazardous Materials, B112, 55– 62.
9. Daneshvar, N., Oladegaragoze, A. and Djafarzadeh, N., (2006). Decolorization of basic dye solutions by electrocoagulation: an investigation of the effect of operational parameters, Journal of Hazardous Materials, B129, 116–122.
10. Ghernaout, D., Ghernaout, B., Saiba, A., Boucherit, A. and Kellil, A., (2009). Removal of humic acids by continuous electromagnetictreatment followed by electrocoagulation in batch using aluminium electrodes, Desalination, 239, 295–308.
11. Hernandez, I.L., Diaz, C.B., Morales, G.R., Bilyeu, B. and Nunez, F.U., (2009). Influence of the anodic material on electrocoagulation performance, Chemical Engineering Journal, 148 (1), 97–105.
12. Hiemenz, P.C. and Rajagopalan, R., (1997). Principles of colloid and surface chemistry, 3rd edition, Marcel Dekker Inc., New York.
13. Holt, P.K., Barton, G.W., Wark, M. and Mitchell, C.A., (2002). A quantitative comparison between chemical dosing and electrocoagulation, Colloids and Surfaces A: 211, 233–248.
14. Ihara, I., Kanamura, K., Shimada, E. and Watanabe, T., (2004). High gradient magnetic separation combined with electrocoagulation and electrochemical oxidation for treatment of landfill leachate, Applied Superconductivity, IEEE Transaction, 14: 2, 1558–1560.
15. Ilhan, F., Kurt, U., Apaydin, O. and Gonullu, M.T., (2008). Treatment of leachate by electrocoagulation using aluminum and iron electrodes. Journal of Hazardous Materials, 154, 381–389.
16. Kobya, M., Can, O.T. and Bayramoglu, M., (2003). Treatment of textile wastewaters by electrocoagulation using iron and aluminum electrodes, Journal of Hazardous Materials, B100, 163– 178.
17. Kobya, M., Bayramoglu, M. and Eyvaz, M., (2007). Technoeconomical evaluation of electrocoagulation for the textile wastewater using different electrode connections, Journal of Hazardous Materials, 148, 311–318.
18. Kobya, M., Demirbas, E. and Akyol, A., (2009). Electrochemical treatment and operating cost analysis of textile wastewater using sacrificial iron electrodes, Water Science and Technology, 60: 9, 2261–227.
19. Kobya, M., Demirbas, E., Bayramoglu, M., and Sensoy, M.T., (2011). Optimization of electrocoagulation process for the treatment of metal cutting wastewaters with response surface methodology, Water Air Soil Pollut, 215, 399–410.
20. Kumar, P.R., Chaudhar, S., Khilar, K. and Mahajan, C., (2004). Removal of arsenic from water by electrocoagulation, Chemosphere, 55, 1245–1252.
21. Larue, O., Vorobiev, E., Vu, C. and Durand, B., (2003). Electrocoagulation and coagulation by iron of latex particles in aqueous suspensions, Separation and Purification Technology, 31, 177–192.
22. Matteson, M. J., Dobson, R.L., Glenn Jr., R.W., Kukunoor, N.S., Waits III, W.H. and Clayfield, E.J., (1995). Electrocoagulation and separation of aqueous suspensions of ultrafine particles, Colloids and Surfaces A: 104, 101–109.
23. Meunier, N., Drogui, P., Montane, C., Hausler, R., Mercier, G. and Blais, J.F., (2006). Comparison between electrocoagulation and chemical precipitation for metals removal from acidic soil leachate, Journal of Hazardous Materials, B137, 581–590.
24. Mollah, M.Y.A., Schennach, R., Parga, J.R. and Cocke, D.L., (2001). Electrocoagulation (EC) – science and applications, Journal of Hazardous Materials, B84, 29–41.
25. Montgomery, D.C., (2001). Design and analysis of experiments, 5th edition. John Wiley & Sons, Inc., New York.
26. Ni'am, M.F., Othman, F., Sohaili, J. and Fauzia, Z., (2005). A combined magnetic field and electro-coagulation process for wastewater treatment, Proc. of Brunei Int'l. Conference on Engineering and technology (BICET). August 15–18, ITB Brunei Darussalam, 389–394.
27. Ni'am, M.F., Othman, F., Sohaili, J. and Fauzia, Z., (2006). Combined magnetic field and electrocoagulation process for suspended solid removal from wastewater, Proc. of the 1st International Conference on Natural Resources Engineering & Technology (INRET), July 24-25, Putrajaya, Malaysia, 384–393, 2006.
28. Ni'am, M. F., Othman, F., Sohaili, J. and Fauzia, Z., (2007a) Electrocoagulation technique in enhancing COD and suspended solids removal to improve wastewater quality, Water Science and Technology, 56: 7, 47–53.
29. Ni'am, M. F., Othman, F., Sohaili, J. and Fauzia, Z., (2007b). Removal of COD and turbidity to improve wastewater quality using electrocoagulation technique, The Malaysian Journal of Analytical
30. Sciences, 11: 1, 198 – 205.
31. Ni'am, M. F., Othman, F., Sohaili, J. and Fauzia, Z., (2008). Electrocoagulation technique for removal of COD and turbidity to iImprove wastewater quality, Ultrapure Water Journal, 25: 3, 36– 42.
32. Othman, F., Sohaili, J. and Zularisham, (2001). Application of magnetic field to enhance wastewater treatment process, The 8th Joint MMM-Intermag Conference, January 7-11, San Antonio, Texas: IEEE.
33. Othman, F., Ni'am, M. F., Sohaili, J. and Fauzia, Z., (2005). Removal of suspended solid from wastewater by combined magnet and electrocoagulation, Proc. of

MSA–International Science Congress (ISC), August 3 - 6, Malaysia.

34. Othman, F., Sohaili, J., Ni'am, M.F. and Fauzia, Z., (2006). Enhancing suspended solids removal from wastewater using Fe electrodes, Malaysian Journals of Civil Engineering, 18: 2, 139–148.

35. Shaw, D.J., (1970). Introduction to colloid and surface chemistry, 2nd edition, Butterworth & Co., London.

36. Sohaili, J., (2003). Effect of magnetic fields on the settling of suspended particles in sewage, University of Technology Malaysia, Ph.D Thesis.

37. Sohaili, J., Othman, F. and Zularisham, (2004). Effect of magnetic fields on suspended particles in sewage, Malaysian Journal of Science, 23, 141–148.

38. Tchamango, S., Njiki, C.P.N., Ngameni, E., Hadjiev, D. and Darchen, A., (2010). Treatment of dairy effluents by electrocoagulation using aluminium electrodes, Science of the Total Environment, 408, 947–952.

39. Tsouris, C., DePaoli, D.W., Shor, J.T., Hu, M.Z.C. and Ying, T.Y., (2001). Electrocoagulation for magnetic seeding of colloidal particles, Colloids and Surfaces A: 177, 223–233.

40. Zhu, B., Clifford, D.A. and Chellam, S., (2005). Comparison of electrocoagulation and chemical coagulation pretreatment for enhanced virus removal using microfiltration membranes, Water Research, 39, 3098–3108.

41. Zongo, I., Maiga, A.H., Wethe, J., Valentin, G., Leclerc, J.P., Paternotte, G. and Lapicque, F., (2009). Electrocoagulation for the treatment of textile wastewater with Al or Fe electrodes: compared variations of COD levels, turbidity and absorbance, Journal of Hazardous Materials, 169, 70–76.

CHAPTER 20 INTERMITTENT AERATION IN BIOLOGICAL TREATMENT OF WASTEWATER

H. Doan and A. Lohi

Keywords: BOD, cyclic aeration, wastewater, packed column, energy conservation

INTRODUCTION

Wastewater treatment plants are usually classified as primary, secondary or tertiary treatment, depending on the purification level to which the plants provide[1] . Primary treatment uses physical processes such as screening and sedimentation to remove large solid particles and pollutants that will settle or float. Secondary treatment uses the physical processes similar to the primary treatment and also provides a biological oxidation of the waste. Tertiary treatment is highly specialized since its use is dependent upon the pollutants to be removed [2].

In secondary treatment, microorganisms convert the organic wastes into stabilized compounds. Typical biological treatment processes make use of trickling filters, rotating biological contactors, aerated lagoons and activated sludge. Activated sludge is the most widely used biological treatment process because the recirculation of the biomass allows microorganisms to adapt to changes in the wastewater composition by a relatively short acclimation process. In an activated sludge process, oxygen is usually supplied to the wastewater by diffusing aeration or mechanical aeration[3]. In diffusing aeration, a diffuser is used to introduce small air bubbles of 1.8-2.5 mm in diameter near the bottom of the tank. The air bubble size is kept small for an efficient oxygen transfer to water. A mechanical aeration system utilizes agitators or mixers to create and entrap air bubbles in water[4] . Aeration of the wastewater enhances the removal of the Biological Oxygen Demand (BOD), which is the amount of oxygen consumed by the microorganisms in oxidation of the pollutants in wastewater. Aeration is used in different units of a wastewater treatment plant: Preaeration, homogenization, biological treatment, nitrogen removal and aerobic sludge treatment. The cost of aeration comes to more than 50% of the total energy consumption of the entire treatment process and that could be reduced substantially by using intermittent aeration[5-7]. Equipment for aeration may be classified into four categories: waterfall aerators, diffusion or bubble aerators, mechanical aerators and pressure aerators. In waterfall aerators, water is spread into droplets or thin films to increase the contact area between water and air. Common waterfall aerators are spray aerators, multiple tray aerators, cascade aerators, cone aerators and packed columns[2]. In the present study, an industrial wastewater from a polymer ecoating process was treated using a packed column as an aerator since it provided a very high surface area of the packing where water and air came into contact and oxygen was transferred from air to water efficiently. The contacting time between air and water in a packed column is an important parameter for obtaining a required level of pollutant removal. Long contacting time between air and water results in a higher amount of oxygen transferred from air to water in the column. On the other hand, the biological oxidation is a relatively slow reaction process. The biological oxidation of the wastewater could be oxygen-diffusion controlled under a high concentration of biodegradable organics initially and becomes reaction controlled in later stage of the treatment. It may thus be unnecessary to aerate the wastewater continuously in the packed column. The aeration process consumes energy. The cost of the aeration contributes a major portion of the overall operating cost of the treatment process. Intermittent aeration thus may entail in some energy saving, i.e., a reduction in the operating cost. Several researchers have investigated

Citation: H. Doan and A. Lohi, Intermittent Aeration in Biological Treatment of Wastewater, ISSN 1941-7020.

intermittent aeration in various biological treatment systems[8-13] . However, intermittent aeration in a packed column had not been investigated. Therefore, the objective of the present study was to determine an optimum on/off aeration cycle in a packed column aerator so as to achieve an acceptable BOD5 removal at a minimum operating cost and maximum energy conservation possible.

MATERIALS AND METHODS

In the present study, a PVC column filled with 15 mm stainless steel Pall rings was used as an aerator. The ratio of the bed length to the column diameter was 4.5 and the ratio of the bed diameter to the packing size was 5. Wastewater from a polymer e-coating process containing an initial BOD_5 of 70 mg L^{-1} was pumped from a holding tank to the top of the packed column. A countercurrent air stream was introduced at the bottom of the column. Pall rings were used to provide a large contact area for the wastewater and the air where oxygen was transferred from air to liquid. In intermittent aeration, the liquid pump was turned on and off automatically at different preset intervals using a timer. During the on aeration cycle, a constant wastewater flowrate of 28 kg m^{-2} sec^{-1} to the column was used. Water exiting the packed column flowed back to the holding tank where it was recirculated to the column. In order to keep the wastewater at a constant temperature of 25°C, a heat exchanger was installed between the pump discharge and the liquid stream entering the packed column as shown in Fig. 1. Air flowrate was maintained at 0.082 kg m^{-2} sec^{-1} for all experiments. Rotameters (Hedland Inc., Racine, WI, USA) were used to measure air and water flowrates to the column. During the off aeration cycle, water remained in the holding tank. Intermittent aearation was done so to determine an optimum air-wastewater contacting time in the column, which provided an adequate aeration of the wastewater for the BOD5 removal and yet some energy saving over continuous aeartion would be realized. Therefore, experiments were performed at varied contacting times between air and water in the column. The contacting time, t_C, is defined as:

$$t_C = \frac{H}{u} \cdot \frac{Q}{V} \cdot t_{op} \qquad (1)$$

Where:

t_{op} = The total aeration time per 24 h of treatment

H = The packing height

u = The superficial velocity of liquid in the column

Q = The volumetric flowrate of liquid

V = The total liquid volume in the tank

H/u = The residence time of liquid in the column

It is relevant to note that for a recirculating system as the one used in the present study, the contacting time is independent of the liquid flowrate as shown in Eq. 2. In addition, the solubility of oxygen in water is also independent of liquid flowrate in the column. Therefore, the liquid flowrate should be set as low as possible in order to reduce the energy consumption for pumping while the liquid rate is still adequate to provide good wetting of the packing in the packed column. It is essential to maintain good wetting of the packing so to ensure a sufficient oxygen transfer from air to water.

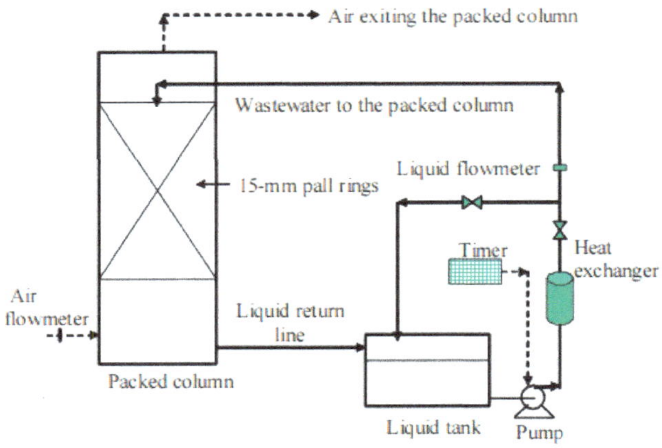

Figure 1: Schematic diagram of the experimental set-up

Table 1: Contacting time of air-water in the packed column

Pump-on time (h)	Pump-off time (h)	No. of cycles 24 h^{-1}	Total on-time 24 h^{-1}	Air-water contacting time (min)	Percentage of energy consumption
4	2.0	4	16.0	34.8	66.7
24	0.0	Continuous	24.0	52.3	100.0
2	2.0	6	12.0	26.1	50.0
2	4.0	4	8.0	17.4	33.3
5	1.0	4	20.0	43.6	83.3
3.5	4.5	3	10.5	22.9	43.8
4	4.0	3	12.0	26.1	50.0
6	2.0	3	18.0	39.2	75.0
10	2.0	2	20.0	43.6	83.3
4	8.0	2	8.0	17.4	33.3
0	24.0	0	00.0	00.0	00.0
1	5.0	4	4.0	8.70	16.7

Using the relationship between the volumetric flowrate and the superficial velocity of liquid, Eq. 1 can be simplified as:

$$t_C = \frac{S \cdot H}{V} \cdot t_{op} \qquad (2)$$

where, S is the cross-sectional area of the column.

Accordingly, the percentage energy consumption can be used to evaluate the economical aspect of each on/off aeration cycle (or contacting time). The percentage energy consumed by the pump for various pumping cycles is defined as:

$$\%P = \frac{t_{op}}{24} \cdot 100 \qquad (3)$$

where, %P is the percentage energy consumption. The 100% energy consumption represents continuous aeration. Twelve different on/off aeration cycles were used in the experiments while all other parameters were kept constant. Table 1 shows the contacting time of air and water and the corresponding percentage energy consumption for different experiments.

In each experiment, water samples were taken and tested at 0, 24, 48 and 72 h of treatment time. The amounts of dissolved oxygen in water at the sampling time and after an incubating period of 5 days at 20°C were measured using a dissolved oxygen meter (Model 52 CE, YSI, Yellow Springs, Ohio, USA). The BOD5 of the water samples were determined accordingly to the standard methods[14]. Polyseed® (InterBio, Inc., Texas) was used to provide microorganisms to the wastewater samples in the BOD bottles for the BOD5 measurements. The seed solution was made up of one capsule of Polyseed® in 500 mL distilled water. 10 mL of the seed solution was added to each BOD bottle of the wastewater.

RESULTS

Effect of aeration intervals on the BOD_5 removal

The variation of the percentage BOD_5 removal with the treatment time for various contacting times is plotted in Fig. 2. When the contacting time between air and water was reduced, the BOD_5 removal decreased accordingly. Nevertheless, for a long treatment time of 72 h, the BOD_5 removal did not decrease significantly when the contacting time was reduced by about 17% (aeration for 20 h per 24 h of treatment). It was also noted that the rate of the BOD_5 removal generally decreased with increases in the treatment time.

The BOD_5 removal appeared to follows the firstorder kinetics that exhibited an exponential decay of the BOD with the treatment time. The first order kinetics of the BOD removal can be expressed as below:

$$\frac{dBOD}{dt} = -k \cdot BOD \tag{4}$$

where, k is the BOD_5 removal rate constant and t is the treatment time.

From Eq. 4, the fractional BOD remaining can be written as:

$$\left(\frac{BOD}{BOD_i}\right) = \exp(-kt) \tag{5}$$

where, BOD_i is the initial BOD of the wastewater.

As indicated by Eq. (5), the rate constant, k, can be obtained from the slope of a plot of ln (BOD/BOD_i) vs. the treatment time. The values of the rate constant obtained for varied aeration cycles are plotted in Fig. 3. From the data in Fig. 3, an empirical relationship between the rate constant and the contacting time ranging from 8.7-52.3 min was also obtained below:

$$k = 2.3 \times 10^{-4} t_C^{1.17} \tag{6}$$

with a coefficient of determination $r^2 = 0.93$ and a RMS (root-mean-square error) of 0.00078 between the values predicted from Eq. 6 and the experimental data.

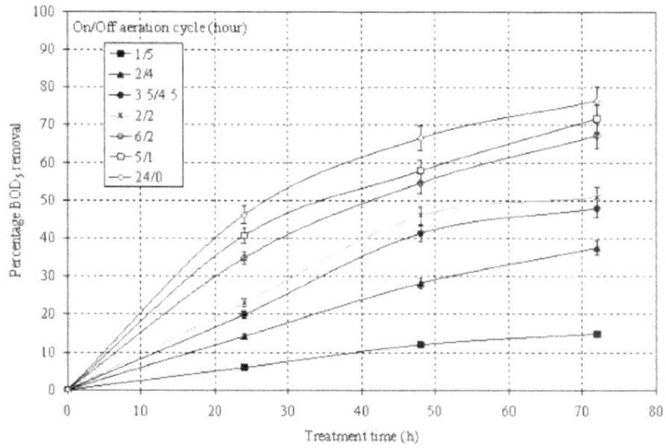

Figure 2: Variation of the BOD removal with the treatment time at different On/Off aeration cycles

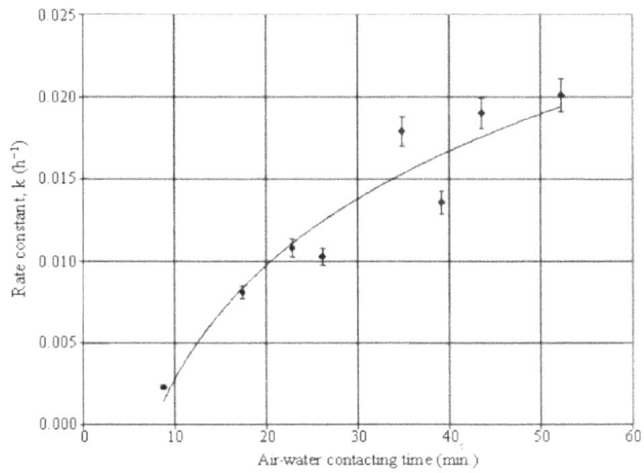

Figure 3: Variation of the rate constant k with the airwater contacting time in the column

The variation of the percentage BOD5 removal with the aeration time for different treatment durations was also investigated and the results obtained are plotted in Fig. 4.

The driving force to oxygen transfer to water was also estimated. Using a standard equation for mass flux, the transfer rate, R_A, of oxygen from air to water in the packed column can be written as below:

$$R_A = V_B \cdot k_L \cdot a \cdot (C^* - C) \tag{7}$$

Where:

V_B = The volume of the packed bed

k_L = The mass transfer coefficient of oxygen in the water film on the surface of the packing

a = The effective interfacial area of the packing

C^* = The saturated concentration of oxygen in water at the gas-liquid interface

C = The concentration of oxygen in the bulk liquid

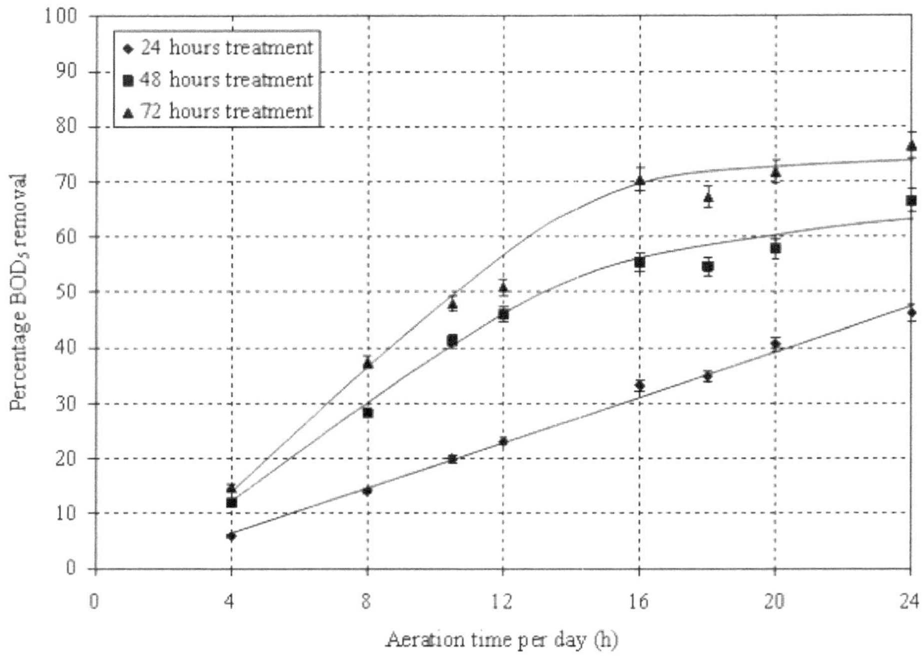

Figure 4: Effect of aeration time on the BOD5 removal for various treatment times

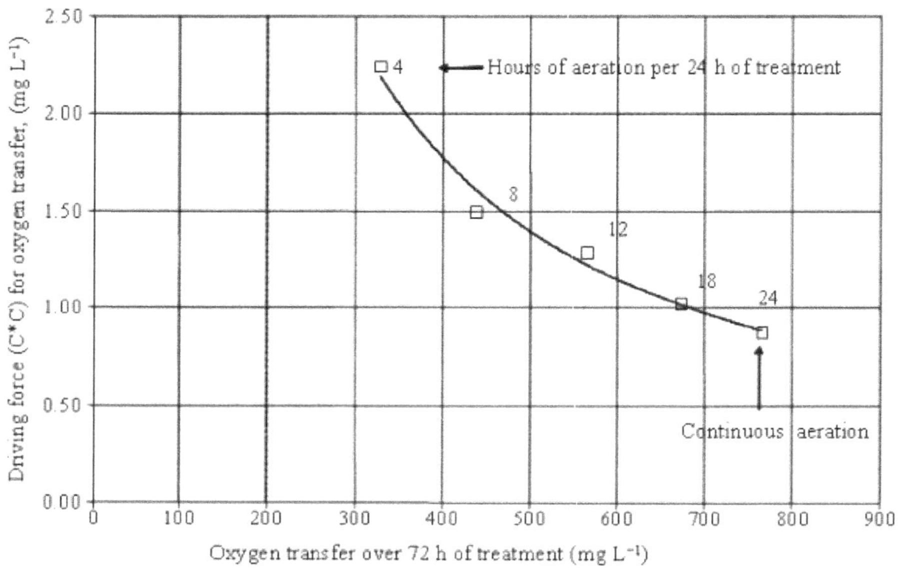

Figure 5: Estimated driving force for oxygen transfer to water at various air-water contacting times in the column

From Eq. 7, the driving force for oxygen transfer (C* -C) can be estimated using the mass transfer coefficient, kL and the effective interfacial area of the packing, a, from correlations given in the literature[15] and the oxygen consumed by both the BOD and the COD removal measured in the experiments under various air-liquid contacting times. The results obtained are presented in Fig. 5. The driving force for the oxygen transfer was within the range of 0.8-2.5 mg L^{-1}. On the other hand, the measured Dissolved Oxygen (DO) in wastewater was substantially higher as shown in Fig. 6. The concentration of DO was above the critical concentration for the BOD removal (10% of the saturation concentration [16]) during 5 h off cycle.

From the profile of DO in Fig. 6, the portion of the total BOD_5 removal by suspended microorganisms in the liquid holding tank over off-aeration cycles can be estimated from the reduction of DO.

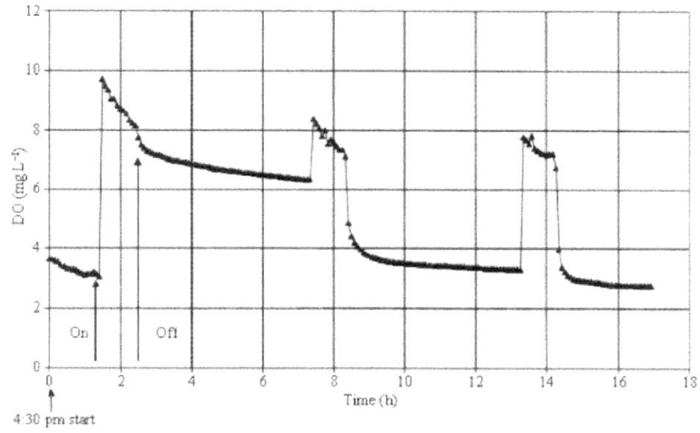

Figure 6: Profile of Dissolved Oxygen (DO) in wastewater for 1 h On/5 h Off aeration cycles

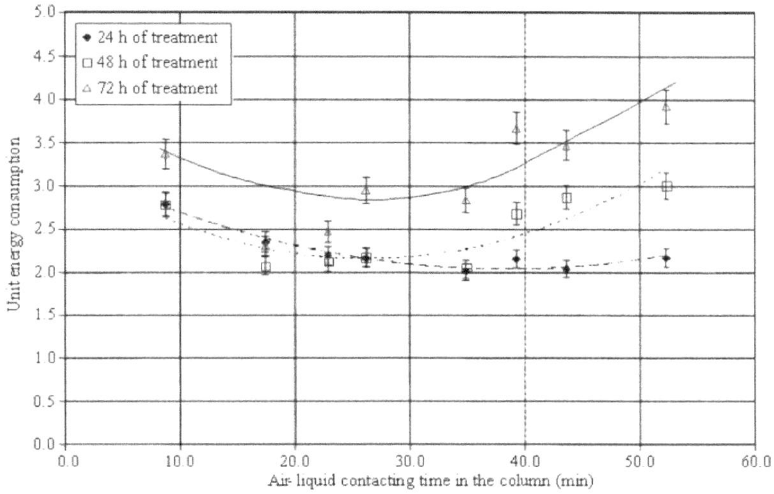

Figure 7a: Optimum aeration time for the treatment of industrial wastewater

It was observed that about 12% of the total BOD_5 removal was due to the suspended microorganisms. In addition, the microorganisms attached to the packing in the column aerator contributed to about 10% of the organic removal during on-aeration cycles while the stripping effect was accounted for about 20% of the total BOD_5 removal. On the overall, about 50% of the total BOD_5 removal was due to suspended microorganisms in the holding tank during on-aeration cycles.

In order to determine the optimum aeration time for the lowest energy consumption, the unit energy consumption was used. The unit energy consumption was defined as the ratio of the time-averaged percent energy consumption and the percentage BOD_5 removal. The results obtained are plotted in Fig. 7a and b.

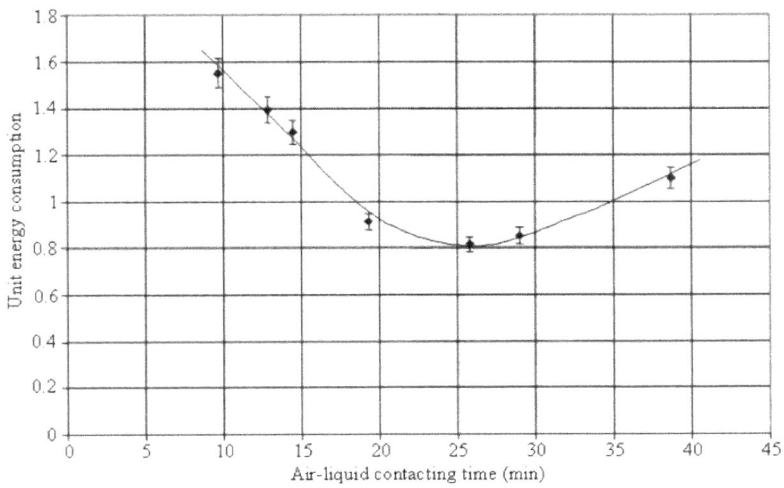

Figure 7b: Optimum aeration time for the treatment of ethylene glycol solution

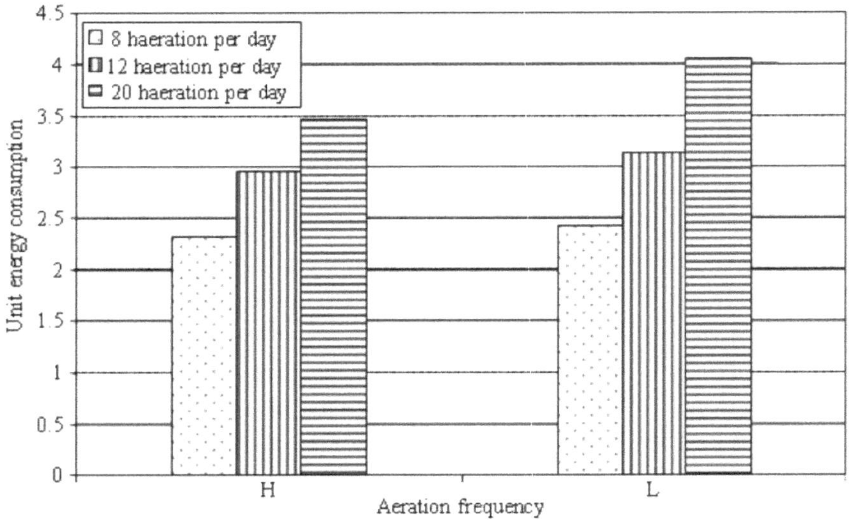

Figure 8: Effect of aeration frequency on the unit energy consumption for various aeration times, L: Low frequency, H: High frequency, twice the number of aeration cycles of L

Those aeration times are equivalent to cases of 33, 50 and 83% energy consumption, respectively. For 8 h aeration, 2 h On/4 h off and 4 h On/8 h off cycles were used. Similarly, 2 h On/2 h off and 4 h on/ 4 h Off cycles were used for the case of 12 h aeration and 5 h On/1 h Off and 10 h On/2 h Off cycles were used for 20 h aeration. For the 8 h aeration, the 2 h On/4 h off aeration cycle was designated as high frequency aeration and used to compare with the 4 h On/8 h off cycle as low frequency aeration. For the same air-liquid contacting time in the column, the aeration frequency (number of aeration cycles) of the 2 h On/4 h off cycle was twice that of the 4 h On/8 h off cycle.

However, in order to have a full and comprehensive picture of the effect of the aeration frequency, a series of experiments with the number of aeration cycles day^{-1} varied from 1-16 (1 h On/0.5 h Off, 2 h On/1 h Off, 4 h On/2 h Off, 8 h On/4 h Off and 16 h On/8 h Off) was carried out. The results obtained for 72 h of treatment are plotted in Fig. 9.

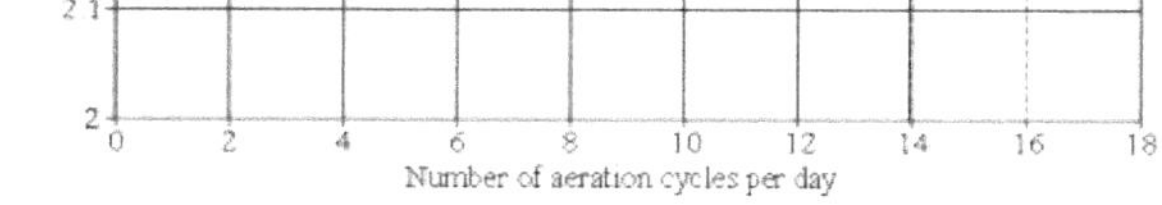

Figure 9: Optimum number of aeration cycles per day for 16 h aeration

DISCUSSION

Wastewater aeration represents one of the most energy intensive operations in a wastewater treatment system, consuming between 50-90% of the total energy cost of a typical municipal treatment facility. Since the solubility of oxygen in water is low, the time required to fully saturate water with oxygen is short. On the other hand, the rate of consumption of oxygen by biological oxidation is rather slow. Therefore, on/off aeration deems to be beneficial. On-Off aeration has been recommended as the most efficient way to reduce the cost of aeration. In the present study, a series of experiments were performed at varied pumping cycles so to obtain an adequate contacting time between air and water while to maintain the same amount of BOD removal.

The BOD5 removal followed the first-order kinetics of the organic concentration that exhibited an exponential decay with the treatment time. This indicates that less oxygen is needed for the biological oxidation in the later period of the treatment. Therefore, it was not necessary to aerate the wastewater continuously for prolonged treatment time. However, for the contacting time less than 23 min (i.e., aeration for less than 10.5 h in 24 h of treatment), the BOD_5 removal tends to follow a zero-order kinetics rather than the widely accepted first order kinetics with respect to the organic concentration. There seems to be a constant rate of BOD_5 removal throughout the duration of the treatment. For these cases, the dissolved oxygen might become the limiting reactant of the biological oxidation process.

As shown in Fig. 4, for 24 h of treatment the BOD_5 removal varies linearly with the contacting time while an exponential decay trend is observed for 48 and 72 h of treatment. For the first 24 h of treatment, the organic concentration in the wastewater was relatively high; hence, more oxygen was needed for the biological oxidation of the organics. The concentration of the dissolved oxygen in water might thus be low during the off cycle of aeration (the pump was off) and become the controlling factor of the BOD_5 removal. The BOD_5 removal was thus increased steadily with increases in the aeration time (or the contacting time). For longer treatment times (48 and 72 h), however, the BOD_5 removal increased rapidly with the contacting time and leveled off at the contacting time of about 35 min. Over the first 24 h of treatment, a substantial reduction of the organic content might have occurred provided an adequate aeration (contacting time) was maintained. The oxygen requirement would then become lower in later stages. The dissolved oxygen concentration might remain at an adequate level during the off pumping cycle, specially for high contacting times of 35 min or higher (the percentage energy consumption of 75% or higher), hence, the BOD_5 removal exhibits a gradual change with the contacting time. At a very low contacting time, i.e., 8.7 min or 16.7% energy consumption (4 h aeration per 24 h of treatment), there might be a lack of dissolved oxygen in the wastewater during the Off cycle. The BOD_5 removal was very low and insignificantly different for all treatment times of 24, 48 and 72 h. Under on-aeration cycles, well-mixed condition as well as high turbulence in the liquid tank was observed. These conditions might have enhanced the transfer of the organics from the bulk liquid to the suspended microorganisms; hence, the BOD_5 removal was higher than that during the off-aeration cycles.

The lower BOD_5 removal at a lower contacting time may indicate that the biomass on the surface of the packing in the column also plays a significant role in the overall bio-oxidation of the organic materials in the wastewater during the treatment process. When the liquid was cascaded down the packed column, the organic materials were transferred from water to the biomass on the surface of the packing where the biological oxidation occurred. When the wastewater was in the holding tank, organic substances might be oxidized only by the suspended biomass in the liquid phase, which was not in a large amount. When the wastewater stayed stagnant in the holding tank, i.e., no pumping, the BOD_5 removal was about 8% after 72 h of treatment compared to 75% BOD_5 removal under continuous aeration. On the other hand, about 40% of the BOD_5 removal was observed for the case with pumping the wastewater continuously to the column without aeration (no airflow)[17]. The removal of the organic substances might thus occur mainly in the packed column in a similar way as that in a trickle bed filter. However, when the pumping time reached 18 h over 24 h of treatment, no further significant improvement in the BOD removal was observed.

The effect of aeration time on the energy consumption is shown in Fig. 7a and b. The air-liquid contacting time of 52.3 min is for the case with continuous aeration. For 24 h of treatment, the lowest unit energy consumption is at about 34 min of air-liquid contacting time (16 h aeration per day). In other words, the biological treatment of the wastewater was most cost efficient at the contacting time of about 35 min, indicating a 27% energy saving. For 48 and 72 h of treatment, the optimum unit energy consumption falls at the region of 25 min of air-liquid contacting time. This is equivalent to 12 h of aeration per 24 h of treatment. Thus, 50% energy saving is realized. Similar optimum air-liquid contacting time was also obtained with experiments using Ethylene Glycol (EG) solution as a simulated wastewater (Fig. 7b). It is interesting to note that for 72 h of treatment, both ethylene glycol solution and industrial wastewater had similar optimum contacting times in the region of 25 min although the experiments of these two wastewaters were operated under different flowrates and aeration times. For the EG solution, 25 min of air-liquid contacting time is equivalent to 16 h of aeration per day while for the industrial wastewater, this is equivalent to 12 h of aeration per day. This indicates that the unit energy consumption would be a good and representative parameter for determination of the optimum aeration time for different aeration operations.

The time required to fully saturate water with oxygen in the packed column was quite short because the solubility of oxygen in water was low (about 8.0 mg L^{-1}) while the oxygen transfer efficiency in the packed column was high. On the other hand, the rate of consumption of oxygen by the biological oxidation of the wastewater was rather low. An adequate oxygen concentration in the wastewater could thus be maintained for several hours during the off-aeration cycle. Under continuous aeration, the amount oxygen transferred to water could be eventually negligible once the concentration of dissolved oxygen in water was close to saturation since the driving force for the oxygen transfer from air to liquid is very low, resulting in a substantial amount of energy wasted under continuous aeration It is also noted that the optimum contacting time was shortened with the long treatment time. At the beginning of the biodegradation process, the concentration of the organic matters in the wastewater was relatively high; hence, the biodegradation occurred at a relatively fast rate, which resulted in a high oxygen demand. The BOD removal process might thus be oxygen-diffusion-controlled for the first 24 h of treatment. More aeration (i.e., longer contacting time) was needed so to provide sufficient oxygen for the BOD removal. The optimum energy consumption thus fell in a longer contacting time interval. At later stages, i.e., 48 and 72 h of treatment, however, the biodegradation proceeded at a lower rate, hence, less oxygen was required. Therefore, on the average less aeration was needed for a longer treatment time; hence, the optimum point was in a shorter contacting time interval. The results obtained indicates that there existed an optimum contacting time between air and water, which would allow a low aeration cost while an adequate BOD removal was maintained. Significant amount of energy could be saved in biological treatment processes that take advantage of intermittent aeration.

Figure 8 shows the unit energy consumption based on the BOD removal after 24 h of treatment for all three cases of aeration cycles. It appears that the frequency only affected the BOD removal slightly for the case that was either has a substantially low aeration time (8 h) or a moderate aeration time (12 h). On the other hand, for the case of 20 h aeration the BOD removal increased noticeably with high aeration frequency, as reflected by the lower unit energy consumption. For the case with low aeration times of 8 and 12 h, it seemed that the treatment process was inefficient because of the low air-liquid contacting time in the column for oxygen transfer from air to water. For 20 h aeration per day at the high aeration frequency, the dissolved oxygen in the wastewater was probably maintained at a higher level since the off aeration time was only 1 h (the shortest off aeration), resulting in a better unit energy consumption. Therefore, the unit energy consumption (or the BOD removal) didn't change significantly with the aeration frequency. On the mass transfer standpoint, the higher aeration frequency would facilitate a higher amount of oxygen transferred to water over the course of the treatment because with a shorter aeration cycle the driving force for mass transfer would be higher. This might be the reason for the improvement in the BOD removal due to the higher aeration frequency for the case of 12 h aeration. Moreover, it can be seen clearly that the energy consumption was reduced significantly when the number of aeration cycle was increased from 1-8 cycles per day as shown in Fig. 9.

CONCLUSION

The results obtained in the present study indicate that a substantial amount of energy saving up to 50% could be achieved using intermittent aeration for a moderate treatment time of 48-72 h. Under this condition, the BOD_5 was reduced from the initial concentration of 70 mg L^{-1} to about 28 mg L^{-1}. At 16 h of aeration per day, further reduction of the BOD_5 to 20 mg L^{-1} was obtained. However, no further significant increase in the BOD_5 removal was achieved with increases in the aeration time to 18 h or higher.

The dissolved oxygen in wastewater at the end of 5 h off cycle (no aeration) was about 3.0 mg L^{-1}, which was still adequate for biodegration of organics in the wastewater. The BOD_5 removal was found to increase with the aeration frequency that was at a shorter On/Off cycle time and hence, a higher number of cycles per day. For the total aeration time of 16 h per day, the optimum point was obtained with 8 aeration cycles (2 h On/1 h Off). Further increase in aeration frequency to 16 cycles per day didn't lead to any significant change in the BOD_5 removal.

ACKNOWLEDGEMENT

The researchers would like to acknowledge the help of C. Lam and G. Thuraisingham on the experimental research of the present study.

REFERENCES

1. Drinan, J.E., 2000. Water and Wastewater Treatment: A Guide for the Nonengineering Professional. 1st Edn., Technomic Publishing Company, Inc., Pennsylvania, ISBN: -10: 1587160498, pp: 336.
2. American Water Works Association, 1997. Water Treatment Plant Design. 3rd Edn., McGraw-Hill, New York, USA., ISBN: -10: 0070016437 pp: 806.
3. Reynolds, T. and P. Richards, 1995. Unit Operations and Processes in Environmental Engineering. 2nd Edn., PWS Publishing, New York, USA., ISBN: -10: 0534948847, pp: 816.
4. Minett, S., 1995. The optimal conditions for aeration-the views of two leading German researchers. Eur. Water Pollut. Control, 5: 17-19.
5. Ferrer, J., M.A. Rodrigo, A. Seco and J.M. Penyaroja, 1998. Energy saving in the aeration process by fuzzy logic control. Water Sci. Technol., 38: 209-217. DOI: 10.1016/S0273-1223(98)00463-6
6. Hamamoto, Y., S. Tabata and Y. Okubo, 1997. Development of the intermittent cyclic process for simultaneous nitrogen and phosphorus removal. Water Sci. Technol., 35: 145-152. DOI: 10.1016/S0273-1223(96)00890-6
7. Schuyler, R.O., R.W. Ramey, M. Wallace, G. Vance and A. Torres, 1997. On/off aeration saves money and improves effluent quality. Fuel Energ. Abst., 38: 443.http://www.ingentaconnect.com/content/els/01406701/19 97/00000038/00000006/art82383
8. Bicudo, J.R. and I.F. Svoboda, 1995. Effect of intermittent-cycle extended-aeration treatment on the fate of carbonaceous material in pig slurry. Bioresour. Technol., 54: 53-62.
9. Ouyang, C.F. and C.T. Juan, 1995. A study of modified process for the intermittent cycle extended aeration system. Water Sci. Technol., 31: 173-180. http://www.iwaponline.com/wst/03109/wst031090173.htm
10. Yang, P.Y. and Z. Wang, 1999. Integrating an intermittent aerator in a swine wastewater treatment system for land-limited conditions. Bioresources, Technol., 69: 191-198. http://cat.inist.fr/?aModele=afficheN&cpsidt=1761740
11. Chang, J.S., A.C. Chen, M.Y. Hung and M. Shih, 1999. The operational strategy of intermittent cycle extended aeration system in treating sewage. J. Chem. Technol. Biotechnol., 74: 688-692. http://cat.inist.fr/?aModele=afficheN&cpsidt=1909404
12. Park, K.Y., Y. Inamori, M. Mizuochi and K.H. Ahn, 2000. Emission and control of nitrous oxide from a biological wastewater treatment system with intermittent aeration. J. Biosci. Bioeng., 90: 247-252. DOI: 10.1016/S1389-1723(00)80077-8
13. Luo, A., J. Zhu and P.M. Ndegwa, 2002. Removal of carbon, nitrogen and phosphorus in pig manure by continuous and intermittent aeration at low redox potentials. Biosyst. Eng., 82: 209-215. DOI: 10.1006/bioe.2002.0071
14. APHA., AWWA., WPCF., 1998. Standard Methods for the Examination of Water and Wastewater. 20th Edn., American Public Health Association, New York, USA.
15. Billet, R. and M. Schultes, 1999. Prediction of mass transfer columns with dumped and arranged
16. packings. Chem. Eng. Res. Des., 77: 498-504. http://cat.inist.fr/?aModele=afficheN&cpsidt=1995442
17. Bailey, J.E. and D.F. Ollis, 1986. Biochemical Engineering Fundamental. 2nd. Edn., McGrawHill, New York, USA., ISBN: -10: 0070032122 pp: 928.
18. Doan, H.D. and J. Wu, 2002. Biological treatment of wastewater from a polymer coating process. J. Chem. Technol. Biotechnol., 77: 1076-1083. DOI: 10.1002/jctb.653

INDEX